S0-AWV-350

Polymeric Materials for Corrosion Control

ACS SYMPOSIUM SERIES 322

Polymeric Materials for Corrosion Control

Ray A. Dickie, EDITOR
Ford Motor Company

F. Louis Floyd, EDITOR
Glidden Coatings and Resins

Developed from a symposium sponsored by
the Division of
Polymeric Materials Science and Engineering
at the 190th Meeting
of the American Chemical Society,
Chicago, Illinois,
September 8–13, 1985

American Chemical Society, Washington, DC 1986

Library of Congress Cataloging-in-Publication Data

Polymeric materials for corrosion control.
(ACS symposium series, ISSN 0097-6156; 322)

"Developed from a symposium sponsored by the Division of Polymeric Materials, Science and Engineering at the 190th Meeting of the American Chemical Society, Chicago, Illinois, September 8-13, 1985."

Includes bibliographies and index.

1. Corrosion and anti-corrosives—Congresses.
2. Polymers and polymerization—Congresses.
3. Protective coatings—Congresses.

I. Dickie, R. A., 1940- . II. Floyd, F. Louis, 1945- . III. American Chemical Society. Division of Polymeric Materials: Science and Engineering. IV. American Chemical Society. Meeting (190th: 1985: Chicago, Ill.) V. Series.

TA462.P57 1985 620.1'9204223 86-20646
ISBN 0-8412-0998-7

PRINTED IN THE UNITED STATES OF AMERICA

ACS Symposium Series

M. Joan Comstock, *Series Editor*

FOREWORD

The ACS SYMPOSIUM SERIES was founded in 1974 to provide a medium for publishing symposia quickly in book form. The format of the Series parallels that of the continuing ADVANCES IN CHEMISTRY SERIES except that, in order to save time, the papers are not typeset but are reproduced as they are submitted by the authors in camera-ready form. Papers are reviewed under the supervision of the Editors with the assistance of the Series Advisory Board and are selected to maintain the integrity of the symposia; however, verbatim reproductions of previously published papers are not accepted. Both reviews and reports of research are acceptable, because symposia may embrace both types of presentation.

CONTENTS

MATERIALS FOR CORROSION PROTECTION

INDEXES

PREFACE

THERE ARE THREE CRITICAL STEPS in obtaining satisfactory performance from corrosion protective materials: *design* of the device, component, or structure; *selection* of materials for the intended application; and *control* of the manufacturing process or application conditions. The principal concern of the present volume is evaluating and selecting materials for use in corrosive environments. Intelligent materials selection requires a thorough understanding of protection and failure mechanisms and relies on the availability of appropriate test methodology. Papers concerning these topics make up the bulk of this volume. A few additional chapters address unusual materials or unexpected applications of polymeric corrosion-protective materials.

The symposium on which this book is based was organized to provide a forum for discussion of recent advances in the use of polymeric materials in corrosion control. Most of the papers presented in the symposium are included in this volume. Several chapters have been added. These include an introductory overview as well as separate review chapters on how organic coating systems protect against corrosion, on mechanisms of adhesion loss of organic coatings, and on the interfacial chemistry of adhesion loss in aggressive environments.

This volume provides a slice-in-time view of the progress in the science and technology of polymeric materials for corrosion control. The editors hope it will prove thought provoking and will contribute to a continuing discussion within the polymeric materials community on improved methods for achieving corrosion control.

The editors wish to express their appreciation to Ford Motor Company and to Glidden Coatings and Resins Division, SCM Corporation, for support during organization of the symposium and preparation of this volume. Special thanks are due Cathy Ciarrocchi and Diane DeSimone for their secretarial assistance.

RAY A. DICKIE
Ford Motor Company
Dearborn, MI 48121

F. LOUIS FLOYD
Glidden Coatings and Resins Division
SCM Corporation
Strongsville, OH 44136

1

Polymeric Materials for Corrosion Control: An Overview

Ray A. Dickie[1] and F. Louis Floyd[2]

[1]Ford Motor Company, Dearborn, MI 48121
[2]Glidden Coatings and Resins Division, SCM Corporation, 16651 Sprague Road, Strongsville, OH 44136

Polymeric materials are widely used to control the corrosion of metals, both to maintain appearance and to prevent loss of structural integrity. In this chapter, the fundamentals of metallic corrosion are briefly reviewed. Methods of studying corrosion, and of evaluating the performance of polymeric materials used in corrosion protection, are outlined. Factors that influence the corrosion protective performance of polymeric materials are discussed, and some of the research needs and important unsolved problems are highlighted.

The economic costs and environmental impact of metallic corrosion are well known, and need not be discussed in depth here. It has been estimated (1) that the total cost of corrosion in the United States may be as much as 4% of the gross national product, and that about 15% of the total cost might be avoidable through the economic use of available technology. Most studies of corrosion and its effects understandably concentrate on the cosmetic and structural effects of metallic corrosion; most of the papers in the present volume fall into this category. It should be noted, however, that metallic corrosion and the products of metallic corrosion can deleteriously affect the properties of non-metallic materials, particularly at joints between metals and non-metals. There are also environmental degradation phenomena that can affect non-metallic materials such as plastics, composites and glass directly; some of these phenomena resemble metallic corrosion processes in the effects observed on appearance and structural integrity. Several papers in this volume deal with corrosion effects on adhesive joints and non-metallic materials.

The present chapter begins with a brief overview of metallic corrosion and mechanisms of corrosion control. Methods of evaluating polymer performance and electrochemical characterization techniques are discussed. Barrier and adhesion aspects of corrosion control are reviewed, and some critical issues needing further study are outlined.

0097-6156/86/0322-0001$06.00/0

Metallic Corrosion

Metallic corrosion has been the subject of many textbooks and scholarly compendia (e.g., 2-4), and a number of introductory treatments dealing with corrosion and corrosion protection are also available (e.g., 5-7). In this context, the term "corrosion" refers to the chemical degradation of a metal by its environment. The reactions are most often heterogeneous redox reactions and occur at the metal-environment interface. The anodic reaction is typically the oxidation of the metal; the cathodic reaction is reduction of a non-metal, typically oxygen. If the product of the metal oxidation forms a tight and adherent film, the corrosion process may be self-limiting. If the products of the corrosion reaction are soluble in the corrosive medium, or are permeable to it, then metallic corrosion can proceed. Corrosion is often represented in terms of a simple electrochemical model. The anodic and cathodic half reactions of the corrosion cell may occur at adjacent or widely separated sites on the metal surface; the electrical circuit is completed by electronic conduction within the corroding metal and ionic conduction within the aqueous electrolyte. In natural corrosion, it is common for the sites of the anodic and cathodic corrosion reactions to become more or less widely separated. In such cases, the anodic sites tend to become acidic and the cathodic sites tend to become basic. These changes in pH can be large, and can have serious implications for the performance of polymeric materials.

The corrosion of iron is one of the most widespread and technologically important examples of metallic corrosion. In the presence of water and oxygen, the corrosion of iron proceeds to form a complicated mixture of hydrated iron oxides and related species; a complete description is beyond the scope of the present discussion, and the interested reader is referred to the previously cited general references on corrosion as well as to the well known descriptions of electrochemical equilibria in aqueous solution given by Pourbaix (8, 9). Iron is a base metal, subject to corrosion in aqueous solutions. In the presence of oxidizing species, iron surfaces can be passivated by the formation of an oxide layer; if the oxide layer formed is imperfect, rapid corrosion may occur. In simplest form, the reaction of iron to form iron oxide can be written as:

$$4\ Fe + 2\ H_2O + 3\ O_2 \rightarrow 2\ Fe_2O_3 . H_2O$$

The first step in the corrosion process is the dissolution of iron to form ferrous ion:

$$Fe \rightarrow Fe^{++} + 2\ e^-$$

In general, the pH decreases at sites of anodic dissolution due to hydrolysis reactions such as:

$$Fe^{++} + H_2O \rightarrow FeOH^+ + H^+$$

The cathodic reactions commonly observed are the evolution of hydrogen and the reduction of oxygen; hydrogen evolution is kinetically favored under acidic conditions, while oxygen reduction is kinetically favored under neutral and basic conditions.

$H_3O^+ + e^- \rightarrow 1/2\ H_2 + H_2O$ (acid solutions)

$H_2O + e^- \rightarrow 1/2\ H_2 + OH^-$ (neutral/basic solutions)

$1/2\ O_2 + H_3O^+ + 2\ e^- \rightarrow 3\ H_2O$ (acid solutions)

$1/2\ O_2 + H_2O + 2\ e^- \rightarrow 2\ OH^-$ (neutral/basic solutions)

The pH at cathodic sites tends to increase due to the production of hydroxide ion and/or consumption of hydrogen ion.

It is interesting to contrast the behavior of iron with that of aluminum. Aluminum is a very base metal; yet aqueous solutions in the neutral pH range that are quite aggressive toward iron often have little effect on aluminum. In the presence of acid solutions, aluminum dissolves with the formation of Al^{+3} ions, while under alkaline conditions, it dissolves as aluminate ions, AlO_2^-. Over the mid-range of pH from about 4 to 9, a passivating film of aluminum oxide tends to form. The structure and composition of the oxide depend on the conditions under which it is formed, and the corrosion performance of aluminum tends to be dominated by the performance of the oxide layer. Certain solution species, notably chloride, can disrupt the oxide layer and cause localized pitting. Control and modification of the aluminum surface oxide layer has been extensively studied, and is of particular importance in the protection of aluminum substrates.

Mechanisms of Corrosion Control

Corrosion can be controlled by isolation of the metal from the corrosive environment; by suppression of the anodic dissolution of metal; and by suppression of the corresponding cathodic reaction. Isolation of corrosion prone metals from corrosive environments is probably the most general mechanism of the corrosion protection afforded by paint films, sealers, and similar polymer-based materials. Effective isolation requires that polymeric materials have good barrier properties and remain adherent in the presence of water and the products of metallic corrosion. Barrier properties and adhesion aspects of corrosion control are discussed in detail in subsequent sections.

The anodic dissolution of metal can be suppressed by lowering the potential so that oxidation of the metal is thermodynamically impossible; this is the principle of cathodic protection of steel. Cathodic protection relies on either an external source of electric current or coupling of the metal to be protected with a more active metal (e.g., steel is protected by coupling to zinc). With the exception of some zinc-containing organic coatings applied to steel, cathodic protection is not a major mechanism of protection by polymeric materials. The mechanism of action of zinc containing coatings has been the subject of some disagreement. Part of the effectiveness of zinc pigmented coatings may be due to the formation of zinc corrosion products after an initial period of true cathodic protection. The action of the zinc compounds formed has been variously ascribed to a blocking of the pores of the film and to passivation of the surface. The evaluation (using impedance methods, see also Ref.

10) and modification of zinc pigmented coatings are discussed in this volume by Szauer and Miszczyk. The chemistry of zinc-rich and modified zinc-rich coatings were also discussed by Fawcett in the symposium on which this book is based. The paper was not made available for inclusion in this volume, but the preprint version is available (11).

Suppression of the anodic reaction can be achieved by the use of oxidizing inhibitors; essentially, the inhibitor is called upon to form (and maintain) an impervious and passivating oxide film on the surface of the metal. If the oxide film formed is imperfect, however, it is possible for rapid localized corrosion to take place. In the case of aluminum, inhibitors can be used to stabilize the oxide film against hydration, as discussed in this volume by Matienzo et al. Organic coatings designed for corrosion protection of ferrous metals often incorporate metal chromates as oxidizing inhibitors. The use of inhibitors in coatings, and the requirements for an ideal inhibitor, have been discussed by Leidheiser (12). As discussed by Funke both in this volume and elsewhere (13), the usefulness of active corrosion inhibiting pigments is open to question: the binders used for paints containing corrosion inhibiting pigments must be somewhat water permeable for the pigments to work, at least partially vitiating the barrier effect of the coating.

Adsorption inhibitors act by forming a film on the metal surface. The action of traditional oil-based red lead paint formulations presumably involves the formation of soaps and the precipitation of complex ferric salts that reinforce the oxide film. There has been substantial interest in recent years in development of replacements for lead-based and chromate-based inhibitor systems. Adsorption inhibitors based on polymers have been of particular interest. In this volume, Johnson et al. and Eng and Ishida discuss inhibitors for copper; 2-undecylimidazole is shown to be effective in acid media, where it suppresses the oxygen reduction reaction almost completely. Polyvinylimidazoles are shown to be effective oxidation inhibitors for copper at elevated temperatures. Also in this volume, Chen discusses the use of N-(hydroxyalkyl)acrylamide copolymers in conjunction with phosphate-orthophosphate inhibitor systems for cooling systems.

In many industrial coating applications, inorganic conversion coatings are used as surface pretreatments for metals. Such treatments typically result in the formation of an insoluble metal chromate or phosphate on the metal surface. The effectiveness of zinc phosphate conversion coatings has been related to their role in suppressing the cathodic reduction of oxygen (14). Bender et al. (15) have reviewed the literature extensively. The performance of inorganic conversion coating systems is dependent on bath composition and deposition conditions, on the initial condition of the substrate, and on the final rinse or post-treatment used. In this volume, Lindert and Maurer discuss a novel film-forming organic post-treatment for inorganic phosphate conversion coatings. Agarwala discusses a modified chromate conversion coating for aluminum.

Methods of Evaluating Polymer Performance

Performance Tests. The underlying goal of corrosion testing is generally the prediction of service performance, whether directly for a device or system, or indirectly in the design or formulation of a new material or process. The ultimate test is performance in the intended application; a close second is exposure of test panels to the normal service environment. Testing based on natural exposure is, of course, time consuming, and a large number of laboratory test methods have been developed to assess aspects of polymer properties and corrosion protection system performance. Performance tests generally involve exposure of a system, component, or test piece to a simulated or accelerated corrosion environment; evaluation of results is typically based on an assessment of the type and extent of corrosion failure. Property tests generally involve the measurement of a single, isolable, material property, or of a change in material property with exposure to an aggressive environment. Evaluation of results is typically in terms of a correlation with performance tests or field performance data.

The fundamental problems of accelerated performance testing are the selection of appropriate test conditions, and the determination and validation of acceleration factors. Ideally test conditions should be selected to accelerate all the relevant chemical reactions and physical processes equally. The determination of acceleration factors typically requires, and hence poses the same problems as, performance tests under natural exposure conditions.

Organic coatings are commonly evaluated using salt water immersion, salt fog or spray, modified salt exposure tests (e.g., salt fog with added SO_2), and various cyclic exposure tests. Humidity exposure and water immersion, and, for many applications, physical resistance tests (adhesion, impact resistance, etc.) are widely used preliminary tests. Standard methods for most of these tests are given in compilations of standard tests such as the Annual Book of ASTM Standards (16). Test methods have been extensively reviewed (e.g., 17-23).

Despite their long and common use, none of the popular laboratory corrosion tests are entirely satisfactory. The use of standard laboratory tests to establish comparative rankings of the corrosion performance of different materials is especially susceptible to error. In one recent study, for example, the results of exterior exposure and standard laboratory tests were used to compare corrosion inhibitive primers (24); it was found that standard salt fog tests showed substantial differences in paint performance that were not observed under field exposure conditions. Cyclic exposure tests have been proposed that incorporate periods of exposure to humidity, salt water immersion, temperature cycling, and dirt (25, 26); such tests are substantially more complicated than conventional laboratory tests, but in some cases give better agreement with corrosion performance in service. Cyclic exposure tests have also been applied to precoated steels, both with and without paint coatings (27). The mechanism of failure in cyclic exposure testing has been the subject of some discussion. Standish (28) argues that the cyclic test allows corrosion products like those observed in service to form under the coating; such corrosion products are typically not observed in salt spray. Jones (29) has also discussed the formation of a bulky oxide layer. Elsewhere in this volume, Dickie discusses surface analytical results on the cyclic immersion failure of organic coatings on phosphated steel substrates; delamination of the coating is found to be associated with dissolution of the conversion coating. This result is consistent with observations of van Ooij (30) on locus and mechanism of coating delamination on phosphated steel.

Electrochemical Characterization Techniques. Since corrosion is an electrochemical process, it is not surprising that a considerable amount of work has been reported over the years on electrical and electrochemical techniques for the study of the corrosion process. Leidheiser (31) and Szauer (32, 33) have provided good reviews of the principal techniques. Walter has recently provided a review of DC electrochemical tests for painted metals (34). Both AC and DC methods have been employed to study a variety of issues related to corrosion and corrosion protection. DC techniques are especially useful for studying substrate processes, while AC impedance techniques are most useful for studying processes relating to coated substrates and the performance of coatings.

DC techniques include measurement of DC resistance, determination of polarization behavior, and measurement of polarization resistance. Coating resistance has been correlated with corrosion performance by a number of workers. As summarized by Leidheiser (31), the results of several independent investigations suggest that coating resistance below about 10^6 ohm/cm^2 is associated with the formation of visible under-film corrosion. Parallel DC resistance measurements on thin film metal substrates have been used to study the deterioration of coated metals; the technique successfully detected the effects of water after migration to the coating/metal interface (35).

Polarization methods involve changing the potential (or current) of a corroding system in both the anodic and cathodic directions while monitoring current (or potential). By manipulating the resulting information, an understanding of the corrosion process can be obtained. For example, Beese (36) has used the linear polarization technique to develop information related to corrosion in beer and beverage cans that are coated with an organic enamel. Such information was ultimately employed to develop improved coatings for the container. Groseclose et al. (37) employed an anodic polarization technique to quantify the quality and variability of both cold rolled and hot rolled steels. This information was used to accurately predict the relative salt spray performance of the subsequently coated steels, and evaluate the effect of abrasive polishing and sandblasting of the substrate. The polarization resistance method widely used for studying metal corrosion has also been applied to painted metals. In principle, the polarization resistance is inversely proportional to the corrosion rate of the metal. For coated metals, the method is complicated by the correction for ohmic potential drop, diffusion limitations, and changes in film properties under the applied potential (32).

AC techniques are highly varied, but tend to converge upon the use of impedance spectroscopy. In recent publications, Hubrecht et al. (38), Mansfield and Kendig (39), and Kendig et al. (40) have reviewed the application of impedance spectroscopy to coating systems. By examining the AC impedance of the coated system as a function of frequency, useful information is obtained regarding both the barrier properties of the coating and the corrosion susceptibility of the substrate. Under proper conditions, information can be extracted relating to the interfacial layer as well. Information can also be extracted relative to the presence of water and ions in paint films as shown by Lindqvist (41). A good example of the latter has been given by Padget and Moreland (42). In most cases, barrier properties of coatings are ultimately found to be highly important to the prevention of corrosion of the substrate. In this particular case, it was also shown that the presence of a barrier film seems to augment the formation of a strongly passive layer at the interface between the coating and the substrate. An interesting application of electrochemical techniques involves the characterization of zinc rich paints. Fernandez-Prini and Kapista (43) and Lindqvist et al. (44) describe both DC and AC techniques for characterizing zinc rich coatings in such a way that subsequent salt spray testing is rationalized.

The present volume contains a number of papers relating to the issue of electrochemical testing. Morcillo et al. compare the results of AC impedance measurements with accelerated and outdoor exposure test results. Vijayan reports the use of AC impedance testing to study the effects of various components of the phosphating pretreatment process, paint thickness, and test variables on subsequent salt spray results. Moreland and Padget update their work on AC impedance as it pertains to the study of the passive layer which forms between a barrier coating and a steel substrate. Butler and Bartoszek-Loza describe their use of DC open circuit potentials to correlate with salt spray data as a function of post-bake temperature of UV cured coatings. Eden and co-workers describe their studies involving electrochemical noise measurements to study corrosion as it progresses. The authors' position is that the coating breakdown/failure on a steel substrate is accompanied by a change in the electrochemical noise signal, which gives a rapid indication of the state of the coating. Lomas et al. describe their novel work with harmonic analysis, combined with AC impedance testing in an attempt to detect corrosion of thickly coated substrates.

Barrier Aspects of Corrosion Control

The relative importance of the barrier function of organic coatings in corrosion protection has been debated for years. It is clear that, if a metallic substrate could be completely isolated from its environment, no corrosion would occur. The degree to which a protection system bars oxygen, water, and ions from the substrate would seem likely to be a measure of the effectiveness of the system in preventing corrosion.

Historically, a number of different theories regarding the role of the barrier function in corrosion protection have emerged. Studies by Mayne and his co-workers (45-49), Bacon et al. (50), and Cherry (51) indicated that neither the permeability of water nor the permeability of oxygen could be the rate determining factor in corrosion control by organic coatings, since neither was sufficiently low to provide effective isolation of the metallic substrate. Protection was attributed to the high electrical resistance and low ionic permeability of coatings that afforded good protection. To varying extents, Guruviah (52), Bauman (53), and Kresse (54) disagreed with the earlier workers regarding the limits of oxygen and water permeability in films. Haagen and Funke (55, 56) agreed with Guruviah and Bauman that oxygen permeability was the controlling factor; they observed that water permeability was the determining factor for the loss of adhesion, but not for corrosion.

More recently, workers in the field have recognized the probable need for a multiple parameter model to understand the corrosion protection process. Funke (57) proposed a model based on water permeability, oxygen permeability, and adhesion under high humidity conditions. The model was used to rationalize the rank order of salt spray results of seven different electrocoating systems. No mathematical treatment of the data was offered. Floyd et al. (58, 59) introduced a mathematical analysis of a wide range of properties in comparison with salt spray results. A barrier mechanism for corrosion control was postulated. This technique was also applied to Funke's earlier data, with a similar result. The model was further elaborated to take into account the existence of an electrochemical component in the model as a back-up to the primary barrier component. Floyd et al. further observed that no adequate characterization of this electrochemical interaction between paint and substrate existed.

The permeability of polymer systems is influenced by the properties of the polymer, by the presence of pigments or fillers, and by the interaction between polymer and fillers. Hulden and Hansen (60) have recently reviewed water permeation in coatings. Regularity of structure, crystallinity, and low segmental mobility are stated to give low permeability. Cross-link density has also been cited as resulting in reduced permeability, but results presented in this volume by Muizebelt and Heuvelsland suggest that cross-link density may be irrelevant in this respect. As Funke notes elsewhere in this volume, some of the factors that contribute to low permeability may interfere with adhesion; in particular, polar functional groups appear to be essential to achieving good adhesion, but are likely to increase permeability and contribute to water sensitivity.

Pigmentation can have a profound effect on permeability. The use of barrier pigments has been suggested as an alternative to the use of active inhibitive pigments, many of which are objectionable on environmental grounds (13). Flake shaped pigments are particularly effective, but pigment geometry is not the only important factor. If water can accumulate at the pigment-binder interface, as evidently happens in the case of mica, water absorption tends to increase with pigment volume concentration and, although permeability is still reduced by incorporation of the pigment, the effect is much smaller than with, for example, comparable loadings of aluminum flake (13). The influence of inert pigments on permeability and corrosion protective properties has been reviewed briefly by Hulden and Hansen (60), and has been discussed in a number of papers by Kresse (e.g., 61, 62). The mechanism of action of inert barrier pigments is commonly stated to be to increase the diffusion pathway to the substrate; it is also possible that pigments may tend to block or prevent the formation of pathways for direct ionic conduction to the substrate.

Interfacial and Adhesion Aspects of Corrosion Control

Basic Mechanisms of Adhesion: Acid-Base Interactions.

The understanding of polymer adhesion has been greatly advanced in recent years by the recognition of the central role of acid-base interactions. The concept of an acid was broadened by G. N. Lewis to include those atoms, molecules, or ions in which at least one atom has a vacant orbital into which a pair of electrons can be accepted. Similarly, a base is regarded as an entity which possesses a pair of electrons which are not already involved in a covalent bond. The products of acid-base interactions have been called coordination compounds, adducts, acid-base complexes, and other such names. The concept that acids and bases vary in their ability to interact with one another was introduced by Pearson in 1968 (63, 64). He introduced the concept of polarizability of the acid or base unit, describing it on a hard-soft scale. Hard acids are those of high electronegativity and low polarizability. Soft acids, in contrast, are large in size, have high polarizability, and low electronegativity. For the purposes of this book, it is important to remember that hard acids react most readily and form the strongest complexes with hard bases, while soft acids react most readily and form the strongest complexes with soft bases.

Drago and co-workers introduced an empirical correlation to calculate the enthalpy of adduct formation of Lewis acids and bases (65). In 1971, he and his co-workers expanded the concept to a computer-fitted set of parameters that accurately correlated over 200 enthalpies of adduct formation (66). These parameters were then used to predict over 1200 enthalpies of interaction. The parameters E and C are loosely interpreted to relate to the degree of electrostatic and covalent nature of the interaction between the acids and bases. This model was used to generalize the observations involved in the Pearson hard-soft acid-base model and render it more quantitatively accurate.

In 1975, Sorensen (67) used the acid-base interaction concept to rationalize color strength, gloss, and flocculation properties of coating systems having binders of differing acid-base characteristics. Anomalies that appear when using solubility parameter concepts were successfully explained by the acid-base concept. Drago et al. (68) were starting to address the issue of corrections to the solubility parameter concept using this technique at about the same time. A good review of the subject was written in 1978 by Jensen (69). The application of acid-base interactions to the phenomenon of adhesion was discussed by Jensen at an ACS meeting in 1981 (70). Fowkes and co-workers had already been discussing the competitive absorption of polymers onto pigment surfaces in the context of acid-base interactions by this time (e. g., 71).

Manson (72) expanded the concept to the solid state by observing that the strength of composite materials also depended upon the acid-base interaction between continuous and dispersed phases. More directly, Vanderhoff et al. (73) addressed the issue of adhesion of polymeric materials to corroded steel. They synthesized eight corrosion products of iron, and used the interaction scheme developed by Fowkes and Manson first to characterize the iron corrosion products as Lewis acids or bases and then to select polymer vehicles for practical coating systems. Such results were employed to enhance the adhesion of epoxy systems to substrates which were predominantly iron oxide in nature. A good overview of these issues was presented by Fowkes in 1983 (74).

Fowkes and co-workers also clearly demonstrated that the physical interaction of polymers with neighboring molecules was determined by only two kinds of interactions: London dispersion forces and Lewis acid-base interactions (75). Calculations based on this concept were shown to correct many of the problems inherent in the solubility approach. They were also able to use the concept to study the distribution of molar heats of absorption of various polymers onto ferric oxides, and thereby more accurately described the requirements for adequate adhesion to steel substrates (76). In the symposium on which this book is based, Fowkes summarized work showing that the polar interactions between polymers and metal surfaces that are important to adhesion are entirely of Lewis acid-base character. Calorimetric and infra-red spectroscopic methods for determining the E and C constants for polymers and metal oxides were presented. The full manuscript was not made available for publication in this volume, and the reader is therefor invited to consult the preprint manuscript (77).

Role of Adhesion in Corrosion Protection. Many of the theories regarding the mechanism of corrosion failure suggest that the loss of adhesion precedes the onset of corrosion, and is therefore of critical importance in understanding the process; Parker and Gerhart (78) considered adhesion to be crucial to corrosion performance. For organic coatings, the strength of the adhesive bond between coating and substrate does not appear to be the critical issue; what does appear to be important is that during and after environmental exposure the coating should be able to withstand the forces applied to it in its intended application. The adhesion of virtually all coatings is adversely affected by exposure to water or humid environments. Walker (79) found that the adhesive strength of a wide variety of coatings dropped from 20 to 40 MPa to 5 to 15 MPa in a direct pull-off test after exposure to humid environments. The initial (dry) bond strength was not a good predictor of performance. Haagen and Funke (56) observed that good protection was obtained if wet adhesion was good, even if the paint was highly water permeable. The importance of adhesion to corrosion protection is further discussed in this volume and elsewhere (80) by Funke.

Mechanisms of adhesion loss under various exposure conditions have been extensively studied. A survey is given elsewhere in this volume by Leidheiser, and specific examples of adhesion loss are discussed in detail by Thornton et al., Maeda et al., and Troyk et al., among others. Acoustic emission has been used to study coating adhesion and the effects of water immersion on coatings on water (see, e. g., 81-83). In this volume, Callow and Scantlebury discuss the possibility of using acoustic emission as a monitoring tool to investigate corrosion-induced debonding.

Modern surface analytical methods have led to much more detailed understanding of the interfacial chemistry of adhesion loss processes. Surface analytical studies of interfacial chemistry are reviewed in this volume by Dickie; in this paper, as in a recent paper by Castle and Watts (84), it is concluded that no single chemical mechanism adequately accounts for all of the observed behavior. In the simplest cases, loss of adhesion appears to involve displacement by water. Displacement of coatings by corrosion generated hydroxide, chemical degradation of the organic coating, and chemical attack on the underlying substrate surface or conversion coating have also been observed. Further examples of interfacial studies are given in this volume by Maeda et al., and a discussion of the reactions in conversion coatings during corrosion has been given by van Ooij (30).

A related but little studied area of adhesion and corrosion protection involves the chemical effects of metal substrates on coatings and other polymeric materials and conversely of polymeric materials on metals. In the curing of certain air-oxidizing coatings on steel, for example, reduction of ferric to ferrous species in the surface metal oxide, substantial thinning of the oxide, and oxidation of the coating material have been reported to occur in the interfacial region. These phenomena have been studied by infra-red and X-ray photoelectron spectroscopic techniques, and are discussed in this volume by Nguyen and Byrd and by Dickie. The stored components used to make polyurethane foams are subject to long term chemical degradation; the degradation products have been associated with the corrosion of storage containers. Wischmann discusses the problem, and suggests formulation changes for improved performance.

Adhesives and sealers can be an important part of a total corrosion protection system. Structural bonding procedures and adhesives for aluminum, polymer composites, and titanium are well established in the aerospace industry. Structural bonding of steel is gaining increasing prominence in the appliance and automotive industries. The durability of adhesive bonds has been discussed by a number of authors (see, e.g., 85). The effects of aggressive environments on adhesive bonds are of particular concern. Minford (86) has presented a comparative evaluation of aluminum joints in salt water exposure; Smith (87) has discussed steel-epoxy bond endurance under hydrothermal stress; Drain et al. (88) and Dodiuk et al. (89) have presented results on the effects of water on performance of various adhesive/substrate combinations. In this volume, the durability of adhesive bonds in the presence of water and in corrosive environments is discussed by Matienzo et al., Gosselin, and Holubka et al. The effects of aggressive environments on adhesively bonded steel structures have a number of features in common with their effects on coated steel, but the mechanical requirements placed on adhesive bonds add an additional level of complication.

Effects of Polymer Composition on Corrosion Control

Polymer composition poses not one but many critical issues for the development of materials for corrosion control. As outlined in previous sections of this chapter, the elements of molecular design for good adhesion, good barrier properties, and effective use of pigments in organic coatings are often in conflict. There does not appear to be a unifying theoretical basis on which these conflicting factors can be resolved, and an empirical balancing of properties remains an essential part of new product development for corrosion control. It is not surprising that the details of composition critical to performance often remain proprietary or appear only in the patent literature.

In addition to the customary desire for improved material performance, the development of new materials has, in recent years, been shaped by the demand for non-polluting or ecologically neutral materials. Restrictions on solvent emissions from industrial and maintenance paints, and limitations on lead-based and chromate corrosion-inhibitive pigments has had a major impact on corrosion protective material technology. In the field of organic coatings, there has been major emphasis on water-borne and so-called high solids coatings. Hill and Wicks (90) have discussed design criteria for high solids coatings; a recent book on reactive oligomers discusses a number of polymer systems of interest in high solids coatings (91). A variety of water-soluble and water-dispersible resins have been described in the literature (e.g., 90-96); the Technical Committee of the New England Society for Coatings Technology has published a series of articles on the design of waterborne coatings for the corrosion protection of steel (97-99). The development and commercial implementation of anodic and, subsequently, of cathodic electrodeposition coatings binders for electrodeposition have made possible major improvements in the corrosion protection of appliances and motor vehicles. The chemistry of binders for electrodeposition has been reviewed by Schenk et al. (100) and by Kordomenos and Nordstrom (101); the cathodic electrodeposition process has been discussed by Wismer et al. (102).

In the present volume, several papers deal with unique materials or applications: Schreiber describes work on plasma-deposited films from organo-silicone and inorganic (SiN) starting materials. Moreland and Padget discuss studies of a chlorine-containing vinyl acrylic copolymer that is applied as an acidic aqueous formulation and that promotes in situ formation of a protective film. White and Leidheiser discuss coating resins for the protection of steel exposed to sulfuric acid; Hojo et al., the behavior of epoxy and polyester resins in alkaline solution. Dreyfus et al. present results on coatings designed to protect glass in alkaline environments. Ibbitson et al. discuss structure-property relationships in tin-based anti-fouling paints.

The protection of microelectronics from the effects of humidity and corrosive environments presents especially demanding requirements on protective coatings and encapsulants. Silicone polymers, epoxies, and imide resins are among the materials that have been used for the encapsulation of microelectronics. The physiological environment to which implanted medical electronic devices are exposed poses an especially challenging protection problem. In this volume, Troyk et al. outline the demands placed on such systems in medical applications, and discuss the properties of a variety of silicone-based encapsulants.

Critical Issues

A topical symposium provides a forum for the review and updating of work in a given field, and provides an opportunity to identify critical issues. The editors of this volume would like to suggest that the following issues are among those needing additional study:

Systems. Corrosion is usually studied in an isolated fashion in the laboratory, but in practice is clearly the result of interacting systems in the environment. Studies need to be conducted on the way in which the component parts of corroding systems interact under actual environmental conditions, and on the way in which the components of the environment interact with the total corroding system. This would suggest not only design work for corrosion protection systems, but also additional work on the sensing and monitoring of corrosion in real hostile environments.

Methods. To observe that corrosion testing in the laboratory frequently fails to predict what happens in real-world environments is to admit that the mechanisms controlling corrosion in such environments are not understood, even at this late date of study. Mechanism-based test methods for monitoring corrosion are needed that will provide reliable and rapid prediction of service life for corrosion-susceptible systems. It is expected that statistical analysis will play a large role in any such effort, since the failure modes observed often vary widely within a statistical distribution. Such a recognition has led to the development of statistical methods for the description of mechanical failure of materials, and it is suggested that a similar effort would bear fruit in the analysis of corrosion phenomena.

Paints. Epoxies and cathodic electrocoats represent major advances in the field of corrosion control by organic coatings. The performance of these coatings represents a plateau which has not been departed from in over a decade. A great deal of attention has been devoted to interfacial processes in corrosion over the last several years, yet control of these processes remains an elusive goal. It is conceivable that, if means can be found to control interfacial failure processes, a new generation of coatings can be developed that will establish a considerably higher plateau of performance. Conversion coatings and surface treatments may play a vital role in this development.

Plastics. Part of the trend to substitute plastic and composite substrates for metals can be attributed to a desire to avoid the process of metallic corrosion and subsequent failure. Relatively little attention has been called to the possible failure modes of plastics under environments considered corrosive to metals. More extensive work should be conducted on the durability and life expectancy of plastic and composite materials under end-use environments. A further consideration is the potential for polymer degradation by the products of metal corrosion in hybrid structures comprising metal and polymer components. Since it is expected that coatings will continue to be used to protect plastic and composite substrates, ancillary programs need to be conducted on the mechanisms by which coatings can protect such substrates.

Adhesives. In many applications, there are substantial functional and economic reasons to prefer adhesive bonding over mechanical fastening of metals, of plastics, and of mixed-substrate joints. The role of corrosion in the failure of adhesive bonds is therefore becoming an increasingly crucial one. The performance demands placed on adhesive bonds by the combination of mechanical loading and aggressive environments are particularly severe, and it is clear that studies involving combined mode testing need to be greatly expanded. It is anticipated that information developed in the study of adhesives and the study of coatings should inter-relate in a sufficiently strong fashion that both fields will benefit from such studies.

Literature Cited

1. NBS Special Publication 511-1. Economic Effects of Metallic Corrosion in the United States. A Report to Congress by the National Bureau of Standards. SD Stock No. SN-003-01926-7, 1978; NBS Special Publication 511-2. Economic Effect of Metallic Corrosion in the United States. Appendix B. A report to NBS by Battelle Columbus Laboratories. SD Stock No. SN-003-01927-5, 1978.
2. Evans, U. R. "The Corrosion and Oxidation of Metals", St. Martins Press: New York, 1960; ibid., 1st Supplementary Volume,

St. Martins Press: New York, 1968; ibid., 2nd Supplementary Volume, Edward Arnold: London, 1976.
3. Uhlig, H. H. "Corrosion and Corrosion Control", 2nd ed., Wiley: New York, 1971.
4. Shreir, L. L. "Corrosion", 2nd ed., Newnes-Butterworths: London, 1976.
5. Evans, U. R. "An Introduction to Metallic Corrosion", 2nd ed., Edward Arnold: London, 1972.
6. Scully, J. C. "The Fundamentals of Corrosion", 2nd ed., Pergamon: Oxford, 1975.
7. Van Delindes, L. S., Ed. "Corrosion Basics: An Introduction", National Association of Corrosion Engineers: Houston, 1984.
8. Pourbaix, M. "Lectures on Electrochemical Corrosion", Plenum: New York, 1973.
9. Pourbaix, M. "Atlas of Electrochemical Equilibria in Aqueous Solutions", 2nd English ed., National Association of Corrosion Engineers: Houston, 1974.
10. Szauer, T.; Brandt, A. J. Oil Col. Chem. Assoc. 1984, 67, 13.
11. Fawcett, N. C. Polym. Mat. Sci. Eng. 1985, 53, 855.
12. Leidheiser, H., Jr. J. Coat. Technol. 1981, 53(678), 29.
13. Funke, W. J. Coat. Technol. 1983, 55(705), 31.
14. Zurilla, R. W.; Hospadaruk, V. Trans. SAE 1978, 87, 762.
15. Bender, H. S.; Cheever, G. D; Wojtkowiak, J. J. Prog. Org. Coat. 1980, 8, 241.
16. "Annual Book of ASTM Standards", Part 27, "Paint - Tests for Formulated Products and Applied Coatings", American Society for Testing and Materials: Philadelphia, issued annually.
17. Von Fraunhofer, J. A.; Boxall, J. "Protective Paint Coatings for Metals", Portcullis Press: Redhill, 1976.
18. Funke, W. J. Oil Col. Chem. Assoc. 1979, 62, 63.
19. Funke, W. Farbe Lack 1978, 84, 380.
20. Funke, W.; Machunsky, E.; Handloser, G. Ibid., 1979, 84, 498.
21. Funke, W.; Zatloukal, H. Ibid., 1979, 84, 584.
22. Funke, W. in "Corrosion Control by Coatings", Leidheiser, H., Jr., Ed.; Science Press: Princeton, 1979, p. 35.
23. Rowe, L. C.; Chance, R. L. in "Automotive Corrosion by De-icing Salts", Baboian, R., Ed.; National Association of Corrosion Engineers: Houston, 1981, p. 133.
24. Athey, R; Duncan, R.; Harmon, E.; Hartmann, M.; Iszak, D.; Nakabe, H.; Ochoa, J.; Shaw, P.; Specht, T.; Tostenson, P.; Warness, R. J. Coat. Technol. 1985, 57(726), 71.
25. Opinsky, A. J.; Thompson, R. F.; Boegehold, A. L. ASTM Bull. 1953 (Jan), 47.
26. Hospadaruk, V.; Huff, J.; Zurilla, R. W.; Greenwood, H. T. Trans. SAE, 1978, 87, 755.
27. Lambert, M. R.; Townsend, H. E.; Hart, R. G.; Frydrych, D. J. Ind. Eng. Chem. Prod. Res. Dev. 1985, 24, 378.
28. Standish, J. V. Ind. Eng. Chem. Prod. Res. Dev. 1985, 24, 1985.
29. Jones, D. A. Polym. Mat. Sci. Eng. 1985, 53, 470.
30. van Ooij, W. Polym. Mat. Sci. Eng. 1985, 53, 698.
31. Leidheiser, H., Jr. Prog. Org. Coat. 1979, 7, 79.
32. Szauer, T. Prog. Org. Coat. 1982, 10, 157.
33. Szauer, T. Prog. Org. Coat. 1982, 10, 171.
34. Walter, G. W. Corr. Sci. 1986, 26, 39.

35. McIntyre, J. F.; Leidheiser, H., Jr. Ind. Eng. Chem. Prod. Res. Dev. 1985, 24, 348.
36. Beese, R. E.; Allman, J. C. in "Modern Container Coatings", Strand, R. C., Ed.; ACS Symposium Series No. 78, American Chemical Society: Washington, DC, 1978.
37. Groseclose, R. G.; Frey, C. M; Floyd, F. L. J. Coatings Technol. 1984, 56(714), 31.
38. Hubrecht, J.; Vereecken, J.; Piens, M. J. Electrochem. Soc. 1984, 131, 2010; see also Piens, M.; Verbist, R.; Vereecken, J.; in "Organic Coatings: Science and Technology"; Parfitt, G. D.; Patsis, A. V., Eds.; Dekker: New York, 1984; Vol. 7, p. 249.
39. Mansfield, F.; Kendig, M. W. Werkst. Korros. 1985, 36, 473.
40. Kendig, M. W.; Allen, A. T.; Jeanjaquet, S. L.; Mansfield, F. NACE Paper No. 74, presented at National Association of Corrosion Engineers National Meeting, Boston, 1985.
41. Lindqvist, S. A. Corrosion, 1985, 41(2), 69.
42. Padget, J. C.; Moreland, P. J. J. Coatings Technol. 1983, 55(698), 39.
43. Fernandez-Prini, R.; Kapista, S. J. Oil Col. Chem. Assoc. 1979, 62, 93.
44. Lindqvist, S. A.; Meszaros, L.; Svenson, L. J. Oil Col. Chem. Assoc. 1985, 68, 10.
45. Mayne, J. E. O. J. Oil Col. Chem. Assoc. 1957, 40, 183.
46. Mayne, J. E. O.; Cherry, B. W. Intern. Congr. Metallic Corrosion, 1st, London, Engl., Butterworths: London, 1961. p. 539.
47. Mayne, J. E. O.; Maitland, C. C. Off. Dig., Fed. Socs. Paint Technol. 1962, 34, 972.
48. Mayne, J. E. O. Trans Inst. Metal Finishing 1964, 41(4), 121.
49. Mayne, J. E. O.; Cherry, B. W. Off. Dig. Fed. Soc. Paint Technol. 1965, 37, 13.
50. Bacon, R. C.; Smith, J. J.; Rugg, F. M. Ind. Eng. Chem. 1948, 40, 161.
51. Cherry, B. W. Australas. Corr. Eng. 1974, 18(10), 23.
52. Guruviah, S. J. Oil Col. Chem. Assoc. 1970, 53, 660.
53. Bauman, K. Plast. Kautsch., 1972, 19, 455.
54. Kresse, P. Pigm. Resin Technol. 1973, 2(11), 21.
55. Haagen, H.; Funke, W. J. Oil Col. Chem. Assoc. 1975, 58, 359.
56. Funke, W.; Haagen, H. Ind. Eng. Chem. Prod. Res. Dev. 1978, 17, 50.
57. Funke, W. J. Oil Col. Chem. Assoc. 1979, 62, 63.
58. Floyd, F. L.; Frey, C. M. Org. Coatings Plastics Chem. 1980, 43, 580.
59. Floyd, F. L.; Groseclose, R. G.; Frey, C. M. J. Oil Col. Chem. Assoc. 1983, 66, 329.
60. Hulden, M.; Hansen, C. M. Prog. Org. Coat. 1985, 13, 171.
61. Kresse, P. Farbe Lack 1974, 80, 817.
62. Kresse, P. Ibid. 1977, 83, 85.
63. Pearson, R. G. J. Chem. Ed. 1968, 45, 581.
64. Pearson, R. G. Ibid. 1969, 45, 643.
65. Drago, R. S.; Wayland, B. B. J. Am. Chem. Soc. 1965, 87, 3571.
66. Drago, R. S.; Vogel, G. C.; Needham, T. E. J. Am. Chem. Soc. 1971, 93, 6014.
67. Sorensen, P. J. Coat. Technol. 1975, 47(602), 31.
68. Drago, R. S.; Parr, L. B.; Chamberlain, C. S. J. Am. Chem. Soc. 1977, 99, 3203.

69. Jensen, W. B. Chem. Rev. 1978, 78, 1.
70. Jensen, W. B. Rubber Chem. Technol. 1981, 55, 881.
71. Fowkes, F. M.; Mostafa, M. A. Ind. Eng. Chem. Prod. Res. Dev. 1978, 17, 3.
72. Manson, J. A.; Lin, J. S.; Tibureio, A. Org. Coat. Appl. Polym. Sci. Proc. 1982, 46, 121.
73. Vanderhoff, J. W.; Bennetch, L. M.; Cantow, M. J.; Earhart, K. A.; El-Aasser, M. S.; Huang, T. C.; Kang, M. H.; Micale, F. J.; Shaffer, O. L.; Timmons, D. W. Org. Coat. Appl. Polym. Sci. Proc. 1982, 46, 12.
74. Fowkes, F. M. in "Physicochemical Aspects of Polymer Surfaces", Mittal, K. L., Ed.; Plenum: New York, 1983. p. 583.
75. Fowkes, F. M. J. Polym. Sci. Polym. Chem. Ed. 1984, 22, 547.
76. Joslin, S. T.; Fowkes, F. M. Ind. Eng. Chem. Prod. Res. Dev. 1985, 24, 369.
77. Fowkes, F. M. Polym. Mat. Sci. Eng. 1986, 53, 560.
78. Parker, E.; Gerhart, H. Ind. Eng. Chem. 1967, 59(8), 53.
79. Walker, P. Off. Dig. Fed. Soc. Paint Technol. 1965, 37, 1561.
80. Funke, W. J. Oil Col. Chem. Assoc. 1985, 68, 229.
81. Rawlings, R. D.; Strivens, T. A. J. Oil Col. Chem. Assoc. 1980, 63, 412.
82. Rooum, J. A.; Rawlings, R. D. J. Mater. Sci. 1982, 17, 1745.
83. Rooum, J. A.; Rawlings, R. D. J. Coat. Technol. 1982, 54(695), 43.
84. Castle, J. E.; Watts, J. F. Ind. Eng. Chem. Prod. Res. Dev. 1985, 24, 361.
85. Kinloch, A. J., Ed. "Durability of Adhesive Bonds", Applied Science Publishers: London, 1983.
86. Minford, J. D. J. Adhesion 1985, 18, 19.
87. Smith, T. J. Adhesion 1984, 17, 1.
88. Drain, K. F.; Guthrie, J.; Leung, C. L.; Martin, F. R.; Otterbrun, M. S. J. Adhesion 1984, 17, 71.
89. Dodiuk, H; Drori, L.; Miller, J. J. Adhesion 1985, 18, 1.
90. Hill, L. W.; Wicks, Z. W. Prog. Org. Coat. 1982, 10, 55.
91. Harris, F. W.; Spinelli, H. J., Eds. "Reactive Oligomers", American Chemical Society: Washington, 1985.
92. Krishnamurti, K. Prog. Org. Coat. 1983, 11, 167.
93. Hopwood, J. J. J. Oil Col. Chem. Assoc. 1965, 48, 157.
94. Lerman, M. A. J. Coat. Technol. 1976, 48(623), 35.
95. Qaderi, S. B. A.; Bauer, D. R.; Holubka, J. W.; Dickie, R. A. J. Coat. Technol. 1984, 56(719), 71.
96. Woo, J. T. K.; Ting, V.; Evans, J.; Marcinko, R.; Carlson, G.; Ortiz, C. J. Coat. Technol. 1982, 54(689), 41.
97. New England Society for Coatings Technology Technical Committee J. Coat. Technol. 1981, 53(683), 27.
98. Ibid. 1982, 54(684), 63.
99. Lein, M. M.; Brakke, B.; Keltz, G.; Kiezulas, M. P.; Leavy, C. M.; Marderosian, R.; Withington, D. Ibid. 1983, 55(703), 81.
100. Schenk, H. U; Spoor, H.; Marx, M. Prog. Org. Coat. 1979, 7, 1.
101. Kordomenos, P. I.; Nordstrom, J. D. J. Coat. Technol. 1982, 54(686), 33.
102. Wismer, M.; Pierce, P. C.; Bosso, J. F.; Christenson, R. M.; Jerabek, R. D.; Zwack, R. R. J. Coat. Technol. 1982, 54(688), 35.

RECEIVED June 16, 1986

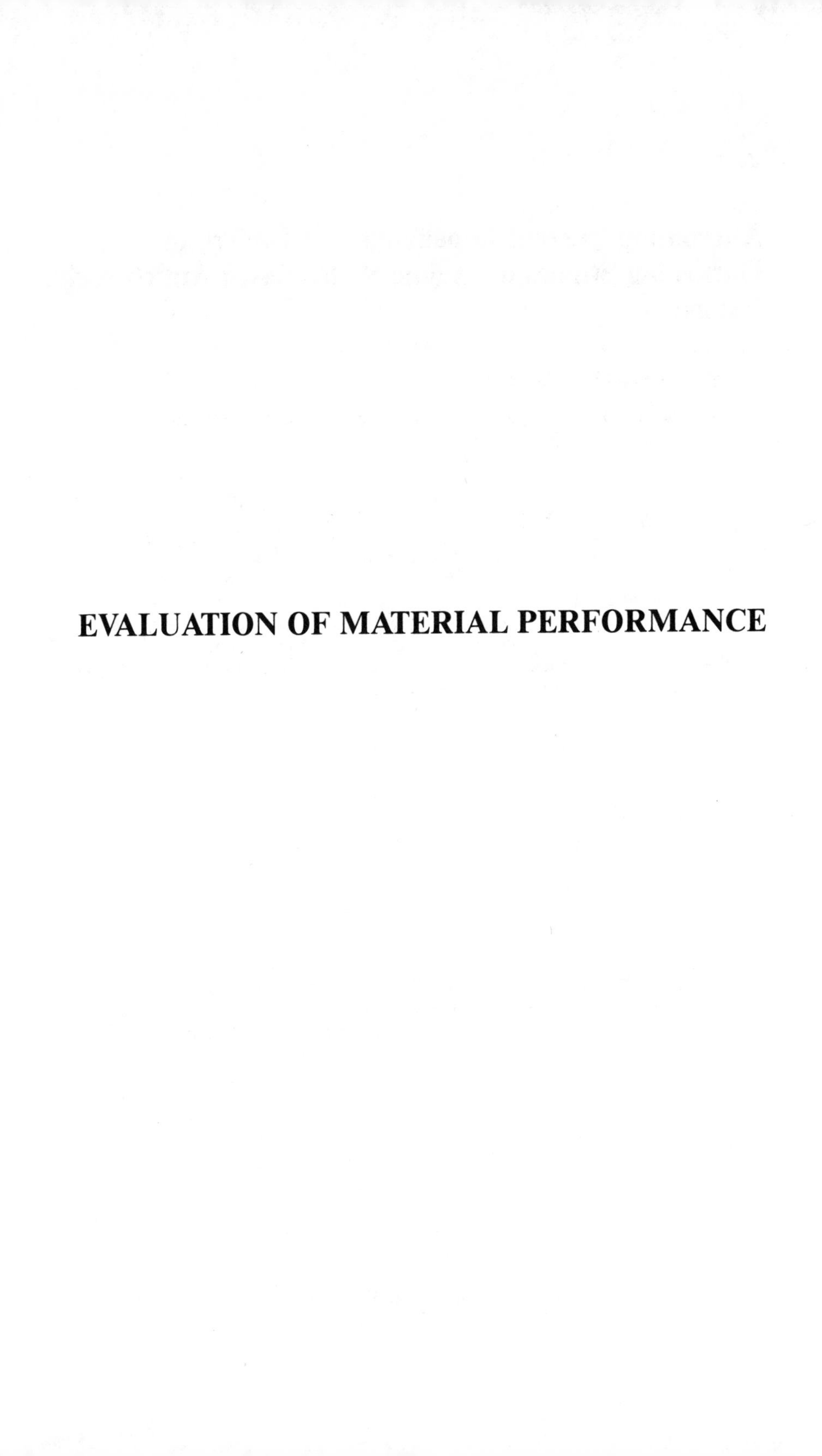

EVALUATION OF MATERIAL PERFORMANCE

2

Alternating Current Impedance and Underfilm Darkening Studies on Acidic Water-Based Anticorrosive Paints

P. J. Moreland and J. C. Padget

Imperial Chemical Industries PLC, Mond Division, Technical Department, P.O. Box 8, The Heath, Runcorn, Cheshire, England

The protective properties and interface reactions on mild steel substrates of an acidically formulated (pH5) water-borne paint based upon a chlorine containing vinyl acrylic copolymer have been examined using a variety of techniques. Traditional electrochemical polarisation curves as well as ac impedance studies were used to investigate the corrosion process in "wet" formulations associated with the occurrence or absence of "flash rusting". Investigation of an underfilm darkening phenomenon observed upon exposure testing of some similarly formulated coatings and associated with excellent long term protective performance are also presented. An arrest in a corrosion process, after some period involving insignificant metal loss was observed, evidenced the formation of a protective interface film. The characterisation of the film showed that its properties were in accord with the recognised protective performance of the coating system.

An acrylate modified vinyl chloride - vinylidene chloride latex copolymer (Haloflex 202) has been developed in our laboratory (1,2) specifically for the preparation of water-borne anti-corrosive primer paints. This carefully designed copolymer, hereafter referred to as a chlorine-containing vinyl acrylic copolymer, exhibits a low water vapour permeability (detached film) of approximately 100 fold less than that of typical acrylic latex polymers intended for the preparation of anti-corrosive primers, and when formulated into paint is capable of giving excellent anti-corrosive performance on smooth and blasted steel.

Chloropolymer latices differ from other latices in that they undergo a dehydrochlorination reaction at alkaline pH; the higher the pH the higher the rate of dehydrochlorination. Thus when such a latex is formulated into a paint at the typical paint pH (7-9)

0097-6156/86/0322-0018$06.00/0

there is a downward drift in pH and an increase in chloride ion concentration in the aqueous phase. Although the rate of dehydrochlorination can be reduced by reducing the chlorine content of the polymer, in our experience this reduction is at the expense of both the barrier properties and the anti-corrosive performance. We have shown that the rate of dehydrochlorination of the high chlorine content copolymer is negligibly small at pH $\leq$ 4.5. Accordingly acidic paint formulations were developed (1,3) (see Appendix), which exhibited very little change in pH or chloride ion concentrations during storage. Such acidic paint fomulations require the dispersed components (ie polymer and pigment particles) to be strongly sterically stabilised if they are to remain colloid stable. Ethylene oxide-propylene-ethylene oxide block copolymers were found to be particularly suitable as they increased the rate and extent of latex particle coalescence due to a surface plasticization effect(4), without downgrading the barrier properties or anti-corrosive performance. We recently reported an ac impedance study on this system (4).
Although acidic paint formulations based on the chlorine-containing vinyl acrylic latex copolymer give excellent anti-corrosive performance, they do exhibit two unusual features not present in the corresponding alkaline formulations :

a) The extent of flash rusting on grit blasted mild steel during the "wet" film condition progressively decreases with decreasing pH, being extensive at pH $\geq$7 and being very slight at pH $\leq$5; this effect is, in our experience, irrespective of latex chemical type.

b) When applied to a mild steel substrate, some acidic paints (eg pH 4.5) exhibit a darkening effect at the interface between the substrate and the dry film. This underfilm darkening is visually observable on detatching the dried film from the substrate. The time taken for this darkening to develop varies according to exposure conditions; being longer for example during natural exposure than during the ASTM salt spray test.

An important point to note is that the underfilm darkening is not associated with any fall off in film performance in terms of adhesion, cross-cut undercutting, blistering or exfoliation.

The objective of the work described in this paper is to obtain an understanding of the above two effects.

The role played by formulation pH during the "wet" film conditions present during flash rusting can be studied using electrochemical techniques such as potential monitoring, dc potentiodynamic scanning and ac impedance since the presence of soluble ions in the paint aqueous phase gives rise to adequate conductivity. The substrate darkening which can take place beneath dry paint films was studied by both an electrical resistance method and an iron pick-up method (to measure interface metal loss), ac impedance and conventional accelerated performance tests (to characterise protective performance of the coatings) and the nature of the underfilm darkening was chemically characterised using a range of modern surface analysis techniques.

Experimental

Substrates, Paints and Coatings. Both flash rusting and underfilm darkening studies exployed a white, pH 4.5, primer formulation (see Appendix), designated as Standard, based upon Haloflex 202, a chlorine-containing vinyl acrylic latex. A zinc phosphate free formulation, designated as Non-Standard, of pH6 was prepared by substitution of zinc phosphate for barytes. The comparison paints were a commercial butyl acrylate-methyl methacrylate water borne primer, formulated at pH 9, and a solvent based chlorinated rubber primer.

The paints were spray applied to a number of substrates. Two coatings gave a final dry film thickness between 85 to 130 microns. For ac impedance studies the coatings were applied to shot (steel) blasted cold rolled steel plates. For interface metal-loss studies by electrical resistance the coatings were applied to 10 micron Fe foils (99.85 purity), adhesively attached by Araldite 2003 epoxy paste on one side to grit blasted glass plates.

Grit (Al_2O_3) blasted mild steel Q panels, spray coated on both sides, were used for iron pick-up studies. Similar panels were used for chemical characterisation.

Measurement Techniques. DC polarisation curves on freshly abraded mild steel in bulk paints were determined using a traditional 3-electrode potentiodynamic technique. A 50 ml cell employed a disc mild steel electrode (area 0.33 cm^2), saturated calomel reference and platinum counter electrode. Polarisation curves were made at a scan rate of 2V/Hr between -950 to -450 mV vs sce.

AC impedance measurements were also made in bulk paints. A Model 1174 Solartron Frequency Response Analyser (FRA) with a Thompson potentiostat developed ac impedance data between 10 KHz and 0.1 Hz at the controlled corrosion potential E_c. The circuit has been described in the literature(5).

In underfilm darkening studies, ac impedance data was obtained for coated grit-blasted steel panels exposed to 3% NaCl using the FRA in a 2-electrode setup(4).

The electrical resistance (ER) method to monitor physical changes in an electrically conducting material is well known(6). 10 cm x 2.5 cm coated iron foils of 10 micron thickness were electrically resistance monitored during exposure to various corrosive environments to follow metal thickness loss at the paint/metal interface. The circuit shown in Figure 1 allowed foil resistance increase ΔR, which is directly related to metal thickness loss Δd according to Equation (1), to be determined within 0.003 microns.

$$\Delta R = - \frac{R_t \Delta d}{d_o} \quad (1)$$

where R_t = foil resistance at time t and d_o = initial thickness.

The resistance was determined by measuring the voltage drop across the foil caused by passage of a brief controlled current of 100 mA.
A micro computer system allowed voltage and current measurements to be synchronised plus data logging and averaging of measurements. Alongside each "measurement" foil was a "control" foil, coated with paint plus a protective epoxy coating, which did not corrode and allowed resistance measurements to be normalised.

Results and Discussion

Flash Rusting (Bulk Paint and "Wet" Film Studies). The moderate conductivity (50-100 ohm-cm) of the water borne paint formulations allowed both dc potentiodynamic and ac impedance studies of mild steel in the bulk paints to be measured. (Table I). AC impedance measurements at the potentiostatically controlled corrosion potentials indicated depressed semi-circles with a Warburg diffusion low frequency tail in the Nyquist plots (Figure 2). These measurements at 10, 30 and 60 minute exposure times, showed the presence of a reaction involving both charge transfer and mass transfer controlling processes. The charge transfer impedance θ was readily obtained from extrapolation of the semi-circle to the 'real' axis at low frequencies. The transfer impedance increased with exposure time in all cases.

At 60 minutes only, dc potentiodynamic curves were determined from which the corrosion current I_c was obtained by extrapolation of the anodic Tafel slope to the corrosion potential. The anodic Tafel slope b_a was generally between 70 to 80 mV whereas the cathodic curve continuously increased to a limiting diffusion current. The curves supported impedance data in indicating the presence of charge transfer and mass transfer control processes. The measurements at 60 minutes indicated a linear relationship between I_c and Θ^{-1} of slope 21mV. This confirmed that charge transfer impedance could be used to provide a measure of the corrosion rate at intermediate exposure times and these values are summarised in Table 1.

Table I shows that the corrosion rate in the Standard pH 4.5 paint was larger than that in the Non-Standard (zinc phosphate free) paint at pH 6, with all rates decreasing with time. The rates decreased with further increase of paint pH to 8. The increased tendency of flash rusting with increase of pH from 6 to 8 was, therefore, associated with lower substrate corrosion currents. Comparison between Non-Standard and Standard paint adjusted to pH 6 with NH_3 showed little difference in corrosion rates implying that pH was more influential than the presence of zinc phosphate at this pH. Adjustment of the Non-Standard paint with H_2SO_4 to an equivalent pH 4.5 of the Standard paint showed good agreement between corrosion rates. This result also indicated pH to be more influential than the presence or absence of zinc phosphate with regard to corrosion currents. The absence of flash rusting at pH 4.5 is therefore associated with higher corrosion currents.

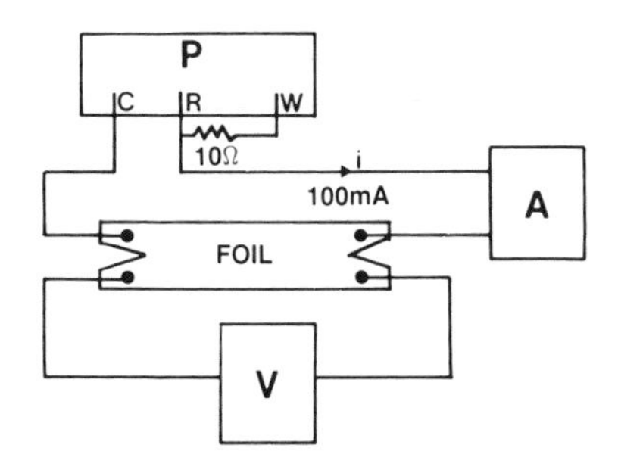

Figure 1. Circuit for electrical resistance measurement of foils.

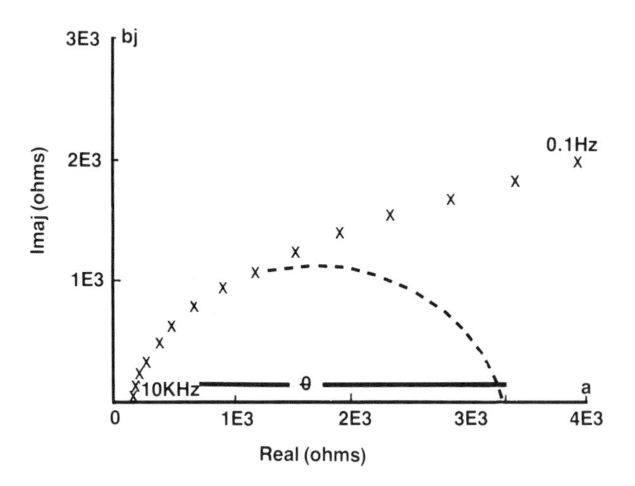

Figure 2. Nyquist plot for mild steel in standard paint.

Table I. Summary of Electrochemical Data in Bulk Paints

Paint	pH	E_c (sce) volt	Corrosion Rate** (inch/yr x 10^3) 10 min	30 min	60 min	Θ* (ohm)	I_c μA	b_a mV
1-STANDARD	4.6	-0.68	14.9	10.6	9.6	3100	6	80
2-NON-STANDARD	6.0	-0.69	7.9	7.1	6.5	4600	3	80
1-pH adjust (NH_3)	6.0	-0.67	8.3	6.2	5.9	5100	4.5	70
2-pH adjust (NH_3)	8.0	-0.71	3.3	3.2	3.1	9500	1	60
2-pH adjust (H_2SO_4)	4.5	-0.67	14.9	10.6	9.9	3000	6	110

*Θ is the charge transfer impedance on 0.33 cm^2 electrode area at 60 minutes.
**Corrosion Rate = B/Θ where B = 21mV and 1μA/cm^2 = $4.7x10^{-4}$inch/yr

In cooperative studies with Sykes and Lewis(7), potential-time measurements were determined on grit blasted steel panels coated with "wet" films of Standard and Non-Standard Paints. At pH5 the active corrosion potential relating to an oxide free condition is rapidly established (Figure 3). However, at pH 7 a passive or "oxide-covered" potential is initially exhibited, which after a slow decay, undergoes a rapid decrease to the oxide-free film potential. The interpretation of these phenomena is that at pH 5 the majority of the surface, originally being oxide covered metal, rapidly suffers general dissolution at a negative film-free potential. At pH 7 the majority of the surface is oxide covered but the embedded iron grit is more active and undergoes local dissolution. This localised dissolution gives rise to isolated rust spots (ie flash rusting) by the process described later. Localised grit particle dissolution eventually appears to lead to reductive dissolution of the surrounding oxide-covered metal and associated rapid decrease of corrosion potential.

It is therefore believed that at pH 6 and greater the corrosion process is localised and large local concentrations of ferrous iron are achieved. At pH 6 the oxidation to ferric iron is very rapid (8) and precipitation of $Fe(OH)_3$ occurs to exhibit localised corrosion or "flash-rust" spots. At pH 5 and below a small but finite uniform dissolution of the iron substrate occurs. However, in this pH range the oxidation of the ferrous dissolution product to ferric ion is considerably slower, by almost 1000 times, and hence "flash rusting" is not observed.

Underfilm Darkening

These studies were conducted to examine the nature and behaviour of substrate discolouration observed on mild steel coated with the Standard formulation paint upon exposure to aggressive environments.

Electrical Resistance - Interface Metal Loss. Figure 4 shows the electrical resistance increase and corresponding foil thickness loss determined for the three coating systems during constant immersion in 3% NaCl. The pH 9 acrylic coating showed a marked foil thickness loss of 4 microns within 20 days which continued with exposure and was accompanied by visual gross deterioration with surface rusting and blistering. However, the pH 4.5 chlorine containing vinyl-acrylic coating exhibited much smaller metal loss, which approached an arrest after the first 10 days exposure. The coating showed some localised rust spots after the longer period of 50 days. The chlorinated rubber coating exhibited no metal loss in the 50 days and no discolouration or rust spots were observed. Alternate immersion exposure gave similar behaviour (Figure 5) to constant immersion (Figure 4).

From the metal loss determinations it was possible to construct the mean corrosion rate - time curves for the acrylic and vinyl-acrylic systems. From Figure 6 it is clear that an initial interface metal loss process in the chlorine-containing vinyl-acrylic system quickly arrests within 12 days exposure, and the overall metal loss at 50 days is only of the order of 0.75 micron. The acrylic system shows no such arrest and indeed the corrosion rate increases with time to destruction of the foil.

The low metal loss figures for the chlorine containing vinyl acrylic system noted above were supported by analytical measurements of iron pick-up in coatings on grit blasted Q panels. Under constant immersion in 3% NaCl the iron pick-up, equivalent to 0.13 micron metal loss, remained virtually unchanged after 14 days over the 63 days exposure. However, the acrylic system had an equivalent metal loss of 0.55 micron at 14 days which increased steadily to 1.43 microns at 63 days.

AC Impedance. The grit blasted panels used in these studies were identical to test panels used to compare coating performance by traditional methods. AC impedance data in the form of Nyquist plots is shown in Figure 7 for the chlorine-containing vinyl-acrylic coating exposed under alternate exposure to 3% NaCl over 51 days. The initial impedance was high at $> 10^8$ ohm on 10 cm^2, but the apparent film resistance decreased upon exposure and exhibited a minimum between 9 and 23 days exposure. The film's behaviour can be modelled as a Randles equivalent RC circuit in which ionic film resistance R decreased and film capacitance C increased with exposure time up to 9 days. The depressed semi circular behaviour at 4 and 9 days indicated a dispersion in the time constant RC for the film. Between 9 and 23 days it is evident that the film resistance increased and continued to do so up to the 51 day measurement. Similar behaviour was obtained for this type of coating under constant immersion conditions (Figure 8) though the recovery to higher impedance at 51 days was not as marked. In all cases, even at minimum impedance values the film resistance was high at $> 5 \times 10^7$ ohms on unit area and compared favourably with the resistance criterion for a protective coating (9).

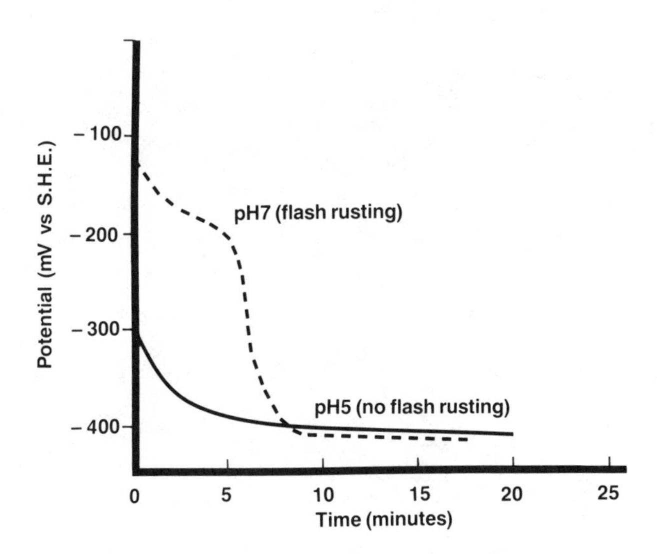

Figure 3. Potential-time behaviour of grit blasted mild steel in drying paint film at pH5 and 7.

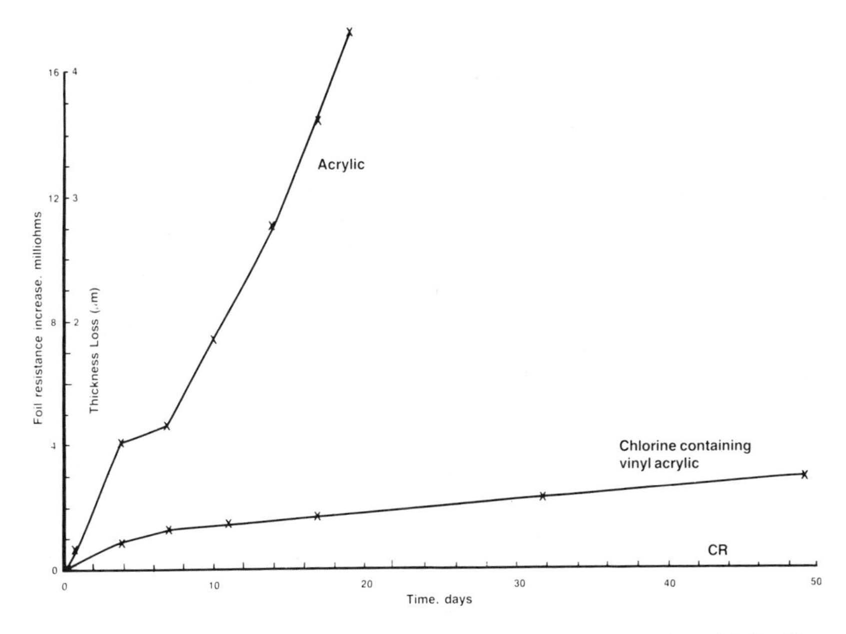

Figure 4. Foil resistance changes-constant immersion 3% NaCl.

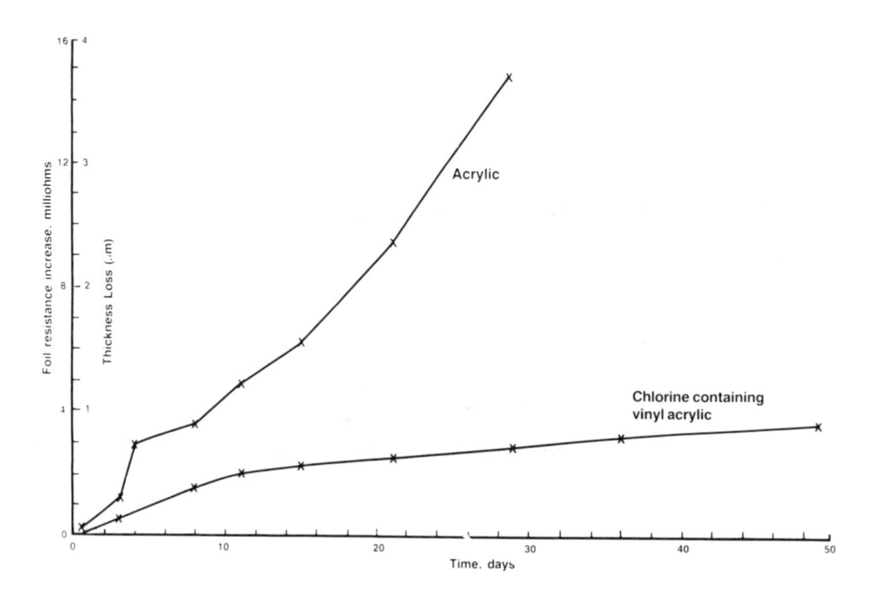

Figure 5. Foil resistance changes-alternate immersion in 3% NaCl.

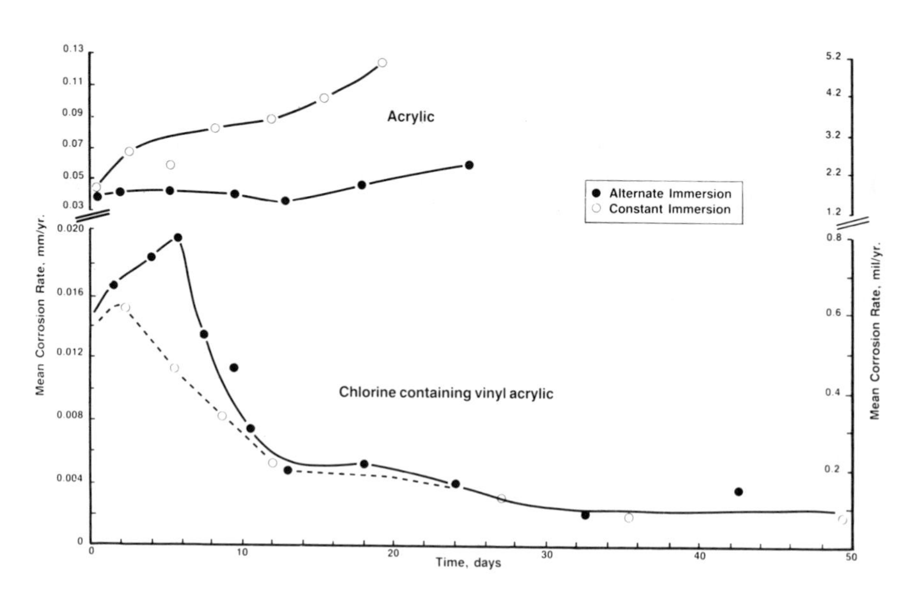

Figure 6. Mean corrosion rate-time curves.

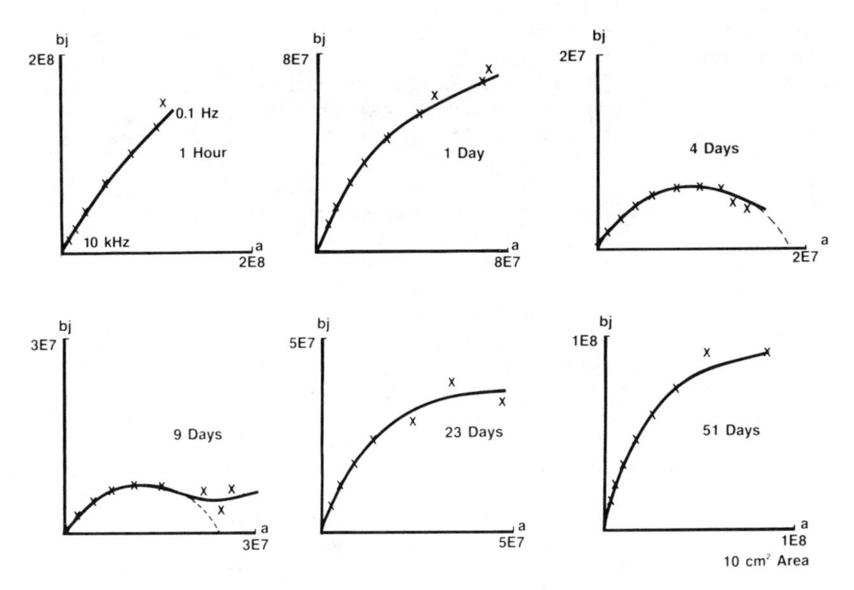

Figure 7. Nyquist plots - chlorine containing vinyl acrylic coating alternatively immersed in 3% NaCl.

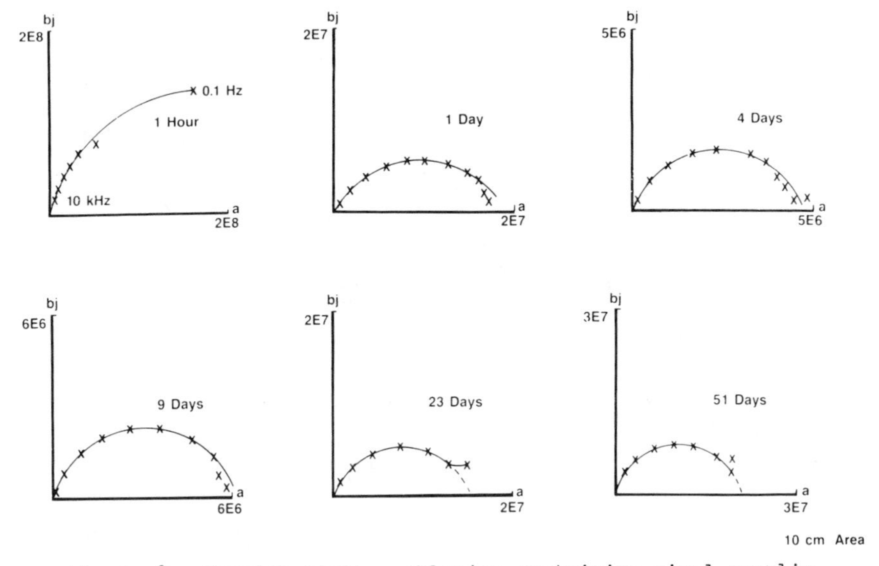

Figure 8. Nyquist plots - chlorine containing vinyl acrylic coating constantly immersed in 3% NaCl.

The initial decrease in ionic film resistance and increase in capacitance can be associated with either NaCl electrolyte or water entry into the film. From ER measurements this period is associated with a metal loss process at the substrate surface. However, between 9 to 23 days the ionic film resistance increases, which is associated with an arrest in metal loss at the substrate surface in ER measurements. It appears, therefore, that with the knowledge of an underfilm darkening phenomenon occurring at the substrate/coating interface, a film of a protective (ie passive or high ionic resistance) nature is produced during exposure.

As shown in Figure 9 the impedance of the acrylic coating immediately showed low values which did not increase. The coating showed marked rusting and exfoliation. Chlorinated rubber coatings maintained a high impedance similar to that of the chlorine-containing vinyl-acrylic coatings though the development of a pinhole after long exposure led to a lower impedance as shown in Figure 9.

The three coating systems were also exposed to hot salt spray. In this case, it appeared that the minimum impedance of the chlorine-containing vinyl-acrylic coating occurred within the first 5 hours exposure and thereafter the impedance remained high (>10^7 ohms). This behaviour is probably due to fast entry of electrolyte and/or water into the film under the more aggressive conditions to form an interface film. As in previous experiments the acrylic coating had low impedance (<1000 ohm) which did not increase, and chlorinated rubber retained a high impedance throughout (>10^8 ohm) unless a pinhole developed.

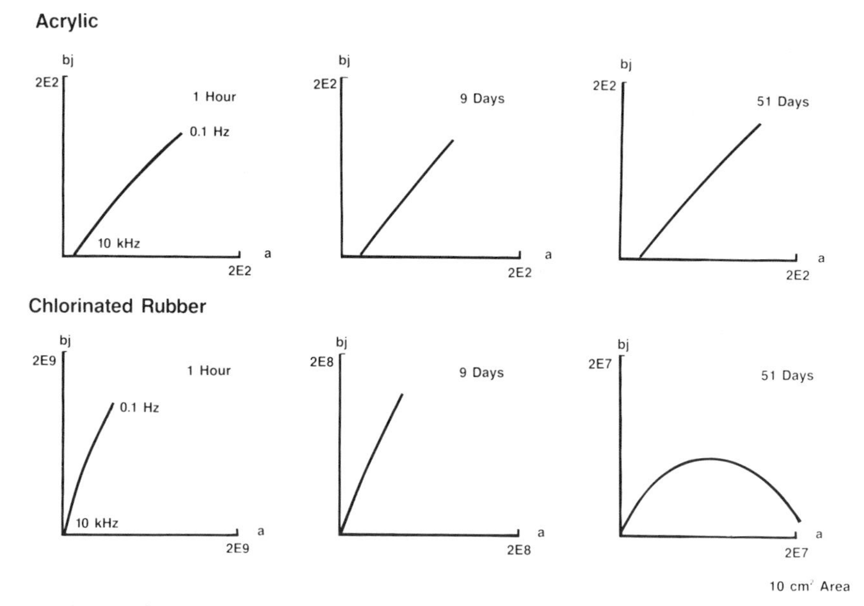

Figure 9. Nyquist - acrylic and chlorinated rubber coatings alternatively immersed in 3% NaCl.

Chemical Characterisation. Chemical characterisation of the underfilm darkening beneath chlorine-containing vinyl-acrylic films was sought using a variety of techniques on the substrate surface as well as the backside of stripped films. The substrates were grit blasted and plain mild steel Q panels exposed up to 98 days in hot salt spray and examined within hours of removal.

XRD identified Fe_3O_4 plus paint constituents of $BaSO_4$, TiO_2 and $Zn_3(PO_4)_2$ together with a less identifiable major phase (7.7Å, 2.68Å, 2.36Å). This phase is now believed to be a member of a class of compounds referred to as the pyroaurite group. These compounds have the general formula:

$$M_x\ R_y\ (OH)_{2x+3y-2z}\ (A^{2-})_z.2H_2O$$

where M is a divalent cation, R is a trivalent cation and A is an anion, commonly CO_3^{2-}, but can be OH^- Cl^- and possibly others (10). The members of this group form hexaganol plate type crystals of which pyroaurite has the formula $Mg_6\ Fe_2\ (OH)_{16}\ (CO)_3\ 3\ H_2O$. The layer structure of such compounds allows the accommodation of a variety of anions and cations (10) and the large number of hydroxyl groups may provide a buffering capacity. This buffering capacity has been recognised as a probable influential factor in the protective property afforded by similar hydrotalcite type films on aluminium in sea water (11). The buffering properties could also constrain localised attack and promote lateral movement of the corrosion process and film formation.

SEM studies supported the above in observation of platelet type crystals containing Fe and Cl (by EDAX). ESCA revealed little detail but SIMS identified a number of hydroxy and oxychloride species including Fe(OH)+, FeO+, FeO-, FeOCl- and $FeCl_3$ to support the presence of chloride in the pyroaurite type film. LIMA indicated a number $Fe_x O_y$ + peaks with x as y as high as 2 or 3 whereas x and y are generally 2 for FeOOH.

Conclusions

Flash rusting exhibited in neutral to alkaline water borne formulations appears to occur through a localised corrosion process probably involving grit "activity" present from blasting, either directly or indirectly, in an electrochemical process. At such pH the rapid oxidation of ferrous to ferric ion produces intense local precipitation of ferric hydroxide evidenced as flash-rust spots. The process can be eliminated by formulating at a lower pH, eg pH 4.5 which gives rise to a uniform corrosion process at the substrate surface. It has been shown that a chlorine containing vinyl acrylic coating can be satisfactorily formulated at this pH. Under this condition, oxidation to ferric ion with subsequent precipitation does not occur and hence flash rusting is not observed.

It appears that under aggressive corrosive conditions electrolyte may enter the film and stimulate corrosion but the chlorine containing vinyl acrylic coating quickly promotes formation of a protective film, as evident by ER, ac impedance and iron pickup measurments, with insignificant overall metal loss. Indeed, ac impedance measurements on typical substrate surfaces indicated film resistances to remain very high even during formation of the interface film. The long term protective field performance behaviour of such coating reflects the protective character of this system. However, water borne systems formulated at pH 7-9 eg conventional acrylics are both capable of producing flash rusting during coating and incapable of producing in-situ protective films in the presence of corrosive environments.

Literature Cited

1 Burgess, A J; Caldwell, D; Padget, J C; JOCCA, 1981, 64, 175.
2 US Patent 4,341,679.
3 European Patent 0,035,316.
4 Padget, J C; Moreland, P J; J Coating Technol, 1983, 55, 39.
5 Piens, M; Verbist, R; In "Corrosion Control by Organic Coatings", NACE, Houston, Texas, 1980; p 163.
6 Dravnieks, A; Cataldi, H A; Corrosion, 1954, 10, 224.
7 Sykes, J; Lewis, G; unpublished work at Oxford University.
8 Minegishi, T; Asaki, Z; Higuchi, B; Konds, Y; Met Trans B, 1983, 14B, 17.
9 Bacon, R C; Smith J J; Rugg, F M; Ind Eng Chem, (1948), 40, 161.
10 Taylor, H F W; Mineralogical Mag, 1973, 39, 377.
11 Austing, C E; Pritchard, A M; Wilkins, N J M; Desalination, 1973, 12, 251.

Appendix

Primer Formulation Based on Chlorine Containing Vinyl Acrylic Latex

Ingredient	% w/w	Ingredient	% w/w
Vinyl Acrylic Copolymer (Haloflex 202)	59.7	Micronised Zinc Phosphate	5.8
Non-ionic block copolymer surfactant	3.1	Micronised barytes	15.6
		Titanium dioxide	2.6
Hydroxy-propyl methyl cellulose	0.2	Water	10.8
De-foamer	0.2	pH	5
Butyl glycol	2.0	Pigment volume concentration	20

The name Haloflex is a trademark, the property of Imperial Chemical Industries PLC.

RECEIVED March 5, 1986

3

New Methods in Electrochemical Assessment of Polymer Coatings on Steel

J. P. Lomas, L. M. Callow, and J. D. Scantlebury

Corrosion and Protection Centre, University of Manchester Institute of Science and Technology, Sackville Street, Manchester, M6O 1QD, England

Harmonic Analysis has been recently studied at UMIST as a corrosion monitoring test following the original theoretical work by Meszaros. This study looks at the possibility of using this technique to enhance the accuracy of Impedance based corrosion rate measurements on painted metal electrodes. A series of 12 chlorinated rubber coated specimens (of 2 different formulations) have been allowed to degrade under immersion for over 3 years and exhibit a variety of defects. This paper aims to show whether a combination of Harmonic Analysis and Impedance tests, can give a realistic ranking of the degree of degradation. Some specimens have remained intact and the Impedance measurements are purely capacitive, such specimens do not exhibit a stable rest potential and no additional information can be obtained from Harmonic Analysis. Those specimens showing certain forms of corrosion attack, give corrosion rates that are determinable using both techniques. In other instances, the results obtained are widely divergant.

The use of the impedance technique in the study of polymer coated steel, has been thoroughly described elsewhere(1). The present paper compares this technique with that of harmonic analysis, originally proposed by Meszaros(2). The authors have presented preliminary data using the latter technique(3) wherein the early stages of polymer breakdown have been studied. The current paper extends this work to polymers which have been immersed for a considerable period of time. The harmonic method gives information not available from the impedance technique in the Tafel slopes and the corrosion current are directly measurable. A brief summary of the harmonic method and the equations used are given below.

A small sinusoidal perturbation is applied potentiostatically to the system under investigation and the resulting current sine wave is analysed in terms of its second and third harmonics (i_2 and i_3), i_1 being the fundamental. The corrosion current is calculated

0097–6156/86/0322–0031$06.00/0

from the following equation:

$$i_{corr} = \frac{i_1^{\;2}}{\sqrt{48}\,\sqrt{2i.\,i_3 - i_2^{\;2}}}$$

and the Tafel slopes a and c from equations of the type shown below.

$$\frac{1}{\beta a} = \frac{1}{2U_o}\;\frac{i_1}{i_{corr}} + 4\,\frac{i_2}{i_1}$$

where U_o = magnitude of the perturbing sine wave.

Experimental

12 chlorinated rubber coated specimens were prepared, 6 with an aluminium flaked pigment and 6 with a grey mica pigment. These were originally 6 panels, cut in half after painting. Details of the preparation and masking leaving an exposed paint area of 34 cm^2, are given elsewhere [1]. The panels were immersed in artificial sea water for 1042 days as part of a separate series of experiments.

Harmonic analysis was carried out on the specimens 7 days after the impedance measurements in order to allow the specimens to settle down again. An Ono Sokki CF 910 dual channel FFT analyser was used in conjunction with a potentiostat (Thompson Ministat 251) to hold the specimen at its rest potential and to provide the low frequency sine wave perturbation. The second channel was used to measure the harmonic content of the resulting current. The Ono Sokki produces a digitially generated high purity sine wave at a chosen frequency, in this instance, 0.5 Hz. The total harmonic content of the input sine wave was less than 0.45% measured over 10 harmonics. Only the first 3 harmonics are used to calculate the corrosion current.

For each specimen an average over 122 cycles were taken, but because of the high impedances present, a perturbing voltage of 115 mV was used. This is somewhat higher than the theoretical value arrived at by Meszaros[2] but necessary because of the low level of harmonic signal from the specimens. The visual appearance of each electrode was noted for comparison with the electrochemical data.

Results and Discussion

For certain specimens, no stable rest potential could be measured and for these electrodes, the potential was held at -610 mV (SCE) designated by @.

It can be seen from Table 1 that all those specimens which could not support a stable rest potential, also showed impedance plots that were of a purely capacitive type. This impedance was also obtained from other specimens of both pigmentation, which exhibited stable rest potentials (A12 & G6). The diameters of the semicircles (given as R_{sc}) are all large, classically indicative of highly resistive films and no corrosion reaction. Regardless of whether the plot is purely capacitive or shows semicircular behaviour, the capacitance values may be attributed to that resulting from the polymer film itself. With the exception of G4, which gave an anomalous value of 320 pF/cm^2.

Table 1. Impedance and Rest Potential Data

Specimen		Potential (SCE)	Impedance plot	(Ohms cm2) Rsc	Capacitance pF/cm^2
Al	1	-390	 1E8	5.6E7	18
	2	-505	 1E7	-	-
	3	-590	 3E7	1.8E7	35.5
	4	-640	 2E7	2.2E6	-
	5	-631	 4E5	1.1E5	67
	6	@-610	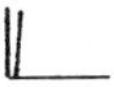 1E7	-	36.6
Grey	1	@-610	 1E7	-	35.8
	2	@-610	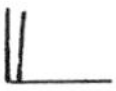 1E7	-	31.7
	3	@-610	 1E7	-	44.6
	4	-550	 2E5	1.6E5	320
	5	@-610	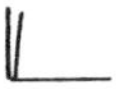 1E7	-	53.3
	6	-680	 1E7	-	-

It can be seen that it was again difficult to obtain results from specimens where no stable rest potential could be measured. The harmonic currents in all cases were low and for certain specimens were of the same order as the distortion resulting from the input sine wave. The Tafel slopes obtained were in general anomalously high and the corrosion rates varied over several orders of magnitude.

As the impedance and harmonic analysis techniques gave different types of data from each other, a direct comparison between Tables 1 and 2 is difficult. However, it should be anoted that G4 gave both semicircular behaviour and a reasonably high corrosion rate. This similarity is also true for Al4 and Al5. Al1 and Al3 showed smaller but still measurable corrosion rates, together with semicircular impedance behaviour and the absolute values measured varied similarly as before.

Any electrochemical technique that could be employed as a means of rapid coating degradation assessment should ideally be able to correlate with the actual observable degradation. The visual appearance of the specimens is outlined in Table 3. This appearance has been relatively stable for the past 600 days.

It may be seen from this table that of the Al flake pigmented

Table 2. Harmonic Analysis Data

Specimen		Potential (SCE)	Fundamental (First harmonic (mV)	Second Harmonic (mV)	Third Harmonic (mV)	i_{corr} (μA)	b_a	b_c
Input sine wave		-	117	.01	.01	-	-	-
Al	1	-390	15.9	.58	.17	.18	275	204
	2	-505	6.57	.2	.15	.005	171	145
	3	-590	87.15	.24	.12	4.69	724	666
	4	-640	5.58E3	181	15	135	706	393
	5	-630	366	.4	.2	17.6	1025	986
	6	@ 610	-	-	-	-	-	-
Grey	1	@ 610	-	-	-	-	-	-
	2	@ 610	-	-	-	-	-	-
	3	@ 610	27	1	1	.16	133	114
	4	-550	1.003E3	1.8	1.4	30.9	742	569
	5	@ 610	-	-	-	-	-	-
	6	-680	161.4	.42	.11	7.3	984	907

Table 3. Visual Appearance of Specimens

Specimen		Description
Al	1	1 x 1 mm + 1 x 3 mm diameter orange spots
	2	2 x 1 mm diameter black spots
	3	2 x 2 mm diameter orange spots
	4	Orange "rosette" 1.5 cm total diameter
	5	no corrosion
	6	"
Grey	1	"
	2	"
	3	"
	6	"
	5	"
	4	Paint blistered, very rusty

specimens, Al4 exhibited extreme breakdown in the form of a corrosion rosette, whilst G4 had blistered and suffered considerable corrosion attack. The remaining specimens either showed no observable degradation or exhibited small, stable and therefore relatively insignificant point site attack.

Comparison between the impedance and harmonic data shows that the rosette was detectable in the impedance data by the emergence of Warburg type behaviour at low frequencies although no quantifiable data as to the rate of this reaction could be obtained. In the harmonic data, the corrosion rate was the highest of those measured, at 135 μA.

Conclusions

It can be seen that for severely degraded specimens, both the harmonic analysis and impedance techniques are capable of detecting the presence of gross corrosion. The harmonics method provides a reasonable estimation of the corrosion rate when the impedance data exhibits Warburg type behaviour. For less severely degraded specimens, especially those exhibiting blister attack, the impedance method is not as successful as the harmonic analysis technique. Where very little corrosion attack has occurred, neither method is capable of providing reliable quantitative data. The non-applicability in certain instances of the impedance technique as a monitoring tool has been reported previously[4]. Further experience with the harmonic analysis technique may be capable of refining the results obtained.

Acknowledgments

Drs. Callow and Lomas would like to thank International Paint plc for financial support. All authors would like to thank Dr. R.P.M. Procter for provision of laboratory facilities.

Literature Cited

1. Callow, L.M. and Scantlebury, J.D., J.O.C.C.A., 64, 83 (1981).
2. Gill, J.S., Callow, L.M. and Scantlebury, J.D. Corrosion, 39, 61 (1983).
3. Devay, J. and Meszaros, L. Acta Chim. Acad. Sci., Hungary, 104, No. 13. 311 (1980).
4. Lomas, J.P., Callow, L.M., Scantlebury, J.D. and G.A.M. Sussex. Presented at 160th Meeting of Electrochem. Soc., Denver, Col. (Oct 1981).

RECEIVED January 22, 1986

4

Application of Electrochemical Noise Measurements to Coated Systems

D. A. Eden[1], M. Hoffman[1], and B. S. Skerry[2]

[1]Corrosion and Protection Centre, University of Manchester Institute of Science and Technology, Sackville Street, Manchester, M60 1QD, England
[2]Sherwin Williams Company Research Center, 10909 South Cottage Grove Avenue, Chicago, IL 60628

This paper describes the application of novel electrochemical techniques to studies of paint films on steel substrates exposed to aqueous environments.

Simultaneous monitoring of the self-generated electrochemical potential and current noise using analogue and digital techniques has been evaluated as a tool for monitoring coating performance. These data obtained have been compared with those from a.c. impedance techniques.

Laboratory measurement procedures used for electrochemical data acquisition and analysis during the monitoring exercise are outlined, and particular emphasis is placed on the electrochemical noise techniques. Electrochemical current noise has been monitored between two identical electrodes and the potential noise between the 'working' electrodes and a reference electrode.

Digital noise measurements have been obtained by use of a microcomputer controlling the sampling rate of a sensitive digital voltmeter employed to measure the potential or current fluctuations. The subsequent analysis of the derived time records is described.

Analogue noise measurements have been made using high gain amplifier/ filter circuits which permit examination of low frequency fluctuations on a 'real-time' basis.

Electrochemical noise monitoring techniques have been used previously in studies of corrosion processes occurring on metals in a variety of environments. Initially, work was directed towards the monitoring of potential noise fluctua-

0097-6156/86/0322-0036$06.00/0

tions, and was used particularly in the identification of the onset of localised attack (i.e. pitting or crevice type attack) [1-3]. Current noise measurements have been used in the studies of electrocrystallisation [4] and pitting [5] with the specimens being held under potentiostatic control.

Recent work [6,7] has been directed towards the simultaneous monitoring of potential and current noise, where the current noise signal is generated by coupling two nominally identical electrodes with a zero resistance ammeter (ZRA), and the potential noise of the couple is monitored with respect to a reference electrode. In this manner no externally applied signal is required.

The potential noise signal provides information pertaining to the type of attack, whereas the current noise provides data which indicate the rate of corrosion and the type of attack. When used in parallel, the two noise measurements may be used to estimate the polarisation resistance of the interface being examined.

When applied to coated metals, the fluctuations observed in the current noise signal are generally low in magnitude with the baseline of detection essentially being limited by the sensitivity of the electronic interface. For the studies cited, the lower limit of the current noise signal is some 10 pico-amps.

For the purposes of this study the responses of a variety of intact and defective coatings were monitored and the results are compared with a.c. impedance data.

The a.c. impedance technique is useful for monitoring changes occurring in coated systems, and the various types of response may be summarised briefly as follows:

a) Intact coatings (no pores) very high impedance produces almost purely capacitive response, difficulty in estimating d.c. component of resistance.
b) Intact coatings (as (a)), with water uptake capacitance increases due to dielectric constant changes.
c) Coatings with minor defects usually produce well defined response with resistive as well as capacitive components.
d) Coatings with major defects show response in which complex behaviour is observed, the coating response moving to higher frequencies due to smaller values of resistance, and in addition, charge transfer and diffusion effects may become evident.

Instrumentation

Digital electrochemical noise. The digital instrumentation used for the noise studies comprised the following:

A Hewlett Packard HP85 Microcomputer
A Hewlett Packard 3478A Digital Voltmeter
A "custom built" multiplexer

A schematic diagram for the experimental set up is illustrated in Figure 1.

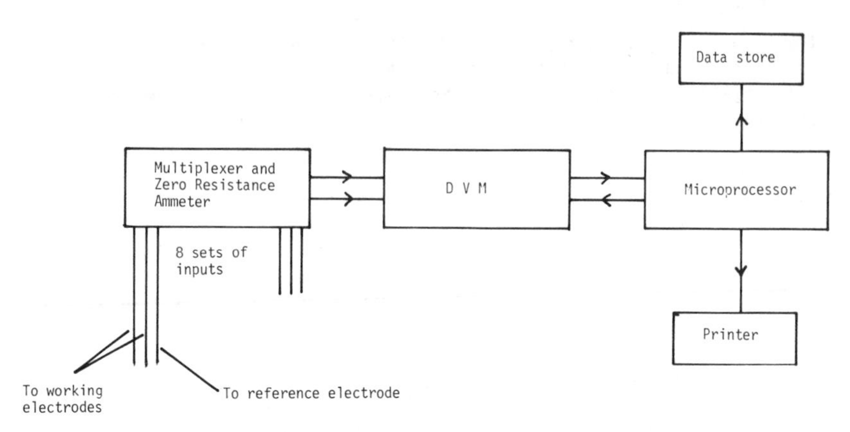

Figure 1. Schematic diagram for digital noise measurements using multiplexed electrodes.

The input multiplexer was designed to allow multi-channel capability and was configured to monitor both potential and current noise fluctuations sequentially on a maximum of eight pairs of samples.

The sampling rate of the digital voltmeter (DVM) was controlled by the microprocessor and channel selection for monitoring was obtained by utilising a pulse output from the DVM.

Time records of the coupling current and potential for the respective samples were obtained and stored for further analysis.

Analogue electrochemical noise. Analogue instrumentation monitoring low frequency noise in a specified bandwidth was used for the analogue measurements. The schematic diagram (Figure 2) illustrates the basic configuration of the instrumentation. The rms values of the noise signals were logged and sent as a 0 - 10V signal to a conventional chart recorder. The signal sensitivity corresponding to the full scale

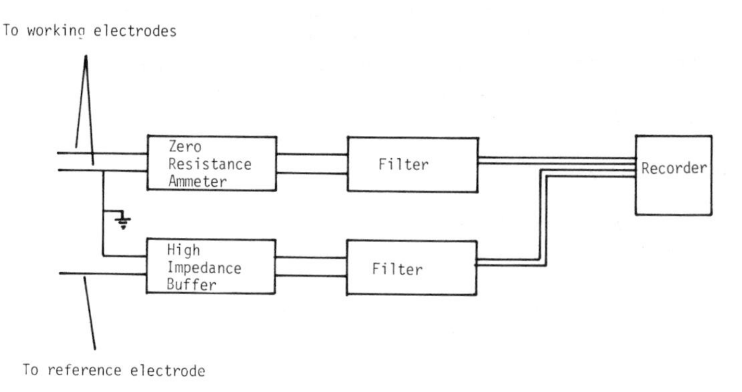

Figure 2. System for analogue current and potential noise monitoring.

response of the chart recorder was 1μV minimum (0V output) to 10mV (10V output) covering four decades at 2.5V per decade.

The current noise signal was monitored by using a sensitive, low noise zero resistance ammeter (ZRA) to couple pairs of identical electrodes; the ZRA acting as a current to voltage converter. This derived potential signal was then fed into a potential noise monitor.

A.c. impedance. Impedance measurements were made using a Solartron 1250 frequency response analyser under computer control using a Hewlett Packard HP85 microcomputer and commercially available software. The coatings were studied in the three electrode mode using a Thompson Ministat. Figure 3 illustrates schematically the experimental arrangement.

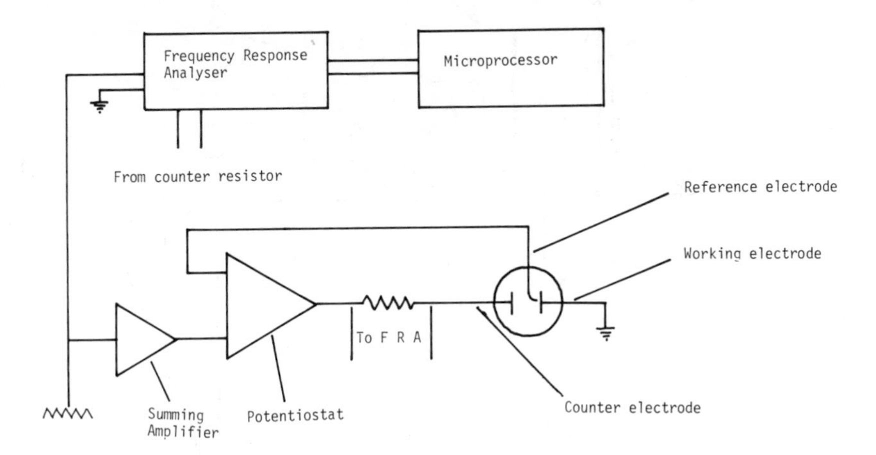

Figure 3. Working arrangement for 3 electrode electrochemical impedance studies.

Ssmple preparation

For the purposes of this study a variety of coatings applied to mild steel substrates were used. The coatings were chosen to provide a range of protection from poor to excellent. The coatings studied were:

1.	Polyurethane (unpigmented)	1 coat ~ 40μm
		2 coats ~ 80μm
2.	Polyurethane (pigmented)	1 coat ~ 45μm
		2 coats ~ 90μm
3.	Bitumen	1 coat ~ 20μm
3.	Bitumen over zinc rich paint	1 coat ~ 20μm

Experimental

Plastic cells of dimensions 5 x 5 x 7.5cms were fixed to the coated specimens using silicone rubber sealant. The silicone rubber was allowed to cure for at least two days prior to filling with electrolyte. Three per cent sodium chloride solution in demineralised water was added to the cells which were prepared as identical pairs. Coupling between the pairs of electrodes was achieved using a sodium chloride/agar salt bridge. Potentials of the specimens were monitored using silver/silver chloride reference electrodes. A platinum counter electrode was introduced into individual cells when monitoring the a.c. impedance response. A typical cell arrangement is shown in Figure 4.

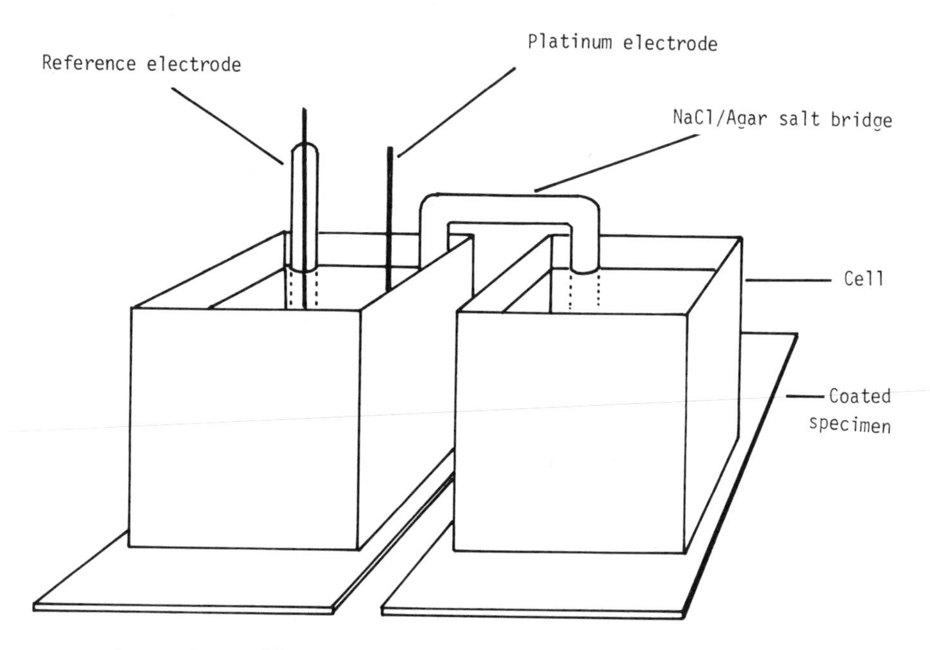

Figure 4. Cell arrangement for electrochemical studies.

During the period of immersion of the samples in sodium chloride electrolyte, electrochemical noise measurements were made using the electronic apparatus previously described. The time records obtained were analysed using statistical techniques to derive mean, standard deviation and coefficient of variance.

The derived value of polarisation resistance was evaluated from the ratio of the standard deviation of the potential noise signal to the standard deviation of the current noise signal, i.e.:

$$\frac{\sigma V}{\sigma i}$$

Data is presented graphically to illustrate the variation in d.c. potential (Figure 5) mean d.c. coupling current, $\bar{i}$ (Figure 6) and $\frac{\sigma V}{\sigma i} \approx Rp$ (Figure 7) with time.

Typical analogue noise traces are illustrated in Figures

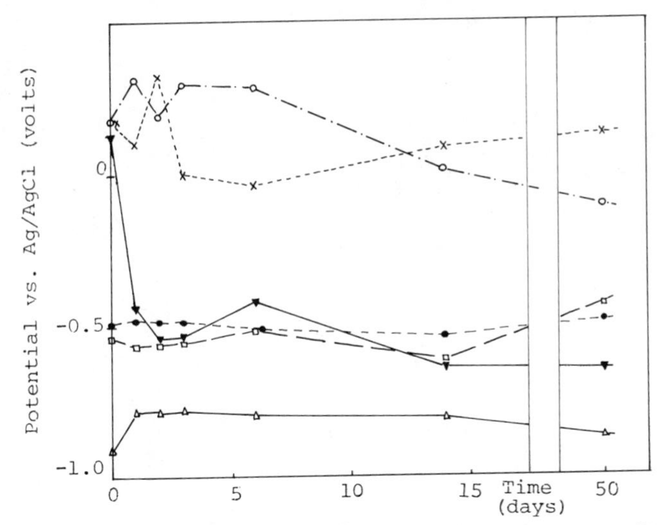

Figure 5. Potential vs time for coated specimens in 3% NaCl. Key: ●, bitumen; △, Zn rich + bitumen; □, polyurethane, one coat (unpigmented); ▼, polyurethane, two coats (unpigmented); O, polyurethane, one coat (pigmented); and X, polyurethane, two coats (pigmented).

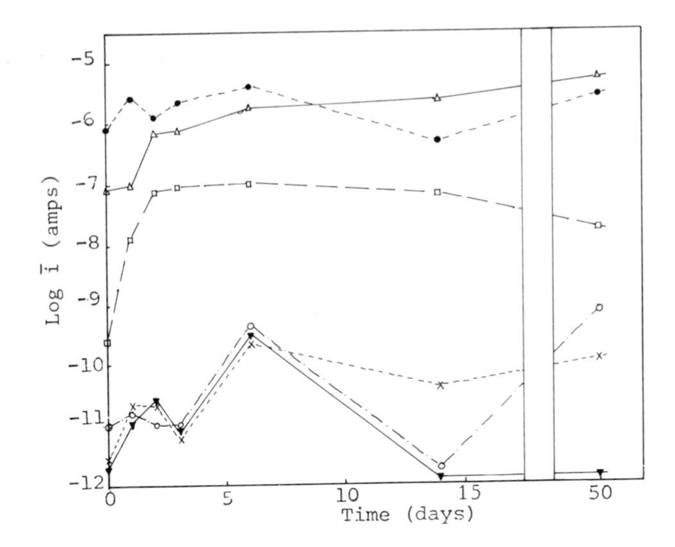

Figure 6. Log $\bar{i}$ vs time for coated specimens in 3% NaCl. Key: same as for Figure 5.

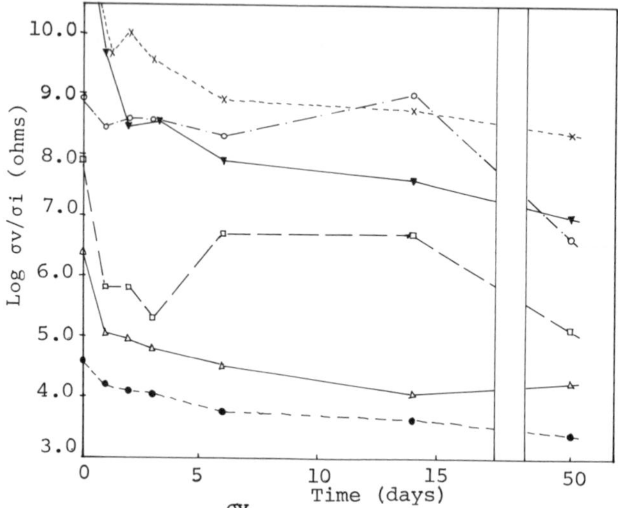

Figure 7. Log $\frac{\sigma v}{\sigma i}$ vs time for coated specimens in 3% NaCl.

8 and 9. Impedance data for typical cells are presented in Figures 10, 11 and 12.

Discussion

From the data obtained for the different specimens it can be seen that there is significantly different behaviour between the poor, porous coatings (bitumen) and the polyurethane paint samples. Of the polyurethane samples only one showed any evidence of corrosion beneath the coating during the duration of the test and this was an unpigmented single coat specimen.

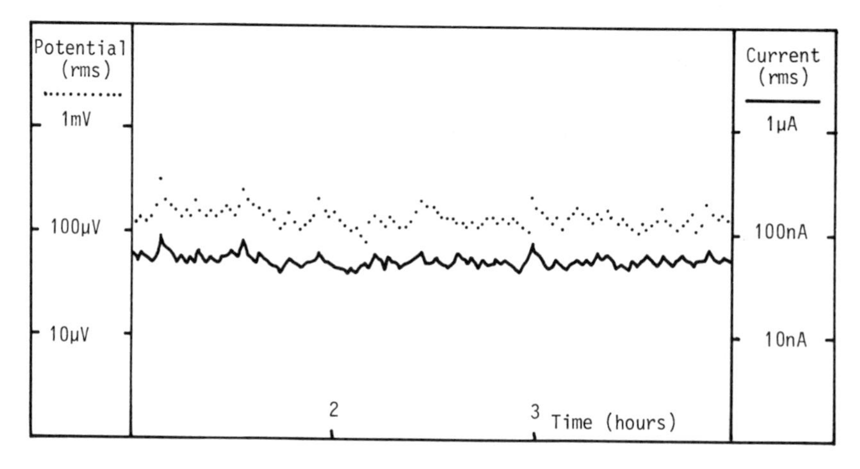

Figure 8. Analogue potential and current noise traces for bitumen on mild steel. Day 1.

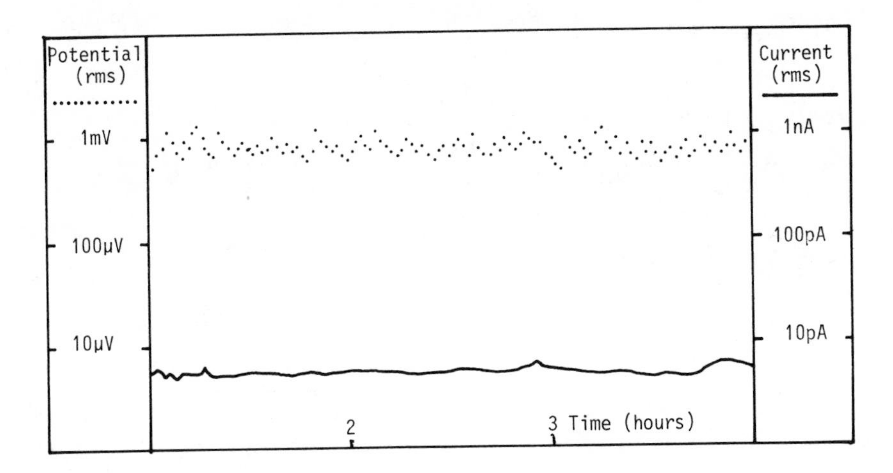

Figure 9. Analogue potential and current noise traces for polyurethane 2 coats unpigmented. Day 2.

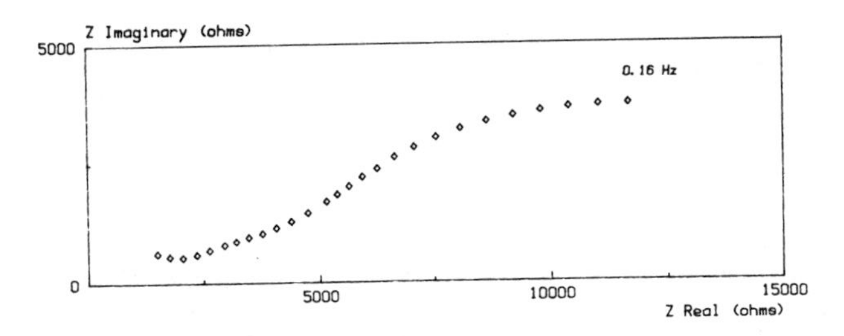

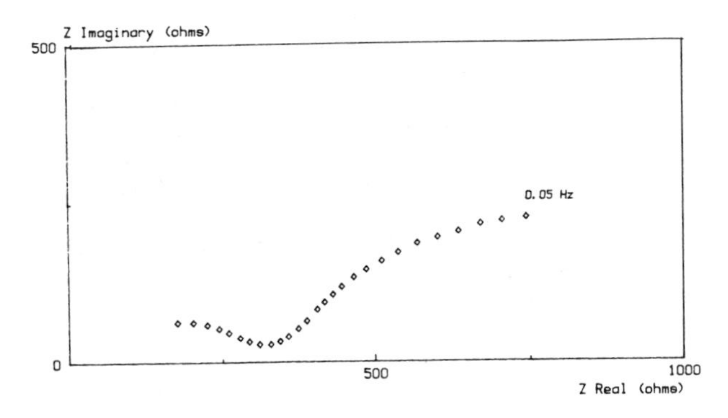

Figure 10. Nyquist plots for bitumen coated mild steel. Day 0 and Day 50.

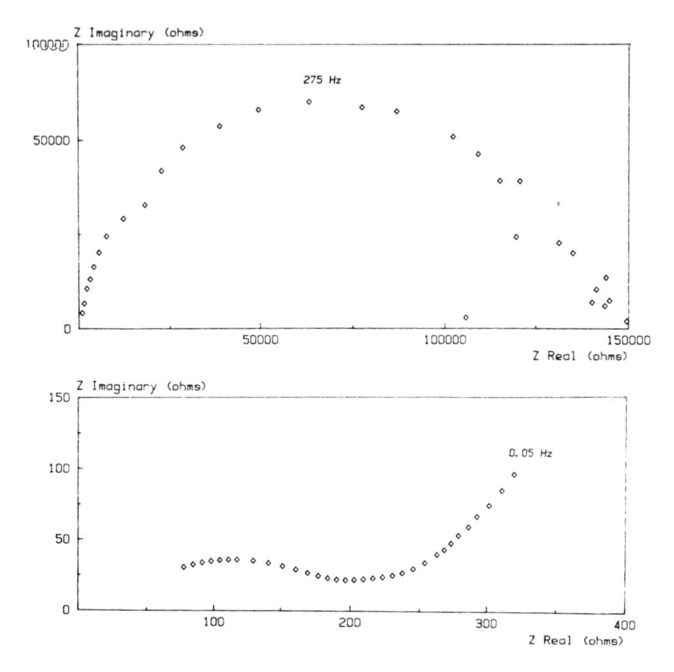

Figure 11. Nyquist plots for zinc rich and bitumen on mild steel. Day 0 and Day 50.

Even so, the low frequency impedance as derived from the noise measurements was still two to three orders of magnitude higher than the bitumen coated system. The better coatings exhibited low frequency impedances some four to five orders of magnitude higher than the bitumen.

The d.c. potentials, however, only indicated whether the material being studied was in a corrosion regime, both the bitumen and unpigmented single coat polyurethane assumed very similar potentials over the period of the test, even though the corrosion rates were grossly different.

The impedance data illustrated in Figures 10, 11 and 12 have been chosen to illustrate the widely differing behaviour of the different coating systems.

In Figure 10, which shows the impedance behaviour of the bitumen coating on mild steel, it is apparent that at day 0, the coating is immediately showing signs of major defective areas, with the impedance response being governed by what appear to be diffusive effects. The response at high frequencies probably being due to the coating itself. The resistance of the system at this stage is greater than 15000 ohms. After 50 days' exposure, the impedance response has changed to one indicating charge transfer and diffusion effects with a resistance greater than 1000 ohms. In comparison, the behaviour of the bitumen coating on zinc rich paint (Figure 11) indicates that at day 0, the coating is showing

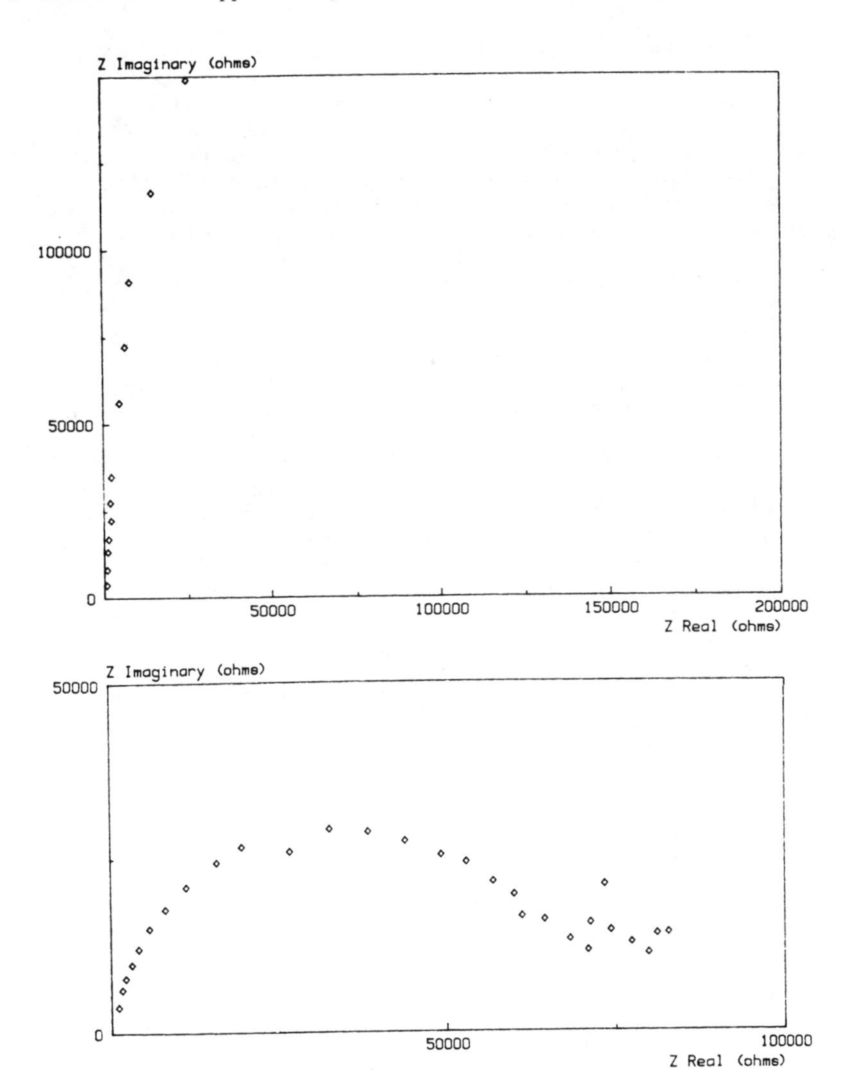

Figure 12. Nyquist plots for polyurethane (1 coat) unpigmented system after 50 days, illustrating coating breakdown on Panel B.

minor defects with an impedance of some 150,000 ohms typical of a coating response, although exhibiting a free corrosion potential (-806mV), which is indicative of a porous coating. After exposure for 50 days this system is showing totally different behaviour with charge transfer and diffusion effects becoming much more evident.

Figure 12 illustrates the difference in impedance behaviour between the two samples of unpigmented polyurethane (applied at ~ 40μm) after 50 days exposure. Panel A gave a response indicative of a good, intact coating (almost purely capacitive), whereas panel B was showing signs of breakdown, with an estimated resistance of some 75,000 ohms.

With the noise techniques, both analogue and digital, no externally applied signal is required, and measurement of the fluctuations around the free corrosion potential provides all the information. The noise technique is useful in that it allows a fairly rapid estimation of the electrochemical impedance of the system being studied, whereas, with for instance, a.c. impedance techniques, very often the minimum frequency studied is still not low enough to provide sufficient information to allow an accurate estimation of the impedance.

With electrochemical noise measurements the d.c. potential of two coupled identical electrodes is governed by the sample with the lowest impedance. It is this lower value of impedance which is monitored by the noise technique, i.e. that of the worst coating of the pair.

For the systems studied, it is interesting to note that the mean level of coupling current also appears to be very useful as a means of studying high impedance systems, but this can cause problems if the current fluctuates around zero and changes polarity. Generally, it would appear to be a better approach to utilise the value of standard deviation of the current signal as a measure of corrosion rate. The coefficient of variance for the current signal gives some indication of the stability of the d.c. coupling current.

If we consider the analagous noise equations derived for electronic components at the low frequency end of the spectrum, one of the equations used to describe the noise is:

$$V_\eta = K_1 \sqrt{\frac{1}{f}} \cdot I_{d.c.} \cdot Rs \qquad (1)$$

where:

$I_{d.c.}$ = d.c. current flowing through device
Rs = source resistance
K1 = constant
f = frequency

Correspondingly the equation for the current noise is:

$$I_n = K_1 \sqrt{\frac{1}{f}} \cdot I_{d.c.} \qquad (2)$$

If we utilise the above equations to describe the low frequency noise signals observed with electrochemical systems, it is apparent that the potential noise signal will provide information pertaining to the value of the Stern Geary constant since:

$$i_{corr} = \frac{B}{R_p}$$

where:

i_{corr} = corrosion current
B = Stern Geary constant
R_p = polarisation resistance

and hence: $Vn = K_1\sqrt{\frac{1}{f}} \cdot B$

whereas the current noise signal will provide information relating to the corrosion rate. It is therefore, not surprising that the low frequency potential noise signals only tend to vary over a few decades, whereas the current noise signals may vary over many orders of magnitude.

Conclusions

1. Electrochemical noise measurements have shown great promise as a monitoring tool in studies of corroding metals in a variety of environments.
2. The application of these sensitive techniques to evaluate the performance of coated specimens would appear to be appropriate for the study of slow corrosion processes and also for the monitoring of coating breakdown/degradation. Since the noise signals are generated by the specimens themselves coating failure is accompanied by a change in the electrochemical noise signal which gives a rapid indication of the state of the coating. Statistical analysis of the data provides a rapid method of assessing the noise levels without the necessity for transposition of the data into the frequency domain by, for instance, FFT techniques.
3. Simultaneous monitoring of current and potential noise and derivation of low frequency values of impedance allows, in some instances, direct comparison with polarisation resistance values derived from, for example, a.c. impedance techniques.

Literature Cited

1. Hladky, K., and Dawson, J.L., Corr. Sci 22, p317 (1981).
2. Hladky, K., and Dawson, J.L., ibid, 23, p231 (1982).
3. Dawson, J.L, Hladky, K., and Eden, D.A., Paper presented at "On line Monitoring of Continuous Process Plant", London, June 1983.
4. Bindra, P., Fleischmann, M., Oldfield, J.W. and Singleton, D., Discussion of Faraday Soc. 56 (1974).
5. Williams, D.E., Westcott, C., Fleischmann, M., Passivity of Metals and Semi Conductors, p217-228, Elsevier Science publishers, Ed. M. Froment.
6. Farrell, D.M., Cox, W.M., Stott, F.H., Eden, D.A., Dawson, J.L., and Wood, G.C., High Temperature Technology Vol. 3, No. 1, February 1985.
7. John, D.G., Hladky, K., Eden, D.A., and Dawson, J.L., Paper presented at Research Sciences Symposium NACE/Corrosion 84, New Orleans, April 1984.

RECEIVED March 6, 1986

5

Electrochemical Characterization of Photocured Coatings

R. Bartoszek-Loza and R. J. Butler

Research and Development Laboratory, The Standard Oil Company, 4440 Warrensville Center Road, Cleveland, OH 44128

Correlations were observed upon electrochemical evaluation of solvent-borne, high nitrile, polymer-based coatings. High nitrile coatings are well known for their barrier properties. The compositions were applied to Bonderite 40 coated steel panels and photocured at room temperature. Dramatic increases in rust rating were observed upon post-thermal treatment of N-vinyl pyrrolidone-based photocured coatings. Furthermore, the open circuit potentials of these coatings correlate with salt spray data over a broad postbake temperature range (25°C-220°C). The same trends were observed for gamma-butyrolactone-based coatings although the extent was not as dramatic. This indicates that electrochemical measurements can be utilized to systematically characterize photocured high nitrile polymer-based coatings.

Metallic corrosion is an electrochemical process associated with the flow of current between surface sites having a difference in electrochemical potential. The assessment and evaluation of organic coatings to prevent metal corrosion has traditionally been accomplished through salt fog testing (ASTM B-117) and long term exposure tests in particular service environments. Electrochemical techniques have often been considered (1), but are not routinely employed in practice.

Absolute correlations between service performance and electrochemical measurements do not appear frequently in the literature. Based on 300 test systems, Bacon and coworkers (2), correlated electrochemical resistance with exposure time. Recently, Mills (3) also observed a correlation between salt fog corrosion and electrochemical resistance. We have found open circuit potential measurements to be extremely useful for the routine evaluation of high-nitrile polymer-based photocured coatings.

Organic coatings function as either inhibitors, sacrificial coatings or barriers (4). While inhibitor and sacrificial coatings protect the substrate from deterioration by preferential corrosion of the coating system, barrier coatings function by isolating the substrate from the corrosive environment. High nitrile polymers are known to possess high resistance to water and oxygen permeation (5).

0097-6156/86/0322-0048$06.00/0

These barrier properties are believed to result from strong, compact -CN dipoles. Close packing of the polymer chain produces powerful intermolecular forces (6).

Metallic corrosion can be characterized by two electrochemical quantities, current and potential. The current associated with a single electrode reaction on a metal surface is related to the potential of the metal by:

$$E = a + b (\log I)$$

where E is the potential (volts), I is the current (amps) and a and b are constants (7). At equilibrium, all anodic and cathodic reactions proceed at an equal, finite rate. The net current flow is zero. The voltage corresponding to this zero net current is the open circuit or corrosion potential. As applied voltage is changed, the resulting current can be recorded to produce a polarization curve (8).

Electrochemical Evaluation of Corrosion Protection by Organic Coatings

In 1979, Leidheiser (9) reviewed the use of corrosion potential measurements with regards to the prediction of corrosion at metal-organic coating interfaces. Wolstenholme had last reviewed this literature in 1970 (10). Work in the 1930-1940's focused on the magnitude of the corrosion potential and how it changed with time (11-14). Negative potentials with respect to uncoated substrates were indicative of corrosion beneath the coating. Positive potentials with respect to uncoated substrates were indicative of the absence of corrosion.

Anomalous cases were noted in which this generalization did not hold. These very empirical measurements were followed by more thorough studies (15). Thin paint films with very low electrical resistance show active corrosion potentials which become more positive as the paint film was increased in thickness. Shapes of the potential/time curves were misleading as a guide to ultimate coating protective properties.

Kendig and Leidheiser (16) electrochemically evaluated thin (9 micron) polybutadiene coatings on steel. They concluded that movement of the corrosion potential in the noble direction was indicative of an increasing cathodic/anodic surface area ratio. Oxygen and water penetrate the coating to produce the cathodic reaction at the metal/coating interface.

Experimental

The electronic components for the measurements consisted of EG&G Model 173 Potentiostat equipped with slow sweep option (0.1 mv/sec) and EG&G Model 376 Logarithmic Current Converter. An EG&G Model 175 Universal Programmer supplied the waveform for running the polarization experiment. The output from the electrometer of the 173 and the log output of the 376 were connected to a Hewlett-Packard Model 7035B X-Y Recorder and the potential plotted versus log current.

Open circuit potentials (potential at zero current) were recorded after a period of 10 minutes at which point the readings were constant for ca. 5 seconds. Sign conventions described in ASTM G5-72 (17) are used to report the data. Polarization curves were obtained 250 mv anodic and cathodic to the open circuit potential.

The electrolyte solution was 0.5M NaCl (distilled water) as described in ASTM G5-72. All potentials are reported relative to a saturated calomel electrode (SCE). A conventional three compartment corrosion cell was used. A polytetrafluoroethylene sleeve inserted through the 24/40 joint holds the test specimen with a known geometric surface area of the specimen exposed to salt solution. An O-ring in the joint eliminates the influence of edge effects. Electrical connection is made to the test sample with a spring loaded wire in contact with the rear (unexposed) side of the specimen. A graphite rod introduced through the top of the cell served as the counter electrode. A saturated calomel electrode, with circuit completed through a Luggin-Haber capillary approximately 1 mm from the sample, served as reference electrode.

All nitrile-based coatings reported in this study were applied to Bonderite 40 coated steel (B40) panels (150-300 mg/ft^2 zinc phosphate pretreatment; The Parker Company). A commercial high nitrile polymer (Barex 210) was employed as the base resin. N-vinylpyrrolidone (Aldrich) and gamma-butyrolactone (Aldrich) were employed as reactive diluents.

All organic coated B40 panels were photocured for the same length of time ensuring the same amount of cure. Panels were then baked at temperatures ranging from 40°C to 220°C (in 10°C increments) for 5, 10 or 15 minutes. Two 5/16" discs were punched from each panel for electrochemical analysis. They were placed in the Teflon cell holder described above. The organic coated B40 panel edges were then masked and placed under the salt fog environment for 24 hours. Corrosion performance was evaluated using ASTM D 610-68 (18). A rating of 10 was given for no appreciable rust. A rating of 0 was given for 100% rusting. The scale is logarithmic between the two extreme endpoints.

Results

Bonderite 40 - Open Circuit Potential Measurements

Twelve different Bonderite 40 coated steel (B40) panels were examined to provide a statistically valid value for the open circuit potential. Their average rest potential was -0.578 V (vs. SCE) with an average deviation of 20 mv. After recording the open circuit potential, polarization curves were obtained (Figure 1).

Corrosion Resistance of the Photocured Coatings

The coating compositions containing 25 weight percent Barex 210 resin (B210) in either N-vinyl pyrrolidone (NVP) or gamma-butyrolactone (GBL) were applied to the B40 panels and photocured at room temperature. The corrosion resistance (18) for these two systems was vastly different. The B40 panels showed 100% rusting (an ASTM rust rating of 0) after 24 hours salt fog exposure. The

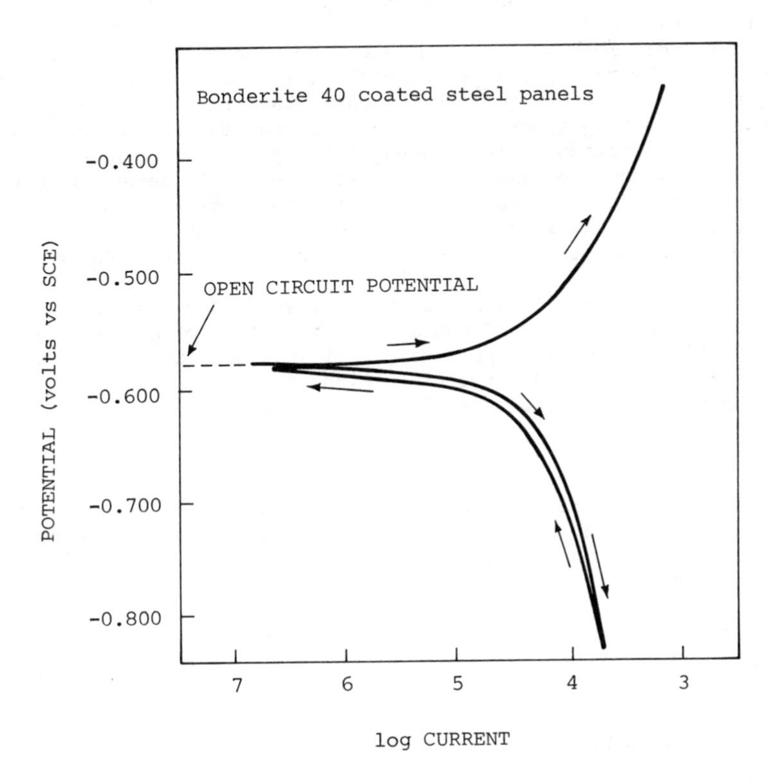

Figure 1. Polarization curve for Bonderite 40 coated steel panels.

B210/GBL system showed no rust (an ASTM rust rating of 10) after 24 hours exposure. However, approximately 50% surface rust (an ASTM rust rating of 1) was observed for the B210/NVP system.

Post-Thermal Treatment of the Photocured Coatings

After the B210 systems were post-thermally treated, they were assessed electrochemically and by 24 hour exposure to the salt fog environment. Post thermally treated B40 panels show no dependence of post thermal treatment temperature on either corrosion performance (all panels showed 100% rusting) or open circuit potential (Figure 2).

For the B210/NVP system, little change in rust rating occurred as the post-thermal treatment temperature increased from 40°C to 90°C (Figure 2). At temperatures greater than 90°C, the rust rating increased, reaching a maximum at 140°C, decreased to a minimum at 170°C, then increased again. When the open circuit potentials of the thermally treated samples was plotted versus temperature, this same behavior was observed. That is, the rest potential tracked the rust rating with the same minimum and maximum. This behavior was observed independent of the time (5, 10 or 15 min) that the sample was heated.

Exposure of the B210/GBL post-thermally treated system to the salt fog environment for 24 hours gave no rusting of any of the panels. All panels had a rust rating of 10. When the open circuit potentials were plotted (Figure 3), minor changes in potential were observed which paralleled the B210/NVP system.

Discussion

In a search for reliable accelerated test methods for determining coating performance, electrochemical techniques have often been explored. The corrosion resistance of a coated steel panel is a composite of the steel quality, its surface finish and the quality of the coating. For this reason, Bonderite 40 coated steel panels were included in our work. They were employed primarily to aid in the interpretation of the electrical measurements for the nitrile-based photocured samples.

Salt Fog Corrosion Resistance

The 24 hour salt fog corrosion resistance for the photocured B210/NVP and B210/GBL systems were vastly different. Using NVP as a diluent, 50% rusting of the sample was seen, while the GBL diluent showed no rusting. This difference is attributed to the inherent water solubility of polyvinylpyrrolidone (19).

Previous work has shown that thermally cured high nitrile polymer coatings have good thermal resistance (20). Thus, the B210/NVP photocured panels were then heated in an attempt to improve their performance. The 24 hour rust rating of these thermally treated test panels varied depending on the temperature of the treatment (Figure 2).

in the salt spray was electrochemically tested. In general, the "good" lot of steel panels exhibited passivation (i.e. nearly constant current) over most of the 250 mv anodic scan. The "poor" lot exhibited loss of passivation. Examination of the respective rest potentials indicated the "poor" steel lot had more noble rest potentials than the "good" steel lot.

Our electrochemical work differed drastically from the Groseclose work in that polymer coated metal samples were employed. Furthermore, we found that coatings can have corrosion resistance when their rest potentials are either more noble (B210/NVP) or less noble (B210/GBL) than the uncoated substrate. Leidheiser (22,23) examined zinc phosphate pretreated panels with automotive primer after 10 days exposure to the salt spray. The start and finish rest potentials of the samples with good paint performance were consistently more negative than those samples with poor paint performance:

	E_{start}(volts)	E_{finish}(volts)
Poor samples	-0.566	-0.757
Good samples	-0.629	-0.770

The effect was more pronounced at the starting potential than at the finish potential. Leidheiser suggested that the best performance is obtained when the cathode/anode surface area ratio is the same as the uncoated metal. Inadequate performance is obtained when the cathode/anode area ratio becomes larger. Our work agrees with Leidheiser's hypothesis. The B210/GBL coatings have rest potentials less noble than the B40 coated steel panels and perform best in the salt fog environment.

Rust Rating - Temperature Correlation

Figure 2 was separated into four regions: Region I (room temperature to 90°C), Region II (100°C-120°C), Region III (130°C-170°C) and Region IV (170°C and above).

In Region I, heat has no effect. In Region II, the enhancement in performance can be explained by changes in polymer matrix crystallinity. Glass transition temperatures for the homopolymer constituents of the B210/NVP matrix range from 86°C to 105°C. This increase in amorphous nature of the matrix should result in a more tortuous path for water and ion permeation and increase the corrosion resistance of the coating.

In Region III, the maximum in the rust rating is observed. The maximum is possible due to the combined changes in the organic coating matrix and the B40 panel. Polyvinylpyrrolidone becomes water insoluble when heated to 150°C due to crosslinking (19). Recent evidence also suggests that polyvinylpyrrolidone itself may act as a corrosion inhibitor (24). As the temperature is increased, further crosslinking may result in the buildup of stress in the coating. If cohesive breakdown were to occur, a pathway for water and ion permeation becomes available.

With regards to the conversion coated substrate, Wittel (25) observed that at temperatures greater than 140°C, tetrahydrate zinc phosphates lose part of their water of hydration. It is likely that the water of hydration liberated in the phosphate recrystallization process has a negative effect on the adhesion of the polymer matrix to the B40 panel.

In Region IV, performance again increases. Crystallographic transformations in phosphate conversion coatings at 180°C are known to adversely affect phosphate crystal adhesion (26). However, nitrile cyclization may be important at these higher temperatures. Cyclization is believed to enhance barrier properties of high nitrile polymers (27,28,29).

Rust Rating - Open Circuit Potential - Temperature Correlation

The open circuit potential data for the B210/NVP system mirrors the behavior of the rust ratings over the temperature range examined. A plausible explanation of the change of the open circuit potential is as follows. As temperature is increased, the composition of the various oxides and hydroxides which make up the zinc phosphate conversion layer and the base iron oxide layer undergo changes. These compositional changes are reflected in the changing open circuit potentials.

In terms of nobility, the open circuit potentials for the B210/NVP system at all temperatures are more positive than the B40 panels. The open circuit potentials for the B210/GBL system at all temperatures are generally more negative than the B40 panels. The open circuit potential trends for the B210/GBL system mimic the B210/NVP system. However, the effects are not as pronounced. We propose that in the GBL system there is a reduction in the cathode/anode area ratio as suggested by Leidheiser (16).

In conclusion, we have shown that a simple, fast, electrochemical measurement, the open circuit potential, can be extremely useful in assessing the effect of diluent for high nitrile photocured coatings. Further work is underway to elucidate the underlying reasons behind the temperature induced changes observed in corrosion performance.

Literature Cited

1. "Corrosion Control by Organic Coatings"; Leidheiser, H. Jr. Ed.; National Association of Corrosion Engineers: Houston, Texas, 1981.
2. Bacon, R. C.; Smith, J. J.; Rugg, F. M. *Ind. Eng. Chem.* 1948, 40, 161.
3. Mills, D. J. *Pitture E Vernici* 1984, 7, 102.
4. Hare, C. FEDERATION OF SOCIETIES FOR COATINGS TECHNOLOGY Units 26 and 27, 1978.
5. Nemphos, S. P.; Salame, M.; Steingeiser In "Encyclopedia of Polymer Technology"; Bikales, N. M. Ed.; Wiley-Interscience: New York, 1976; p. 65.
6. American Cyanamid Co. "The Chemistry of Acrylonitrile"; 1959.
7. Tafel, J. *Z. Physik. Chem.* 1905, 50, 641

8. "Electrochemical Techniques for Corrosion"; Baboian, J. Ed.; National Association for Corrosion Engineers: Houston, Texas, 1978.
9. Leidheiser, H. Jr. In "Corrosion Control by Coatings"; Leidheiser, H. Jr. Ed.; Science Press: Princeton, New Jersey, 1979; p. 143-170.
10. Wolstenholme, J. Corrosion Science 1973, 13, 521.
11. Burns, R. M.; Haring, H. E. Trans. Electrochem. Soc. 1936, 69, 169.
12. Haring, H. E.; Gibney, R. B. Trans. Electrochem. Soc. 1939, 76, 287.
13. Whitby, L. Paint Research Asscn. 1939, Tech. Paper No. 125.
14. Zahn, H. Corrosion 1947, 3, 233.
15. Wormwell, F.; Brasher, D. M. J. Iron Steel Inst. 1950, 164, 141.
16. Kendig, M. W.; Leidheiser, H. Jr. J. Electrochem. Soc. 1976, 123, 982.
17. "Standard Reference Method for Making Potentiostatic and Potentiodynamic Anodic Polarization Measurements", ASTM G5-72.
18. "Evaluating Degree of Rusting on Painted Steel Surfaces", ASTM D610-68.
19. Lorenz, D. H. Encyclopedia of Polymer Technology; Bikales, N. M. Ed.; Wiley-Interscience: New York, 1976; p. 239.
20. Talsma, H; Giffen, M. W. U. S. Patent 4 329 401, 1980.
21. Groseclose, R. G.; Frey, C. M.; Floyd, F. L. J. Ctg. Tech. 1984, 56, 31.
22. Iezzi, R. A.; Leidheiser, H. Jr. Corrosion 1981, 37, 28.
23. Leidheiser, H. Jr. Corrosion 1983, 39, 189.
24. Mostafa, A. El-Khair B. Corr. Prev. and Control 1983, 30, 14.
25. Wittel, K. Ind. Lackier Betrieb 1983, 51, 169.
26. van Ooij, W. J. Proc. 10th Conf. in Organic Coatings Science and Technology, 1984, p. 381.
27. Grassie, N.; Hay, J. N. J. Polym. Sci. 1962, 56, 189.
28. Coleman, M. M.; Sivy, G. T. Carbon 1981, 19, 123.
29. Fochler, H. S.; Mooney, J. R.; Ball, L. E.; Boyer, R. D.; Grasselli, J. G. Spectrochemica Acta, 1985, 41A, 1/2, 271.

RECEIVED January 21, 1986

6

Alternating Current Impedance: Utility in Evaluating Phosphate Coating, Phosphorus–Chromium Rinse, and Paint Performance

C. P. Vijayan, D. Noël, and J.-J. Hechler

Industrial Materials Research Institute, National Research Council of Canada, 75, Boulevard de Mortagne, Boucherville, Quebec, Canada J4B 6Y4

Surface preparation methods are related to impedance behavior of painted aluminum-killed 1006 steel in a 3 weight percent sodium chloride solution. Variation of phosphating time, anodic phosphating, phosphochromic rinse, paint thickness, temperature of test solution and immersion time are studied using two different types of paints. For porous coatings, the beneficial effects of phospho-chromic rinse is confirmed. Decrease in coating resistance and increase in capacitance are observed with increasing time of immersion as well as with increasing test temperature. Progress in deterioration is indicated by the appearance of Warburg-type behavior. Good protection is obtained by phosphating for 5 minutes.

Simplicity and reliability of operation make AC impedance measurements attractive as a technique in the evaluation of coating integrity. As opposed to classical salt spray test, analysis times are shorter with the AC impedance technique and quantitative data are obtained permitting relevant mechanistic information to be derived. Impedance test methods are likely to find many applications in the resolution of unsolved practical problems (1).

Various publications have appeared in recent years regarding the application of AC impedance and polarization resistance test methods in the study of surface modifications such as oxidation, passivation, cathodic deposition, coatings and the study of corrosion reactions (2-8). Analysis of impedance data provides clues regarding the mechanism of reactions likely to be taking place at different interfaces in the system (9,10). Electrical equivalent circuits proposed, incorporating impedance corresponding to electrolyte/surface coating as well as surface coating/metal surface, are of great help in evaluating and improving the nature of coatings (11-17). They are also useful in determining compatible coating/environment combinations.

In the present work, steel surfaces polished, phosphated and painted are studied using AC impedance technique in order to evaluate the protection efficiency of a commercial phosphating solution. The AC impedance behavior of painted metal has been correlated with the immersion time in the phosphating solution and with the desirability of a phospho-chromic rinse (18-23). A comparison of the impedance behavior of two different types of commercial paints is made for various durations of immersion in sodium chloride solution at room temperature, and also for various temperatures at a given duration of immersion.

Mechanistic analyses proposed by Gabrielli et al (25) and Sluyters (26) are made use of in understanding the results obtained in this work.

THEORETICAL

An examination of the theoretical models proposed for metal dissolution and for the general impedance behavior of electrodes enables the rate-determining step of the corrosion reaction to be identified. It is then possible to separately study the rate determining step in order to find a suitable inhibitor or a suitable surface coating.

Bockris et al (24) proposed a two step mechanism for iron dissolution involving an adsorbed intermediate species $(FeOH)_{ads}$.

$$Fe + OH^- \underset{k_{-1}}{\overset{k_1}{\rightleftharpoons}} (FeOH)_{ads} + e$$

$$(FeOH)_{ads} \xrightarrow[k_2]{} (FeOH)^+ + e$$

In the bulk of the solution, the reaction

$$(FeOH)^+ \underset{k_{-3}}{\overset{k_3}{\rightleftharpoons}} Fe^{++} + OH^-$$

takes place but it is not considered to intervene in the electrode kinetics.

This mechanism can be related to the formal dissolution model proposed by Gabrielli et al (25) in terms of faradaic impedance.

If x represents the surface concentration of $(FeOH)_{ads}$ it can be shown that

$$\frac{1}{Z_F} = F\left[k_1'\ (1-\bar{x}) - (k_{-1}'-k_2')\bar{x} - (k_1+k_{-1}-k_2)\ \frac{dx}{dE}\right]$$

$$\bar{x} = k_1/(k_1+k_{-1}+k_2)$$

and

$$\frac{dx}{dE} = \frac{k_1'(1-\bar{x}) - (k_{-1}' - k_2')\bar{x}}{k_1+k_{-1}+k_2+j\omega\beta}$$

where Z_F = faradaic impedance; F = Faraday constant; k = rate constants; k'= derivatives of rate constants with respect to the potential, E; $\omega = 2\pi f$ where f is the frequency in Hz; β = constant coupling the surface concentration of $(FeOH)_{ads}$ and the fractional surface coverage.

The appearance of capacitive or inductive impedance depends essentially on the value of the rate constants. Low frequency loops, in a general case, are all very sensitive to the pH of the electrolyte. The different time constants are attributed to the relaxation of surface coverage by a corresponding number of reaction intermediates.

Rehbach and Sluyters (26) suggest the following general expression for impedance Z (=Z' + jZ'')

$$Z = R_{\Omega} + \frac{R_{ct} + \sigma\omega^{-\frac{1}{2}} - j\left[\omega C_{dl}(R_{ct} + \sigma\omega^{-\frac{1}{2}})^2 + \sigma^2 C_{dl} + \sigma\omega^{-\frac{1}{2}}\right]}{(\sigma\omega^{\frac{1}{2}} C_{dl} + 1)^2 + \omega^2 C_{dl}^2 (R_{ct} + \sigma\omega^{-\frac{1}{2}})^2}$$

with $\sigma = \sigma_o + \sigma_r$; R_{Ω} = ohmic cell resistance; R_{ct} = charge transfer resistance; σ = Warburg coefficient; σ_o, σ_r = separate Warburg coefficients for oxidised and reduced species; C_{dl} = double layer capacitance.

For given values of double layer capacitance C_{dl}, solution resistance R_{Ω} and Warburg coefficient σ, plots of -Z'' versus Z' have been made for selected values of charge transfer resistance, R_{ct} (26). It is observed that at smaller values of R_{ct} (~10 $\Omega\ cm^2$) relaxation due to R_{ct}-C_{dl} and Warburg diffusion behavior are both clearly seen.

The analysis proposed by Gabrielli (25) does not take diffusion effects into consideration. However, this model together with the dissolution mechanism proposed by Bockris show how relative variations in the values of rate constants can give rise to different types of Nyquist diagrams. In other words, it is possible to evaluate rate constants for a particular system by looking at the Nyquist diagram if the experiment has properly been designed.

In systems where diffusion phenomena are of significance, the mechanistic study is facilitated by using the general expression for impedance Z (26). This equation shows for instance how the Warburg coefficient can be evaluated by conducting impedance studies at very low frequencies. These coefficients in turn enable the evaluation of diffusion coefficients for the diffusing species.

Thus it appears that by incorporating parameters such as pore resistance and coating capacitance to the existing theoretical impedance model dealing with metal dissolution one would obtain valuable overall information (14,27). Complemented by results from regular immersion and salt spray tests it should be possible to find satisfactory solutions to corrosion problems of coated metals (9).

A generalised model of electrical equivalent circuit for painted surfaces has been considered in many of the recent publications. Googan (2) used it to study vinyl coatings free of defects and coatings containing defects. Electrocoatings were also evaluated. Musiani et al (27) in their investigation of mild steel coated with electrochemically synthesised polyoxyphenylenes used an electrical equivalent circuit incorporating elements such as coating capacitance C_c and pore resistance R_{pore}. Certain samples showed faster corrosion rates probably due to poor adhesion to the metal substrate. Conversion or pseudo-conversion phosphating which forms a protective layer on the base metal usually improves the final performance of paints by increasing the contact area and thus adhesion and reducing blistering and underskin corrosion.

Piens (9) suggested a similar equivalent circuit slightly modifying the incorporation of electrolyte resistance. Chlorinated rubber pigmented with iron oxide was applied to sand-blasted steel. Impedance measurements were taken after 24 hours of immersion in 0.5M NaCl, at the corrosion potential. Two capacitive semi-circles covering the frequency range 10^5 to 280 Hz and 280 Hz to 0.28 Hz were obtained. The linear section lying between 0.28 Hz and 10^{-2} Hz is characteristic of the diffusion impedance.

The values of C_c and R_{pore} reflect the different aspects of sensibility of a coating to water. Their variation enables one to compare various coatings as water barriers.

The resistance R_{pore} of a coating is often so high that the impedance diagram is no longer a complete semi-circle, but rather an arc intersecting the origin of the axes at high frequencies. In such case, the values for R_{pore} and C_c measured at a given frequency are used to evaluate coatings. These values are influenced by water absorption, the influence being pronounced at lower frequencies.

Parameters such as ageing, influence of pigment concentration and influence of coalescence of emulsion paints can also be studied using AC impedance test methods.

Mansfeld and Kendig (5) evaluated different surface pretreatments of steel and aluminum alloys. They suggest the usage of curve fitting methods such as CIRFIT program in order to overcome the difficulties posed by experimental data scatter. Their study of phosphated and coated steel indicates the appearance of a Warburg-type impedance at longer exposure times. A slope of -1/2 or -1/4 is observed in the Bode plot with a corresponding phase angle maximum of 45° or 22.5°. These behaviors perhaps represent respectively semi-infinite linear diffusion and semi-infinite diffusion in pores.

Padget and Moreland (11) showed that films cast from chloride-containing vinyl acrylic latex copolymers exhibited low uptake of liquid water and ions when the degree of particle coalescence was high. AC impedance measurements supplemented by salt spray and outdoor exposure results showed a relationship between the permeation characteristics and the anticorrosive performance of latex films and latex paints, and showed the advantages to be gained by using low permeability chlorine containing polymer.

Deterioration of the coatings was found to pass through a series of stages each characterized by its own distinctive Nyquist plot. In the first stage, the impedance diagram had little curvature and corresponded to a high impedance value. The plot then became curved due to the ingress of ions into the coating, film resistance fell progressively until rate controlling diffusion processes were apparent and the film had a very low impedance.

Coal tar epoxy and plasticized chlorinated rubber laquer coated on mild steel were studied by Scantlebury et al (28). Impedance plots show a gradual decrease in the value of R_{ct} and the onset of Warburg-type behavior with increasing immersion time in 3 weight percent sodium chloride solution. Appearance of an inductive loop when the coal-tar epoxy had a pin-hole was clearly demonstrated.

The present study deals with two types of paints applied on phosphated steel, the influence of temperatures upto 90°C and immersion times upto 10 days. The results obtained are analyzed in the light of published information briefly described in this section. An electrical equivalent circuit similar to the one used by Musiani et al (27) is considered suitable for the analysis.

EXPERIMENTAL

AC impedance system 368 (EG&G PARC, Princeton, NJ) with Fast Fourier Transform analysis was used along with a potentiostat EG&G model 273. The potentiostat was coupled to an Apple II+ computer through an IEEE-488 interface. A frequency range of 0.01 Hz to 10^5 Hz was used for many of the experiments. The electrolyte used was a 3 weight percent sodium chloride solution and was prepared using ACS certified chemicals and deionized water.

Aluminum-killed 1006 steel sheet (Sidbec-Dosco, Contrecoeur, Québec, Canada) was cut to 1 cm X 1 cm size, soldered to a copper wire, embedded in epoxy, polished to 600 grit, washed respectively in tap-water, methanol and deionized water, dried in a stream of air and was then preserved in a desiccator. Details of phosphating (Oxy-Plus 84 DRS, Laboratoire Brabant Inc., Ville St-Pierre, Québec, Canada), rinsing and painting (Tremclad Rust Paint, Tremco Ltd., Toronto, Ontario, Canada) used in the first set of specimens are shown in Table I. Painted specimens were dried for 2 days at room temperature. The average thickness of single layer paint was 20 µm for these specimens as measured using an instrument NEO-DERM, Model 179-711 (Mitutoyo Mfg. Co. Ltd., Tokyo, Japan). Impedances of coated specimens were measured after 4 hours of immersion in 3% NaCl at the required temperature.

Coated specimens were placed in an open three-electrode electrochemical cell. After 4 hours of immersion at ambient temperature, open-circuit potentials were noted and impedance measurements were made on duplicate samples. Specimens were tested in a salt spray test cabinet (ASTM B117-73) for 1, 17 and 96 hours respectively and their surfaces photographed in order to calculate the percentage of surface covered by corroded spots and blisters (ASTM D610-68).

Table II shows the details of preparation for the second set of 10 painted steel specimens. A polyurethane paint (Marinox SR-2, Mabraco International, 20 des Navigateurs, Québec, Canada) along with an initial coating of an aluminum containing paint (prépolymère d'aluminium, Mabraco International) was the second type of paint system used in this study.

The method of preparation for the second type of paint consisted of mechanical polishing, cleaning, phosphating for 5 minutes with Oxy-Plus 84 DRS solution, drying for 30 minutes without rinsing, application of one layer of precoat followed by air drying for 4 hours at room temperature and application of one layer of Marinox SR-2 paint followed by curing for 8 days at room temperature.

Polarization resistance, R_p, of specimens #25 and #27 was measured as a function of immersion time using Corrosion Console 350A (EG&G PARC, Princeton, NJ).

All specimens anodically phosphated were initially polished to 600 grit, cleaned and immersed in the phosphating solution Oxy-Plus 84 DRS, before applying a potential of +0.8V/SCE. This potential corresponds to the passive zone in the cyclovoltammogram obtained for the test steel in the phosphating solution.

RESULTS AND DISCUSSION

Table I shows the details of surface treatment and coating along with the calculated values of total resistance R and effective capacitance C. For specimens with initial mechanical surface preparation, the Nyquist impedance plot shows the characteristic semicircular behavior with a resistance of the order of 1800 Ω cm^2 and a capacitance of about 40 μF cm^{-2}. As different surface treatments are incorporated on a sequential basis, the complex plane diagram shows a gradual evolution.

Comparison of the capacitance values of specimens phosphated for 5 minutes and 30 minutes respectively and rinsed in phosphochromic solution (specimens #6 and #8) with a non-phosphated specimen (#1) shows values of the same order of magnitude i.e. 20 and 32 μF cm^{-2} as compared to 40 μF cm^{-2}. On the other hand, the specimens #7 and #9 which are phosphated for 5 and 30 minutes respectively but not rinsed, show higher values of capacitance per unit surface in the millifarad range. Since these specimens are not rinsed after phosphating and since the phosphating solution contains a proprietary inhibitor, perhaps a certain amount of the inhibitor is retained on the surface causing this change in capacitance.

Similar trend is observed in specimens coated with one layer of paint (specimens #10, #12, #14, #16) but prepared with and without rinsing. The conclusion holds also for specimens coated with two layers of paint. Resistance values are seen to be higher for specimens that have been washed with phospho-chromic solution after phosphating. Since the rinsing solution is supposed to seal the openings existing between phosphate crystals (19), it is logical that specimens subjected to rinsing show higher resistance values (i.e. less tendency for effecting charge transfer reactions).

On comparing the results of specimens subjected to rinsing with phospho-chromic solution but having same paint thickness (#10 and #14) it is seen that phosphating for 30 minutes does not provide any appreciable benefit. The same is true on comparing specimens #11 and #15.

If one looks at specimens #12 and #16 the conclusions made above regarding non-rinsed specimens hold. The same is true for specimens #13 and #17 which are also non-rinsed but have two layers of paint. These results thus appear to indicate that prolonged phosphating does not provide appreciable additional benefit from the point of view of the corrosion reaction likely to take place at the metal surface. Since paint adhesion on thicker phosphate films is of poor quality and since there is no significant improvement in impedance behavior due to thicker phosphate layers, it appears that only a thin phosphate coating should be used whenever possible.

Double layer paint provides additional protection since such coatings would be less porous than single layer paint. It is also noted that in all specimens that are not rinsed there is a tendency to show inductive loops in the impedance plot. It is not clear if this is due to the adsorption of inhibitor on steel surface or due to the formation of oxides or due to increased porosity (28).

Many specimens (#1, #6, #8, #10 and #12) also show the initiation of Warburg impedance behavior at the lower end of the frequency range covered in this study.

Specimens that were exposed to salt spray testing were withdrawn after 1, 17 and 96 hours of exposure. The appearance of these specimens was rated as per ASTM procedure D610-68 which is a measure of the surface clearly attacked or showing formation of blisters (Table I). Here again it is noted that the conclusions made earlier hold good: phosphated non-rinsed specimens behave poorly as compared to rinsed specimens, thus establishing the necessity of phospho-chromic solution rinse. Salt spray test results do also indicate that phosphating for 30 minutes does not provide any appreciable improvement as compared to 5 minutes phosphating.

Impedance measurements taken on specimens after 96 hours exposure to salt spray show a combination of R-C and Warburg diffusion behavior. This is in agreement with the observation elsewhere (9,11).

Tests conducted on specimens phosphated for 5 minutes but provided with additional paint thickness using Tremclad or Marinox SR-2 are summarized in Table II. Specimens (#18) rinsed with phospho-chromic solution show higher values of resistance and lower values of capacitance with respect to non-rinsed specimens (#20) both having 80 μm of paint thickness.

Specimen #19 tested for impedance after exposure to salt spray tests shows a low charge transfer resistance of 4300 Ω cm^2 and a fairly high double layer capacitance of 316 μF cm^{-2}. This shows the increased tendency of NaCl solution to penetrate the paint film exposed to salt spray test. The capacitance is even higher for specimen #21 prepared with no phospho-chromic rinse.

It is clear on analyzing the results obtained from specimens #18 and #19 as well as from #20 and #21 that film degradation and coating integrity can be followed more efficiently by impedance measurements than by salt spray testing (Table II and Figures 1 and 2).

Results of impedance tests conducted on specimens with the polyurethane paint always showed capacitance in the pF cm^{-2} range, in spite of no rinsing operation being carried out. The resistances are in the 10^7-10^9 Ω cm^2 range.

AC impedance measured using a few specimens anodically phosphated and painted with Tremclad or Marinox paint are shown Table II and Figures 3 and 4. Cyclovoltammetry indicated a notable fall

TABLE I . RESULTS OF AC IMPEDANCE AND SALT SPRAY TESTS FOR DIFFERENT SURFACE TREATMENTS

Spec. #	Details of surface treatment and coating (a)	R (Ω cm^2)	C (F cm^{-2})	rust grade (d)
1	Mechanical polishing + no additional treatment	1800	40×10^{-6}	-
6	5min. phosphating (b) + rinse (c)	8000	20×10^{-6}	-
7	5min. phosphating + no rinse	460	1×10^{-3}	-
8	30min. phosphating + rinse	4000	32×10^{-6}	-
9	30min. phosphating + no rinse	280	0.32×10^{-3}	-
10	5min. phosphating + rinse + 20 μm paint	4.5×10^4	10×10^{-6}	9
11	5min. phosphating + rinse + 40 μm paint	4.5×10^5	1×10^{-6}	9
12	5min. phosphating + no rinse + 20 μm paint	3.0×10^4	20×10^{-6}	1
13	5min. phosphating + no rinse + 40 μm paint	2.2×10^5	4×10^{-6}	4
14	30min. phosphating + rinse + 20 μm paint	5.0×10^4	20×10^{-6}	9
15	30min. phosphating + rinse + 40 μm paint	2.5×10^5	1×10^{-6}	9
16	30min. phosphating + no rinse + 20 μm	8000	50×10^{-6}	1
17	30min. phosphating + no rinse + 40 μm	4.0×10^4	3×10^{-6}	3

a) All specimens are mechanically polished to 600 grit.

b) Specimens are phosphated using a commercial solution, Oxy-Plus 84 DRS, received from Laboratoire Brabant Inc., Ville St-Pierre, Québec, Canada.

c) Dilute solution of phospho-chromic acid mixture is used for the rinsing operation.

(d) A higher rust grade number indicates better protection against corrosion.

TABLE II. RESULTS OF AC IMPEDANCE MEASUREMENTS MADE ON SAMPLES WITH THICKER COATINGS

Spec. #	Details of surface treatment and coating (a)	E_{corr} (V/SCE)	R (Ω cm^2)	C (F cm^{-2})
18	5min.phosphating + rinse + 80 μm paint (Tremclad)	-0.114	$1.5x10^7$	$3.2x10^{-9}$
19	5min. phosphating + rinse + 80 μm paint (Tremclad) + 100h. salt spray test	-0.569	$4.3x10^3$	$316x10^{-6}$
20	5min. phosphating + no rinse + 80 μm paint (Tremclad)	-0.509	$7.0x10^4$	$14x10^{-6}$
21	5min. phosphating + no rinse + 80 μm paint (Tremclad) + 100h. salt spray test	-0.502	$3.5x10^3$	$724x10^{-6}$
22	5min. phosphating + no rinse + 1 precoat + 1 Marinox SR-2 coat (100 μm coating)	-0.126	$8.0x10^7$	$50x10^{-12}$
23	5min. phosphating + no rinse + 1 precoat + 1 Marinox SR-2 coat + 100h salt spray test	-0.173	$1.0x10^9$	$32x10^{-9}$
24	5min. anodic phosphating* + rinse + 32 μm paint (Tremclad)	-0.601	$1.1x10^4$	$40x10^{-6}$
25	30min. anodic phosphating* + rinse + 24 μm paint (Tremclad)	-0.599	$2.3x10^4$	$32x10^{-6}$
26	5min. anodic phosphating* + rinse + 1 precoat + 1 Marinox SR-2 coat (170 μm coating)	-0.170	$8.0x10^7$	$160x10^{-12}$
27	30min. anodic phosphating* + rinse + 1 precoat + 1 Marinox SR-2 coat (160 μm coating)	-0.285	$2.0x10^8$	$316x10^{-12}$

* See text for details.

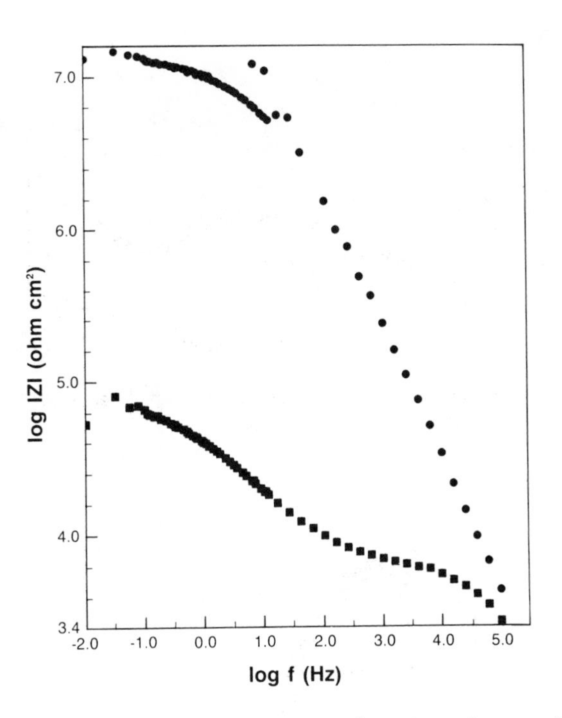

Figure 1. Bode plots for specimens #18 (●) and #20 (■). See Table II for details.

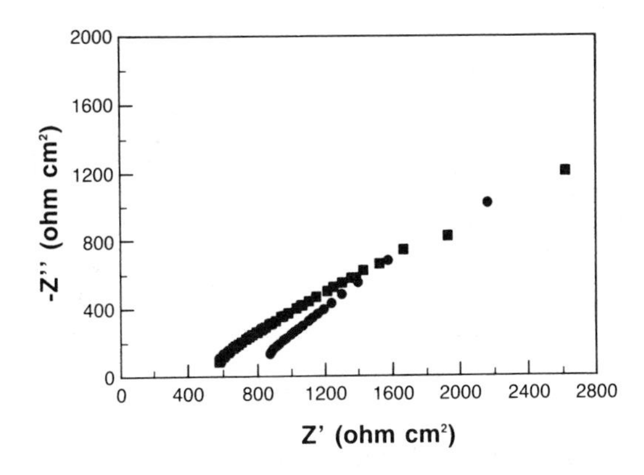

Figure 2. Nyquist plots for specimens #19 (●) and #21 (■). See Table II for details.

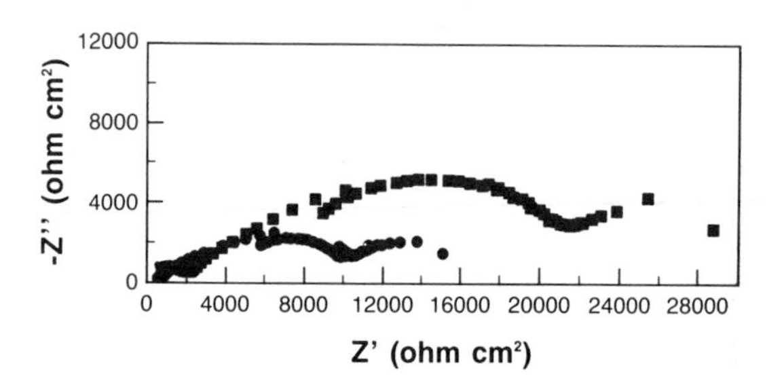

Figure 3. Nyquist plots for specimens #24 (●) and #25 (■). See Table II for details.

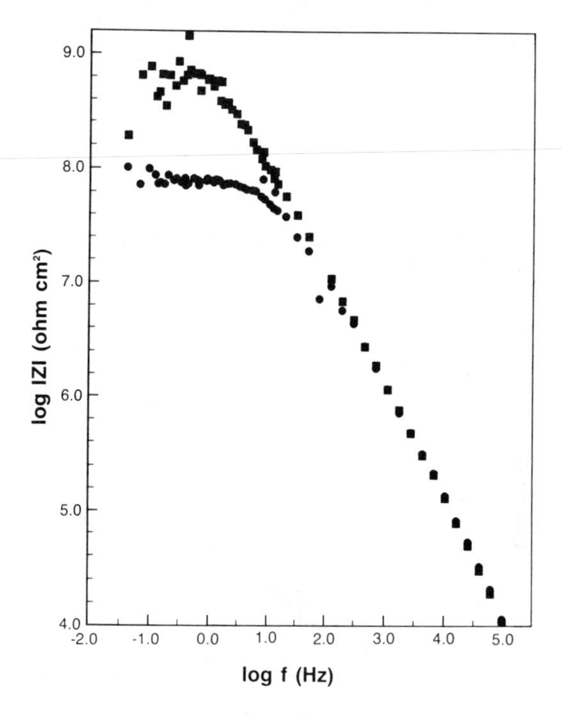

Figure 4. Bode plots for specimens #26 (●) and #27 (■). See Table II for details.

in the anodic current beyond +0.5 V/SCE in the phosphating solution Oxy-plus 84 DRS. However, anodic phosphating at +0.8 V/SCE for 5 and 30 minutes respectively did not produce any improvement in the ultimate performance of painted specimens in spite of the beneficial aspects of this process cited in literature (29,30).

AC impedance measurements taken on the same specimen at different temperatures in the range 25-90 °C are shown in Table III. A specimen with no surface treatment other than mechanical polishing shows $C_{dl} \approx 40\,\mu F\ cm^{-2}$ at 25°C but the value increases appreciably with increasing temperature. The values of R_{ct} for different specimens (#28,#29) show a systematic decrease with increasing temperature whereas the values of C_{dl} show a systematic increase. Figures 5 and 6 show the evolution of impedance plots as a function of temperature. In addition to the variation in the values of R_{ct} and C_{dl}, it is noticed that the Warburg-

TABLE III. AC IMPEDANCE MEASUREMENTS AT DIFFERENT TEMPERATURES

Spec. #	Details of surface Treatment and coating	Temp. (°C)	E_{corr} V/SCE	R (Ω cm^2)	C (F cm^{-2})
28	5min. phosphating + rinse + 40 µm paint (Tremclad) 4h. immersion in 3% NaCl at room temperature	25	-0.588	2.1×10^4	63×10^{-6}
		45	-0.587	7000	166×10^{-6}
		65	-0.580	5000	251×10^{-6}
		90	-0.545	3000	550×10^{-6}
29	5min. phosphating + no rinse + 1 precoat + 1 Marinox SR-2 coat 4h. immersion in 3% NaCl at room temperature (100 µm coating)	25	-0.126	8.0×10^7	0.50×10^{-9}
		45	-0.534	6.5×10^5	2.5×10^{-9}
		65	-0.612	8.0×10^4	16×10^{-9}
		90	-0.708	5.0×10^4	400×10^{-9}
30	5min.phosphating + rinse + 40 µm paint (Tremclad) 4h immersion in 3% NaCl at 45°C.	45	-0.586	1.2×10^4	251×10^{-6}
31	5min. phosphating + rinse + 40 µm paint (Tremclad) 4h. immersion in 3% NaCl at 65°C.	65	-0.438	3500	398×10^{-6}
32	5min. phosphating + rinse + 40 µm paint (Tremclad) 4h immersion in 3% NaCl at 90°C.	90	-0.450	2060	832×10^{-6}

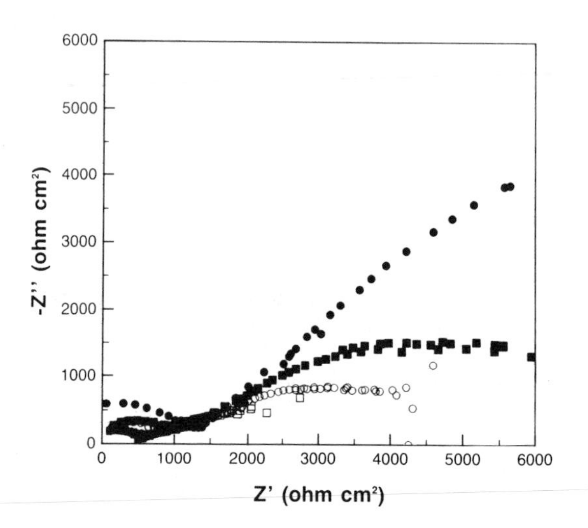

Figure 5a. Nyquist plots for specimens #28 at different temperatures. See Table II for details. Legend: 25°C (●); 45°C (■); 65°C (○) and 90°C (□).

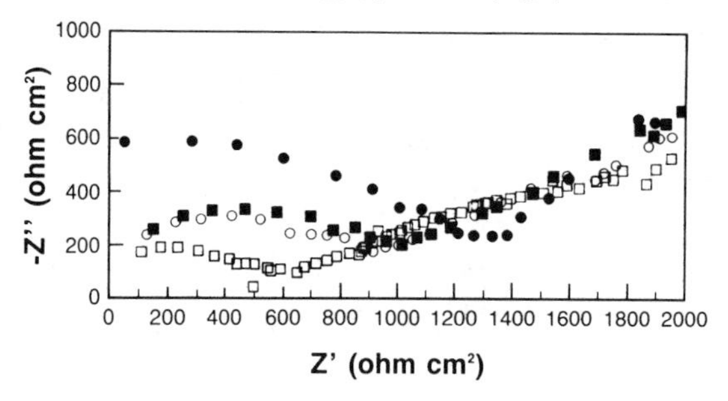

Figure 5b. Enlargement of the initial portion of Figure 5a.

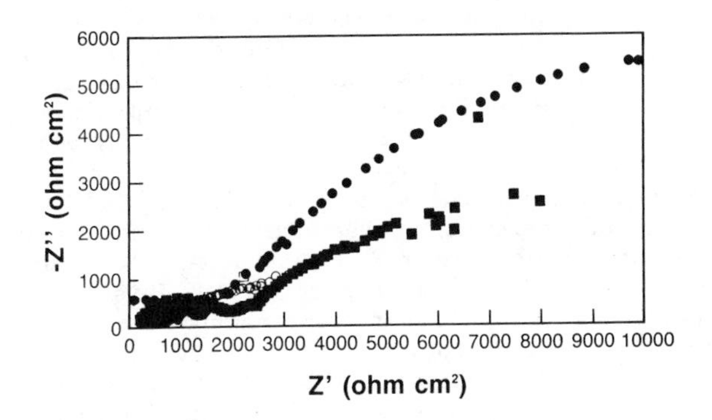

Figure 6a. Nyquist plots for specimens identically prepared but immersed in 3% NaCl for 4 hours at the temperature of impedance measurement. See Table III for details. Legend: specimen #28 (●), # 30 (■); #31 (○) and #32 (□).

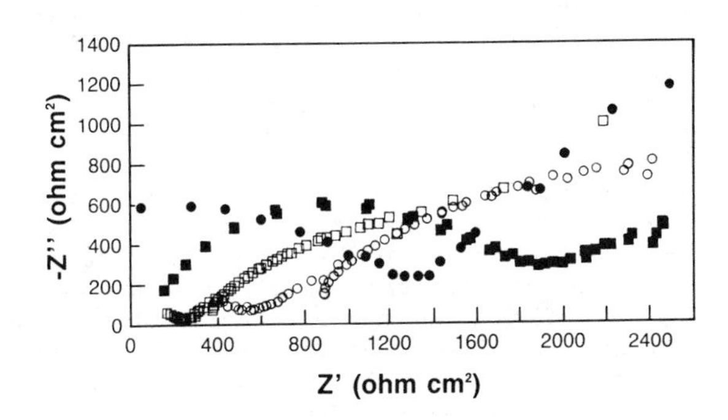

Figure 6b. Enlargement of the initial portion of Figure 6a.

type behavior at low frequencies gradually become more prounonced at higher temperatures.

Figure 5 has been obtained from a specimen (#28) maintained at 25°C for 4 hours in 3% NaCl solution before taking impedance measurements. The temperature was then raised to the desired value for each step.

In the case of Figure 6 separate specimens identically prepared were individually maintained at 25,45,65 and 90°C for 4 hours before impedances were measured at the particular temperature. The variation in impedance as shown in Figures 5 and 6 with temperature appears to be logical. An increase in temperature decreases the resistance but increases the capacitance. In Figure 6, specimen #30 is an exception in that the impedance at 45°C is higher than that at 25°C. This is perhaps caused by unknown errors in manipulation.

It is remarked that specimens maintained at 65°C and 90°C for 4 hours show lower resistances than the resistances measured when the specimen was maintained for 4 hours at 25°C and then the temperature raised to 65°C and then to 90°C. This is also logical because maintenance at a higher temperature for a longer time causes accelerated degradation.

Figure 7 shows the impedance plots as a function of temperature obtained using a specimen having the polyurethane paint coating. The resistance is of the order of 10^5 Ω cm^2 and the capacitance is in the nanofarad range. Higher values of resistance and very low values of capacitance show the better protection offered by this type of paint at all of the temperatures studied.

Apparent activation energies for the degradation reaction was calculated using specimens #28 and specimens #30, #31 and #32. A value of ~7 kcal/mol was obtained. This evaluation was based on the variation of resistance as a function of temperature. A similar value was obtained also with specimen #11 (Table I) after it was exposed to salt spray test (100 hours).

When a thick coating (~100-160 μm) of polyurethane paint is applied (e.g. specimen #29), the activation energy is of the order of 26 kcal/mol. This indicates the higher energy barrier presented by a non-porous coating.

Figure 8 presents the impedance behavior of polyurethane-painted specimen after 4 hours of immersion and after 10 days of immersion. The impedance spectrum does not show any appreciable variation. Further information on the behavior of the Tremclad and the Marinox systems (specimens #25 and #27) for immersion times upto 10 days is given in Table IV. The variation in corrosion potential is not very systematic but the changes in C_{dl}, R_{ct} and R_p appear to follow a logical trend. For the Tremclad coating, the resistance decreases and capacitance increases with immersion time. This indicates incorporation of electrolyte into the paint film. For the polyurethane coating, the resistance stays in the 10^7 Ω cm^2 range and the capacitance in the 400 pF cm^{-2} range indicating resistance to electrolyte incorporation into the film.

The results are in concordance with published information available regarding coating deterioration. It was confirmed that a fall in R_{ct} as well as the appearance of Warburg-type of behavior at low frequencies, especially at higher temperatures, is clearly an indication of lack of protection. A properly prepared surface

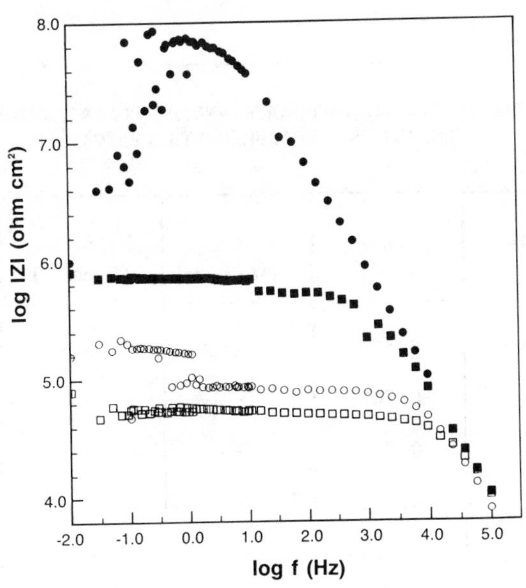

Figure 7. Bode plots for specimen #29: influence of temperature. See Table III for details. Legend: 25°C (●); 45°C (■); 65°C (○) and 90°C (□).

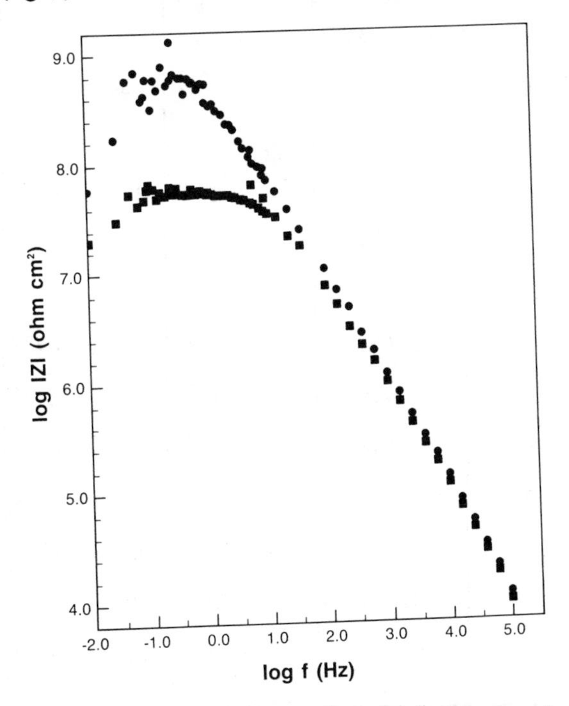

Figure 8. Bode plots for specimen #27: influence of immersion time. See Table IV for details. Legend: 4 h (●); 10 days (■).

TABLE IV. AC IMPEDANCE AND DC POLARIZATION RESISTANCE MEASUREMENTS (25°C)

Spec. #	Immersion time (h.)	E_{corr} V/SCE	R (Ω cm^2)	C (F cm^{-2})	R_p (Ω cm^2)
25	4	-0.599	2.3×10^4	32×10^{-6}	$> 10^5$
	24	-0.708	-	-	7.4×10^4
	48	-0.700	-	-	7.1×10^4
	72	-0.732	-	-	5.4×10^4
	96	-0.734	-	-	4.4×10^4
	168	-0.585	-	-	3.4×10^4
	192	-0.589	-	-	4.5×10^4
	216	-0.574	-	-	4.0×10^4
	240	-0.578	-	398×10^{-6}	3.0×10^4
27	4	-0.122	2×10^8	316×10^{-12}	$>10^5$
	24	-0.234	-	-	$>10^5$
	48	-0.410	-	-	$>10^5$
	72	-0.442	3.4×10^7	320×10^{-12}	$>10^5$
	120	-0.404	4.0×10^7	794×10^{-12}	$>10^5$
	144	-0.382	5.0×10^7	200×10^{-12}	$>10^5$
	168	-0.412	4.0×10^7	631×10^{-12}	$>10^5$
	192	-0.310	1.6×10^7	320×10^{-12}	$>10^5$
	216	-0.290	5.0×10^7	320×10^{-12}	$>10^5$
	240	-0.278	5.0×10^7	251×10^{-12}	$>10^5$

provided with a thick polyurethane coating shows good performance. The economic aspects of the problem and the investment involved in such paint systems need of course to be taken into consideration.

CONCLUSION

1) AC impedance measurements enable the determination of charge transfer resistance and double layer capacitance and other parameters related to coated systems.
2) Decrease in charge transfer resistance and increase in double layer capacitance is observed with increasing time of immersion or with increasing test temperature and gives information on the degree of protection efficiency of a coating.
3) Appearance of Warburg-type behavior shows that diffusion phenomena become predominant in some coatings as their deterioration progresses.
4) Longer phosphating times or anodic phosphating does not provide any appreciable advantage.
5) Phospho-chromic rinse has a beneficial effect on the life of a coated surface.
6) Steel surface phosphated for 5 minutes, rinsed and provided with a polyurethane coating shows good resistance to 3% NaCl solution.

ACKNOWLEDGMENTS

The authors thanks B. Harvey for assistance in conducting some of the tests and in taking pictures of the corroded specimens. Special thanks are expressed to D. Lajeunesse for her help in correcting a few problems related to the electrical system, to D. Simard for his assistance with photographic work. Samples of phosphating solution received from Laboratoire Brabant Canada Inc. (Ville St-Pierre, Québec, Canada) and samples of aluminum-killed 1006 steel sheet received from Sidbec-Dosco (Contrecoeur, Québec, Canada) are gratefully acknowledged.

LITERATURE CITED

1. Leidheiser Jr., H. Corrosion (NACE) 1982, 38, 374-383.
2. Googan, C.G. Proc. UK National Corrosion Conference, London, England, 1982, pp. 13-18.
3. Groseclose, R.G.; Frey, C.M.; Floyd, F.L. J. Coat. Technol. 1984, 56, 31-43.
4. Bombara, G.; Lunazzi, G.C.; Martini, B. Werkst. Korros. 1982, 33, 610-617.
5. Mansfeld, F.; Kendig, M. Proc. 9th Int. Congr. Met. Corros., Toronto, 1984, Vol. 3, 74-84.
6. Macdonald, D.D.; McKubre, M.C.H. In "Electrochemical Corrosion Testing"; Mansfeld, F., Bertocci, U., Eds., ASTM STP 727, American Society for Testing and Materials: Philadelphia, 1981; pp. 110-149.
7. Chao, C.Y.; Lin, L.F.; Macdonald, D.D. J. Electrochem Soc. 1982, 129, 1874-1879.
8. Mansfeld, F. Corrosion (NACE) 1981, 36, 301-307.

9. Piens, M.; Verbist, R.; Vereecken, J. Proc. 9th Int. Conf. Org. Coat. Sci. Technol., Athens, Greece, 1983, p. 137.
10. Hubrecht, J.; Vereecken, J.; Piens, M. J. Electrochem Soc. 1984, 131, 2010-2015.
11. Padget, J.C.; Moreland, P.J. J. Coat. Technol. 1983, 55, 39-51.
12. Epelboin, I.; Gabrielli, C.; Keddam, M.; Takenouti, H. In "Electrochemical Corrosion Testing"; Mansfeld, F.; Bertocci, U., Eds.; ASTM STP 727 American Society for Testing and Materials: Philadelphia, 1981; pp. 150-166.
13. Haruyama, S.; Tsuru, T. In "Electrochemical Corrosion Testing"; Mansfeld, F.; Bertocci, U., Eds.; ASTM STP 727, American Society for Testing and Materials: Philadelphia, 1981; pp.167-186.
14. Rothstein, M., personal communication and Application Note AC-1, EG&G PARC, Princeton, N.J.
15. Kendig, M.W.; Meyer, E.M.; Lindberg, G.; Mansfeld, F. Corros.Sci. 1983, 23, 1007-1015.
16. Mansfeld, F.; Kendig, M.W. Werkst. Korros. 1983, 34, 397-401.
17. Kendig, M., Mansfeld, F.; Tsai, S. Corros. Sci. 1983, 23, 317-329.
18. Gordon, D.C. Corros. Prev. Control Oct. 1984, 7-10.
19. Lorin, G. "La phosphatation des métaux"; Eyrolles: Paris, 1979; p.168.
20. Ottaviani, R.A. Proc. Org. Coat. Appl. Polym. Sci., 1981, 46, 50-55.
21. Dickie, R.A. In "Adhesion Aspects of Polymeric Coatings"; Mittal, K.L., Ed.; Plenum: New York, 1983, pp. 319-327.
22. de Vries, J.E.; Riley, T.L.; Holubka, J.W.; Dickie, R.A. Surf. Interface Anal. 1985, 7, 111-116.
23. Kuehner, M.A. Met. Finish. 1985, 83, 15-18.
24. Bockris, J.O'M.; Kita, H. J. Electrochem. Soc. 1961, 108, 676.
25. Gabrielli, C.; Keddam, M.; Takenouti, H. In "Treatise on Materials Science and Technology"; Scully, J.C., Ed.; Academic: New York, 1983, Vol. 23, pp. 395-451.
26. Sluyters-Rehbach, M.; Sluyters, J.H. In "Comprehensive Treatise of Electrochemistry"; Yeager, E.; Bockris, J.O'M.; Conway, B.E.; Sarangapani, S., Eds.; Plenum: New York, 1984, Vol. 9, pp. 183-191.
27. Musiani, M.M.; Pagura, C.; Mengoli, G. Electrochim. Acta 1985, 30, 501-509.
28. Scantlebury, J.D.; Ho, K.N.; Eden, D.A. In "Electrochemical Corrosion Testing"; Mansfeld, F.; Bertocci, U., Eds.; ASTM STP 727, American Society for Testing and Materials: Philadelphia, 1981; pp.187-197.
29. Rajagopalan, K.S.; Krithivasan, N.; Rajagopal, C.; Janaki, M.E.K. Corros. Maint. 1983, 6, 259-266.
30. Gabe,D.R.; Johal, C.P.S.; Akanni, K.A. Met. Finish. 1985, 83, 41-44.

RECEIVED January 22, 1986

7

Evaluation of Coating Resins for Corrosion Protection of Steel Exposed to Dilute Sulfuric Acid

Malcolm L. White[1] and Henry Leidheiser, Jr.[2]

[1]Center for Surface and Coatings Research, Lehigh University, Bethlehem, PA 18015
[2]Department of Chemistry and Center for Surface and Coatings Research, Lehigh University, Bethlehem, PA 18015

Three types of coatings--a vinyl ester, a polyester and four epoxies--were coated on steel and exposed to 0.1M H_2SO_4 at 60°C. Measurements of corrosion potential, AC conductance, tensile adhesion and weight gain were made on the coated substrates after 1000 hours of exposure, and the values were compared with the observed corrosion of the steel substrate. The best correlation of parameter values with corrosion was found with conductance. Corrosion potential did not show a consistent relationship, and weight gain and tensile adhesion showed no correlation with corrosion. It was concluded that the most important properties for coatings to be used in acid media are low permeability and resistance to degradation by acid. The vinyl ester, a bisphenol A epoxy cured with an aliphatic amine, and a novolac epoxy cured with a mixed aromatic/cycloaliphatic amine provided the best corrosion protection. The saturated polyester and a bisphenol A epoxy cured with a polyamide amine showed significant deterioration in acid and corrosion of the underlying steel. Two novolac epoxies cured with aromatic amines showed intermediate performance.

The mechanism for the initial corrosion of steel in neutral or alkaline solutions is generally accepted as the oxidation of iron from the metallic state to the ferrous ion:

$$Fe \rightarrow Fe^{++} + 2e^- \qquad (1)$$

with the attendant reduction reaction being the formation of hydroxide ion from oxygen and water [1]:

$$1/2O_2 + H_2O + 2e^- \rightarrow 2OH^- \qquad (2)$$

0097–6156/86/0322–0077$06.00/0

A variety of techniques has been developed to measure the condition of a coating so that some evaluation of its protective ability can be made. Many of these are based on electrochemical measurements [2]. The four techniques used in this study are (1) corrosion potential, (2) AC conductance, (3) tensile adhesion, and (4) weight gain.

The corrosion potential is determined by the potential at which the anodic and cathodic reactions occur at the same rate [3]. The AC conductance is a measure of the ease with which charge is transmitted through the coating [4]. The adhesive strength of the coating to the steel surface is affected by reactions occurring at the interface. The weight gain of coatings has been studied by Funke and Haagen [5] who have shown that a weight gain exceeding that of a free film indicates an accumulation of water at the interface [5].

These techniques are frequently used to study corrosion under coatings in neutral solutions and enjoy sporadic success, depending primarily on the type and thickness of coating being studied.

The mechanism of steel corrosion in acid solutions, however, is different from that in neutral solutions in that the reduction reaction is the formation of hydrogen from hydrogen ion:

$$2H^+ + 2e^- \rightarrow H_2 \quad (3)$$

Previous work in this laboratory has established that for epoxy and fluoropolymer coatings exposed to dilute sulfuric acid, there is movement of acid through the coating to the steel surface so that Equation 3 is the predominant reduction reaction [6].

Because of this difference in corrosion mechanism in acid solution, the usefulness of the four evaluation techniques discussed above may be different than in neutral solutions. The purpose of this work was to evaluate these four techniques for predicting the behavior of coating resins in acid solutions. In addition, the ability of several different types of coating resins to protect steel against corrosion in acid solution was evaluated.

Experimental

The following coating resins were used: (1) a vinyl ester (Derakane 470 from Dow Chemical); (2) a polyester (Atlac 382-05 AC from ICI); and (3) four epoxy resin/hardener combinations. The details of the resins and hardeners used are shown in Table I. One of the epoxy/hardener combinations was represented by materials from two sources.

The coatings were applied to one side of a steel substrate by means of a spray gun for the lower viscosity coatings, or by doctor blading with an adjustable Gardner knife for the higher viscosity materials. A casting technique was also used in which a known volume of the coating material was poured into a known area defined by heavy tape and was allowed to spread while on a level surface.

Table I. Coating Materials

Supplier	Designation	Resin	Hardener	Resin: Hardener Ratio	Cured Thickness (mils)	Cured Thickness (mμ)
Vinyl Ester						
Dow Chemical (Derakane 470)	VE	Bisphenol A Vinyl Ester + .15% CoNap*; 36% Styrene	Cumene Hydroperoxide	100:2	7-9	178-229
Polyester						
ICI Americas (Atlac 382-05 AC)	PE	Bisphenol A-Fumarate Polyester + 1% CoNap* + DMA; 50% Styrene	MEKP	100:2	6-12	152-305
Epoxies						
CON/CHEM	BPA/AL	Bisphenol A	Aliphatic Amine	100:20	9-10	229-254
	NOV/AR 1	Multifunctional	Aromatic Amine	100:50	10-12	254-305
Ciba/Geigy	NOV/AR 2	Multifunctional Novolac	Aromatic Amine	100:46	10-11	254-279
Martek	BPA/PA	Bisphenol A	Polyamideamine	2.77:1	8-12	203-305
	NOV/AR/AL	Novolac	Cycloaliphatic/ Aromatic Amine	1.82:1	8-9	203-229

*Cobalt Naphthanate (6%).

The coatings were cured in two steps: first, a room temperature exposure for at least five hours to allow any solvent present to evaporate and/or the resin to gel and, second, an elevated temperature cure. The vinyl ester, polyester and epoxies were baked at 60°C for 4-16 hours. The coating thickness after curing was measured with a micrometer, subtracting the substrate thickness. The thickness of the coatings ranged from 6 to 12 mils (150-300 μm) and is shown for each type in Table I.

The substrates were cold rolled, low-carbon SAE 1010 steel, 32 mils (0.8 mm) thick (Q Panels). They were sandblasted on both sides to a 6 μm profile with silica sand. No further cleaning was done. A circular disk 3.33 cm in diameter was prepared from a larger coated panel with a punch and die set. The disk was placed on a 125 ml widemouth screw-cap polypropylene bottle, using a rubber gasket to make a tight seal, with the coated side facing the inside of the bottle. The disk and gasket were held on the bottle with the cap, from which the central portion had been removed, so the back (uncoated) side of the steel substrate was exposed. The bottle was inverted and half filled with 0.1M H_2SO_4 through a hole drilled in the bottom of the bottle in order to contact the acid with the coating. The inverted bottle was placed in an oven at 60°C.

The corrosion potential was measured by putting a saturated calomel electrode/salt bridge into the solution through the hole in the bottom of the plastic bottle and contacting the back side of the substrate to complete the circuit as shown in Figure 1. A Keithley 600A electrometer was used for the measurement. The AC conductance was determined by inserting a carbon rod into the solution and measuring the conductance at 2 kHz frequency and 200 mv potential with an Extech Model 440 Digital Conductivity Meter. The conductance values were converted to specific conductivity by multiplying by the thickness and dividing by the area (5 cm^2).

The adhesion was measured by fastening a lead anchor of known area (2.84 cm^2) to the coating with a cyanoacrylate adhesive (Loctite 414) and after curing, pulling it off normal to the surface with a Dillon tensile tester. The force to remove the coating was divided by the area of attachment to convert it to a normalized tensile adhesion value.

The weight gain was measured by weighing the coated disk after it was removed from the plastic bottle, following a water rinse and removal of surface water.

The coating and substrate were observed through the hole in the bottle during the exposure to the acid. Since the coatings were transparent, it was possible to observe any visible corrosion occurring on the steel substrate. The corrosion products on the steel were gray or black, except when the coating blistered and some rusting was seen. As corrosion progressed during the acid exposure, the steel surface gradually darkened from the initial light gray of the original sandblasted surface to an almost black surface. The extent of corrosion was estimated from the amount of darkening observed on the steel surface.

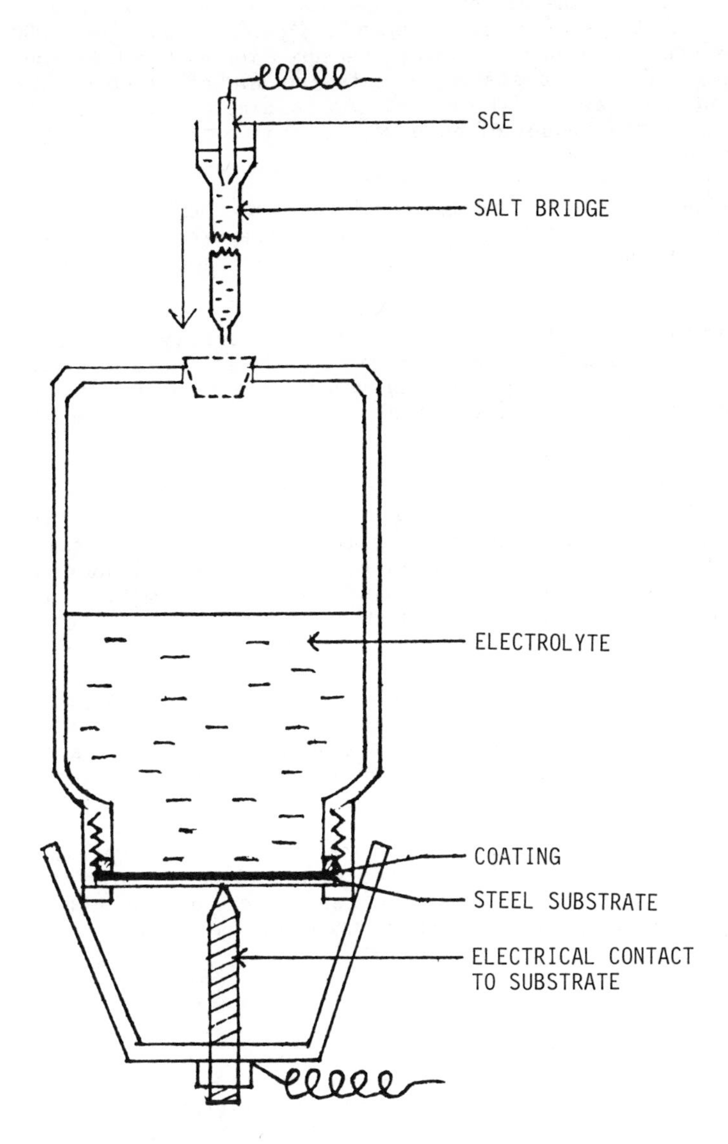

Figure 1. Technique for electrochemical measurements. Reproduced with permission from Reference 12. Copyright 1985, National Association of Corrosion Engineers.

Results

Figure 2 summarizes the values obtained for the four measurements on the seven coatings after exposure to 0.1M H_2SO_4 at 60°C for 1000 hours. The ordinate shows the values measured for each of the four techniques, and the abcissa represents the amount of corrosion observed on the steel under each of the coatings after the acid exposure, with the amount of observed corrosion decreasing from left to right.

The corrosion potentials show a general trend of increasing values with decreasing substrate corrosion, with the exception of the polyester and the novolac epoxy cured with an aromatic/cycloaliphatic amine.

The specific AC conductivity values show a generally decreasing value with decreasing substrate corrosion, with one exception: the novolac epoxy cured with an aromatic/cycloaliphatic amine. This is one of the coatings that also did not fit into the trend for the corrosion potential values.

The tensile adhesion values show no correlation with the extent of corrosion; the bisphenol A epoxy cured with a polyamide amine showed blistering, which represents a complete loss of adhesion. The polyester showed cohesive failure at less than 1000 hours of exposure, so a true adhesion value could not be determined. The other epoxies and the vinyl ester all had values in the 150-200 psi range, with no apparent relationship to the amount of corrosion.

Weight change data were obtained for only four of the seven coatings, and those data showed no correlation with the extent of steel substrate corrosion. The polyester showed a weight loss, rather than a weight gain, probably due to an attack and dissolution of the epoxy by the acid.

It should be noted that the electrochemical measurements (corrosion potential and conductivity) for the two novolac epoxies cured with an aromatic amine from different sources showed good agreement, although the tensile adhesion and weight gain values were not as reproducible.

Discussion

The best performing coatings were the vinyl ester, the bisphenol A epoxy cured with an aliphatic amine, and a novolac epoxy cured with a mixed aromatic/cycloaliphatic amine. The saturated polyester, and a bisphenol A epoxy cured with a polyamide amine showed significant deterioration of the coating material in the acid, and corrosion of the underlying steel. Two types of novolac epoxies cured with aromatic amines showed intermediate performance.

Only one of the four techniques--the conductivity--showed any correlation with the observed extent of corrosion. The lack of correlation of the tensile adhesion values with corrosion is a result of the fact that the method integrates adhesion loss at the substrate

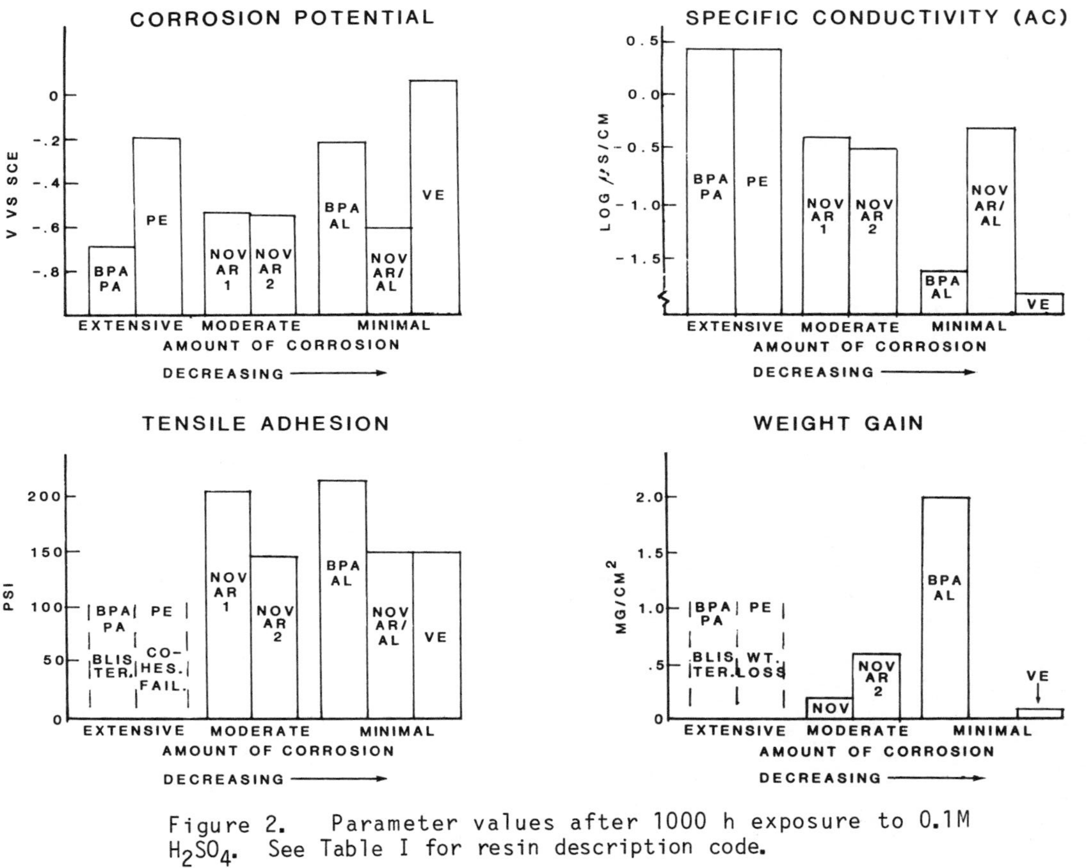

Figure 2. Parameter values after 1000 h exposure to 0.1M H_2SO_4. See Table I for resin description code.

interface and the cohesion loss due to deterioration of the polymer by acid. Also, the corrosion often is localized and the area attacked is a small fraction of the total interfacial area. The weight gain is not a reliable measure of corrosion protection because there may be an attack and solubilization of the coating material itself (as occurred with the polyester), so that the observed change in weight will be the net result of a weight loss from solubilization and a weight gain from water and acid entry into the coating. The only way to separate the two effects is to measure the weight change occurring with a free film of the coating and assume it will be the same as a coating on a substrate; this was the technique used by Funke [5]. There is much more likelihood of attack of coating materials in acid solutions because of increased rates of hydrolysis reactions at low pH's.

The interpretation of corrosion potential has always been difficult. Wolstenholme [3] concluded that the corrosion potential was not an unambiguous indicator of the amouunt of corrosion, and Cerisola and Bonora [7] described the measurement as one with no quantitative relationship to amount of corrosion. The results shown in Figure 2 confirm the questionable value of potential measurements in correlations with the corrosion of an underlying substrate.

The coating conductance, on the other hand, has been reported by a number of people to be related to the extent of corrosion under a coating [8-11]. The conductance, either AC or DC, is a function of the amount of charge that can pass through the coating and this amount of charge is a function of the amount of aqueous phase in the coating that permits charge motion.

Thus, in acid solution it appears that an important property of a coating for corrosion protection is its permeability to acid. This variation in permeability is thought to be the reason for the difference in behavior of coatings observed during exposure to acid environments [12]. The permeability is also affected by the degradation of the coating as caused by reaction with the acid.

Conclusions

Of the four techniques studied for evaluating coatings on steel for corrosion control (corrosion potential, conductance, adhesion and weight gain), the most useful was conductance. Corrosion potential did not show a consistent relationship, and weight gain and tensile adhesion showed no correlation with corrosion.

The best performing coatings studied were a vinyl ester, a bisphenol A epoxy cured with an aliphatic amine, and a novolac epoxy cured with a mixed aromatic/cycloaliphatic amine. A saturated polyester, and a bisphenol A epoxy cured with a polyamide amine showed significant deterioration in the acid and corrosion of the underlying steel. Two types of novolac epoxies cured with aromatic amines showed intermediate performance.

Acknowledgments

The authors are indebted to the Electric Power Research Institute, Palo Alto, California, for supporting this work and to B. C. Syrett of that organization for helpful discussions during the study.

Literature Cited

1. Fontana, M. G.; Greene, N. D. "Corrosion Engineering"; McGraw-Hill: New York, 1978; 2nd ed.
2. Leidheiser, H. Jr. Prog. Org. Coatings 1979, 7, 79-104.
3. Wolstenholme, J. Corr. Sci. 1973, 13, 521-30.
4. Mansfeld, F.; Kendig, M. W.; Tsai, S. Corrosion 1982, 38, 478-85.
5. Funke, W.; Haagen, H. Ind. Eng. Chem. Prod. Res. Dev. 1978, 17, 50-53.
6. White, M. L.; Paper given at ACS Meeting, Philadelphia, PA, August 1984. Submitted to Ind. Eng. Prod. Res. Dev. for publication in 1986.
7. Cerisola, G.; Bonora, P. L. Mater. Chem. 1982, 7, 241-48.
8. D. J. Mills. In Coatings and Surface Treatment for Corrosion and Wear Resistance; Strafford, K. N., Ed.; Horwood: Chichester, England, 1984; pp. 315-30.
9. Rajagopalan, K. S.; Guruviah, S.; Rajagopalan, C. S. J. Oil Col. Chem. Assoc. 1980, 63, 144-48.
10. Touhsaent, R. E.; Leidheiser, H. Jr. Corrosion 1972, 28, 435-40.
11. Vertere, V.; Rozados, E.; Carbonari, R. J. Oil Col. Chem. Assoc. 1978, 61, 419-26.
12. White, M. L.; Leidheiser, H. Jr., Materials Performance 1985, 24, 9-16.

RECEIVED January 27, 1986

8

Comparison of Laboratory Tests and Outdoor Tests of Paint Coatings for Atmospheric Exposure

M. Morcillo[1], J. Simancas[1], J. M. Bastidas[1], S. Feliu[1], C. Blanco[2], and F. Camón[2]

[1]Centro Nacional de Investigaciones Metalúrgicas, Ciudad Universitaria, 28040–Madrid, Spain

[2]Instituto Nacional de Técnica Aeroespacial "Esteban Terradas," Torrejón de Ardoz, Madrid, Spain

Many accelerated laboratory tests have been devised to determine the susceptibility of paint films to breakdown by atmospheric weathering, however, the demand for a generally applicable test exists. In this study different typical paint systems have been subjected to various natural environments and laboratory tests (DEF-1053 Method No. 26, ASTM G53-77, Salt Spray and Electrochemical Impedance Measurements). The results indicate that electrochemical impedance measurements provide a satisfactory correlation with the behaviour of paint coatings as evaluated by visual examination. In addition, it appears that, in certain cases, data obtained by this technique will allow prediction of the metallic corrosion underneath the paint coating when no changes in the appearance of the coating can be externally observed.

In practice, most paint coatings exhibit some reduction in protective properties during their service life. Paint coatings exposed to the atmosphere undergo a progressive degradation which ultimately results in complete loss of protective action. This degradation is attributable to a number of environmental factors, the most important being atmospheric contamination, ultra violet light, moisture, and temperature fluctuations.

It is extremely important to predict coating service life with some degree of certainty. Clearly, the best

0097–6156/86/0322–0086$06.00/0

assesment of a coating's protective efficiency is prolonged exposure to the proposed service conditions combined with regular inspections at predetermined time intervals. However, as the useful life of the organic coatings under atmospheric exposure is often in excess of ten years, it is not always realistic to test new developments until they have failed in service. Consequently accelerated test are necessary to determine the susceptibility of paint films to breakdown by weathering or other climatic factors. Many accelerated laboratory tests have been devised for this purpose.

Artificial Weathering Tests. Progress in this area has been achieved as chemists increased their understanding of the degradation mechanisms of polymeric systems. Since a direct relationship between the intensity of sunlinght and the rate of deterioration of paint seemed to exist, an intense light source was considered to be the primary prerequisite to simulate or accelerate the degradation of polymers in the laboratory.

Presently, there is a great number of apparatus and methods for accelerating weathering in the laboratory (DEF - 1053 No 26, ASTM G23 Carbon-arc type, ASTM G26 xenon-arc type, ASTM G53 fluorescent UV-Condensation type, etc.). However, even though these devices exist, it is difficult to obtain reproducible results. Hence, the rather pessimistic statement of Papenroth (1): "A general accelerated weathering method which is valid in all cases does not exist today and there will never be one".

Accelerated weathering tests are performed in special test chambers all of which are similar in that a coated specimen is exposed in a controlled environment containing a high intensity ultraviolet light source and a facility for spraying water onto the test surfaces.

Accelerated Corrosion Tests. There are as many as a dozen methods (salt fog, Kesternich, etc.) that are currently being used to investigate corrosion resistance of coating systems and a need to develop a better and more dependable method to predict in-use service. A severe drawback of all these tests is that their results often compare unsatisfactorily with practical experience. One reason for the discrepancies is assumed to be the variability of natural exposure conditions. Accordingly, cyclic testing procedures have been developed with which exposure conditions, especially temperature and humidity, are sistematically varied. Results obtained with such cyclic tests agree much better with practical performance than those from constant condition test. However at present there is no single accelerated test which correlates satisfactorily with actual long term performance in any environment; the demand for a generally applicable corrosion test exists (2).

Electrochemical Corrosion Tests. Electrochemical procedures are very well established for investigating the corrosion behaviour of pure metals, alloys and metallic coatings. The application of these techniques to paint coatings is, however, far less common. It was disappointing to find that up to 1973 (3) electrochemical tests did not provide much useful information. While this situation was fully justifiable a decade ago when d.c. electrochemical methods were almost exclusively used, the picture seems to have changed radically when recent results obtained by a.c. impedance measurements are examined. The advantage of these impedance measurements over the classical tests mentioned is its non destructive nature; in addition, corrosion and coating damage may be determined prior to its visual manifestation. This coincides with the present trend towards the development of methods which enable early prediction of the paint coating performance, even before the occurrence of any substantial changes in its appearance.

Experimental

Test samples have been prepared from a 3 mm hot-rolled steel sheet which showed an intact mill scale (Degree A of Swedish Standards, SIS-055900). Thereafter they were shot blasted with S-280 to reach the ASa3 standard, prior to the application of the paint coating. Surface preparation B St 2 was obtained by wire brushing a steel sheet of Grade B obtained in turn by oxidation of Grade A sheet in a contaminant free atmosphere. In "Table I.", the characteristics of the paints used in this study are shown.

To study the effect of contaminants (chlorides and sulphates) at the interface metal/coating, a set of panels (surface A Sa 3) was prepared and dosed with solutions of NaCl and $FeSO_4$ in distilled water and methanol. Subsequently, two paint systems (chlorinated rubber and polyurethane) were applied on these contaminated surfaces.

Table I. Characteristics of paint systems.

Paint System	System Designation	Primer	Undercoat	Topcoat	Total Thickness (μm)
Oil/Alkyd	O/A	INTA 164103	--	INTA 164218	90–120
Alkyd	A	INTA 164201	--	INTA 164218	80–110
Chlorinated Rubber	CR	INTA 164705	INTA 164701A	INTA 164704A	80–100
Vinyl	V	INTA 164604	INTA 164602A	INTA 164603A	100–120
Polyurethane	P	RENFE 03.323.125			120–150
Epoxy/Polyur.	E/P	MIL C-82407	RENFE 03.323.125		90–120

The following tests were used:

1. Natural weathering tests involving different environments: rural, urban, industrial and marine. At present only results for two years of field exposure are available.
2. Artificial weathering tests in climatic chambers in accordance with the following specifications:

2.1. DEF-1053 Method No. 26. Alternative running of carbon arcs in 6-hour periods, with a chamber temperature of about 40ºC. In between carbon arc operational periods, deionized water spraying with a resistivity over 300.000 Ω.cm^2. Testing period: 3500 hours.

2.2. ASTM G53. Condensation (44ºC) and cooling (65ºC) cycles were used of 4-hour duration each, with a 0.5 hour in between. An ultraviolet radiation was continuously produced. Testing period: 2300 hours.

3. Salt Spray (fog) test (ASTM B 117) with a testing period of 3200 hours.
4. Impedance diagram technique. The polarization cell consisted in a transparent plastic tube that was adhered to the paint surface by means of a silicone sealer. The tube contained distilled water and a 25 cm^2 platinized titanium sheet, which was used as auxiliary electrode. Measurements were made with

the double electrode technique (4) and the help of a H.P. 4800-A Vector Impedance Meter, over a frequency range of 50 KH_z to 5 H_z.

Results and Discussion

Regular inspections of rusting and blistering grades shown on painted surfaces were carried out. Failure was evaluated according to ASTM D610 and ASTM D 714 standards. The results obtained are given in "Tables II and III." As is apparent, no visible damage was observed on ASa 3 steel panels ("Table III".) after two years' test exposure to the various atmospheres. The salt fog test, enabled us to arrange the different paint systems according to the exposure time elapsed before the panel reached a particular rating level. We have chosen rust Grade 8 (0.1% rust) since it is a reasonably accepted indicator in this type of studies. It was seen that systems P and E/P remain unaltered after 3200 hours testing. The remaining systems fail at shorter test times.

With regard to the ASTM G 53 and DEF-1052-26 accelerated weathering tests, no close agreement was observed between them. Whereas in the ASTM G53 test the CR and V systems show rust points at 800-1000 hours, in the DEF-1053-26 test both systems keep in almost perfect condition. However, in this test the O/A and A systems are the only ones which show a slight corrosion at 1900 hours. In either of the two cases only a slight corrosion is involved which does not reach Grade 8 in the ASTM range.

Of the BSt2 steel panels under atmospheric exposure ("Table II".) only the CR system shows early rust spots (6-12 months). After the first 3-6 months, rust blisters appear on the paint surface. Later these blister breaks due to accumulation of corrosion products inside them. The rust Grade is higher as is the atmospheric corrosivity.

With this paint system (CR) laboratory tests correlate quite well with the results in the atmosphere. So, in the salt fog chamber the worst paint performance is shown by this system, which is also the only one that shows slight rusting in the ASTM G 53 test. The DEF-1053 test promotes blistering in this system, as well as in the O/A, A and V systems.

When making an overall examination of "Tables II. and III." to correlate the results obtained in the laboratory with those in the natural environments, we are faced with the problem of the short time elapsed in the

Table II(A). Outdoor and laboratory tests of paint coatings applied over ASa 3 steel panels.

System Designation	Outdoor (2 years)	Salt Spray (hrs. to rust Grade 8)	ASTM G53 (2300 hrs.)	DEF-1053(26) (3500 hrs.)
O/A	NF	800	NF	1900 hrs.: slight rusting
A	NF	1800	NF	1900 hrs.: slight rusting
CR	NF	2100	800 hrs.: light rusting	NF
V	NF	2400	1000 hrs.: pinholes in film and rusting	160 hrs.: slight blistering
P	NF	$>$3200	NF	NF
E/P	NF	$>$3200	NF	NF

NF = No failures in time specified

Table II(B). Outdoor and laboratory tests of paint coatings applied over BSt 2 steel panels.

System Designation	Outdoor (2 years)	Salt Spray (hrs. to rust Grade 8)	ASTM G53 (2300 hrs.)	DEF-1053(26) (3500 hrs.)
O/A	NF	1400	NF	1900 hrs.: slight blistering and rusting
A	NF	800	NF	1900 hrs.: slight blistering
CR	6–12 months: rust coming through blisters. Rust Grade increases with atmospheric corrosivity	500	1000 hrs.: slight rusting	1900 hrs.: blistering
V	NF	>3200	NF	160 hrs.: slight blistering
P	NF	>3200	NF	NF
E/P	NF	>3200	NF	NF

NF = No failures in time specified

Table III(A). Outdoor and laboratory tests of paint coatings applied over ASa 3 and BSt 2 steel panels. Paint system: Chlorinated Rubber.

Surface Condition	Outdoor (2 years)	Salt Spray (hrs. to rust Grade 8)	ASTM G53 (hrs. to slight rusting)	DEF-1053 (26) (hrs. to rust)
ASa 3	NF	2100	800	NF
BSt 2	6-12 months: rusting	500	1000	NF
ASa3, 20 mg/m^2 NaCl	NF	700	800	NF
ASa3, 100 mg/m^2 NaCl	NF	700	1200	NF
ASa3, 500 mg/m^2 NaCl	12-18 months: blistering and rusting	500	1100 (heavy rusting)	NF
ASa3, 250 mg/m^2 $FeSO_4$	NF	700	1500	NF
ASa3, 500 mg/m^2 $FeSO_4$	NF	700	1500	NF
ASa3, 1000 mg/m^2 $FeSO_4$	NF	500	1500	NF

NF = No failures in time specified

Table III(B). Outdoor and laboratory tests of paint coatings applied over ASa 3 and BSt 2 steel panels. Paint system: Polyurethane.

Surface Condition	Outdoor (2 years)	Salt Spray (hrs. to blister Grade 6*)	ASTM G53 (hrs. to blister)	DEF-1053(26) (hrs. to blister)
ASa 3	NF	1800	NF	NF
BSt 2	NF	400	NF	NF
ASa3, 20 mg/m^2 NaCl	NF	1600	NF	NF
ASa3, 100 mg/m^2 NaCl	NF	800	NF	NF
ASa3, 500 mg/m^2 NaCl	12–18 months: slight blistering	150	NF	NF
ASa3, 250 mg/m^2 $FeSO_4$	NF	2400	NF	NF
ASa3, 500 mg/m^2 $FeSO_4$	NF	2400	NF	NF
ASa3, 1000 mg/m^2 $FeSO_4$		2400	NF	NF

NF = No failures in time specified.
* = Blister Grade according conversion Table of ASTM D714 to a numerical scale (5).

latter in order to reach significant differences among the different paint systems. Among the various tests carried out, the salt spray test is the one that best correlates with atmospheric exposure. Apparently it is the most aggresive of all tests, and the only accelerated corrosion test in comparison with the remaining ones considered specifically as accelerated weathering tests.

In the accelerated weathering tests some of the systems showed a certain degradation (changes in colour and/or brightness, chalking, etc.) which have not been included in tables since they were not considered of interest for our purpose. However, such tests provide perhaps the most useful information for evaluating these type of degradation. Accelerated weathering tests are perhaps the most suitable for indicating the behaviour of the coating system itself, but such tests do not always offer an accurate indication of the corrosion protection provided by the coating. These methods are not by themselves corrosion tests and are only likely to cause significant surface deterioration of the paint coating (6,7).

Electrochemical impedance measurements were used to determine the effect of accelerated weathering of the paint coating on the corrosion of the steel substrate. Figures 1 and 2 give the impedance diagrams at different times for the various paint systems carried out on non weathered coatings (N.W.) and on coatings after weathering by means of the ASTM G 53 and DEF-1053- 26 tests for 2300 and 3500 hrs, respectively. In accord with the visual observations, the CR and V systems in the ASTM G 53 test and O/A and A systems in the DEF-1053 -26 test, show a diagram in a semicircle form, which indicates the existance of paths or channels of electrolyte penetration across the film permiting the movement of the chemical species to and from the metal substrate (corrosion, diffusion, etc.) (8). The other systems indicate a paint film in perfect condition and the impedance diagram takes the shape of a straight line forming a certain angle with the imaginary axis. This response is similar to that of a capacitor, with phase angles near to 90º and very high values of the impedance modulus. Thus, this measuring technique reveals quite satisfactorily the corrosion damage observed by the naked eye on painted surfaces.

Effect of Pollutants on the Metal/Paint Interface. "Table III." gives the results of different tests carried out on CR films applied to rusted steel surfaces (BSt 2) or surfaces contaminated with NaCl and $FeSO_4$. In the case of rust

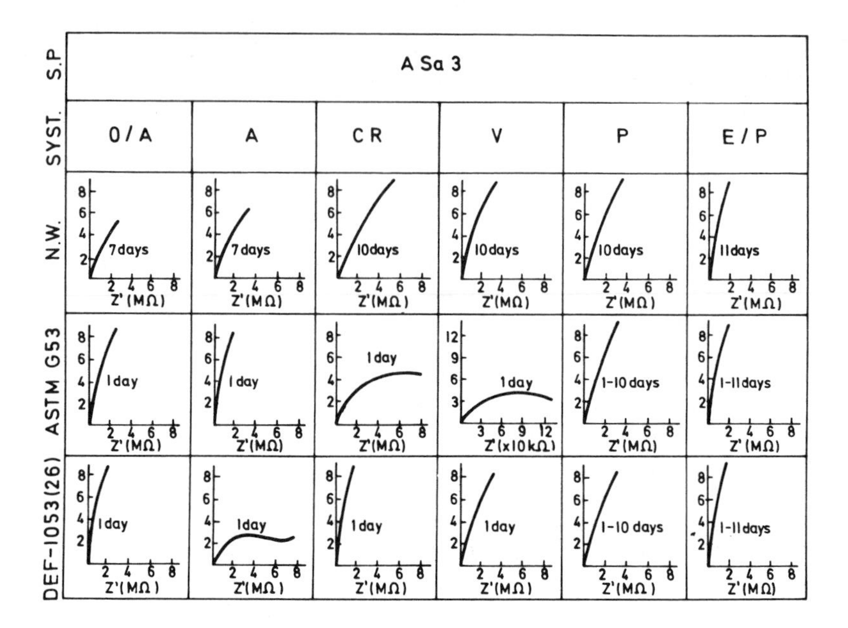

Figure 1. Impedance diagrams of not weathered (N.W.) and weathered coatings in accelerated chambers.

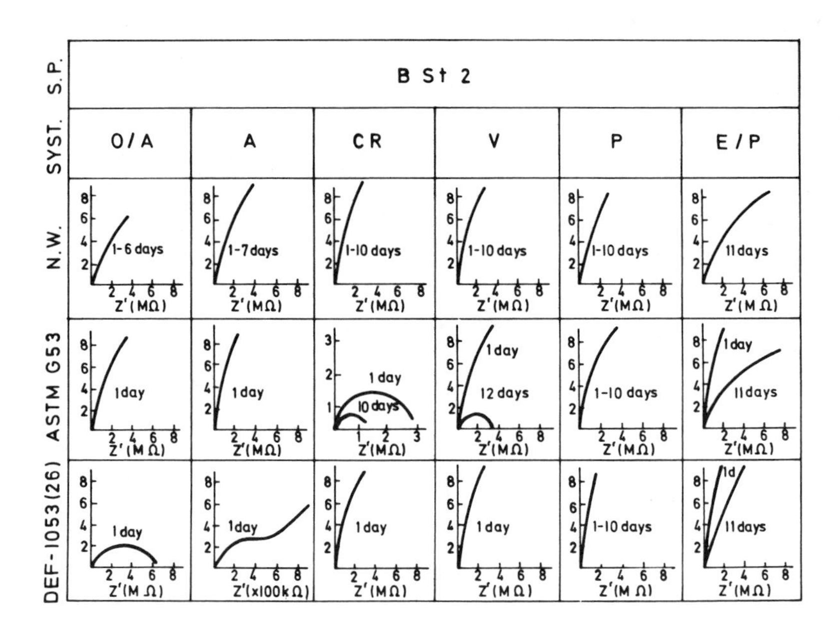

Figure 2. Impedance diagrams of not weathered (N.W.) and weathered coatings in accelerated chambers.

or NaCl contaminants the salt fog test results, in particular "hours to rușt Grade 8", show a quite acceptable relationship with the outdoor exposure results. The dangerous effect of the presence of rust or NaCl on the metal/paint interface can be detected quite early. However, the presence of sulphates in the salt fog test promotes an important corrosion of the substrate unobserved in the natural exposure tests, at least, during the first two years of experimentation.

The ASTM G 53 test also shows, though not so clearly, the effect of the chloride contaminant on the integrity of the paint. The DEF-1053 -26 test however does not seem to be sensitive to the presence of this metallic corrosion accelerator.

The impedance diagrams (Figure 3) show quite well the presence of the oxide which was detected visually on test panels exposed to the effect of the ASTM G 53 test. These diagrams move away from the almost capacitive behaviour and show, in particular, when chloride contamination is involved, a smaller semicircle diameter, the higher the saline contamination on the interface.

In all the paint systems tested in the atmosphere, the presence of the chloride contaminant at its highest concentration, causes the appearance of small blisters. Such blisters may in time burst due to the oxide built-up inside them, as was seen in the case in the CR system ("Table III.") In the P system ("Table III."), the blistering seen with the third level of chlorides continues after two years' outdoor exposure, without rust coming through the coating. This same effect is also seen in salt fog test panels, but it does not appear in those test specimens subjected to accelerated weathering tests.

A very important question is whether the impedance measurements yield information about the deterioration process sooner than is obtained by visual examination. With this in mind it is interesting to comment upon some of the results obtained (8) with the CR system and shown in Figure 4. Together with the impedance diagrams for the various contamination levels studied (0, 20, 100 and 500 mg/m^2 NaCl), photographs are included in it showing the condition of the test panels after the tests in distilled water (231 days) and in a rural atmosphere (2 years). The diagram for the highest chloride level (500 mg/m^2) shows the most advanced stage (smaller semicircle diameter) which corresponds to marked oxidation in the test panels. Diagrams corresponding to lesser amounts of contaminants, although they are found at lower

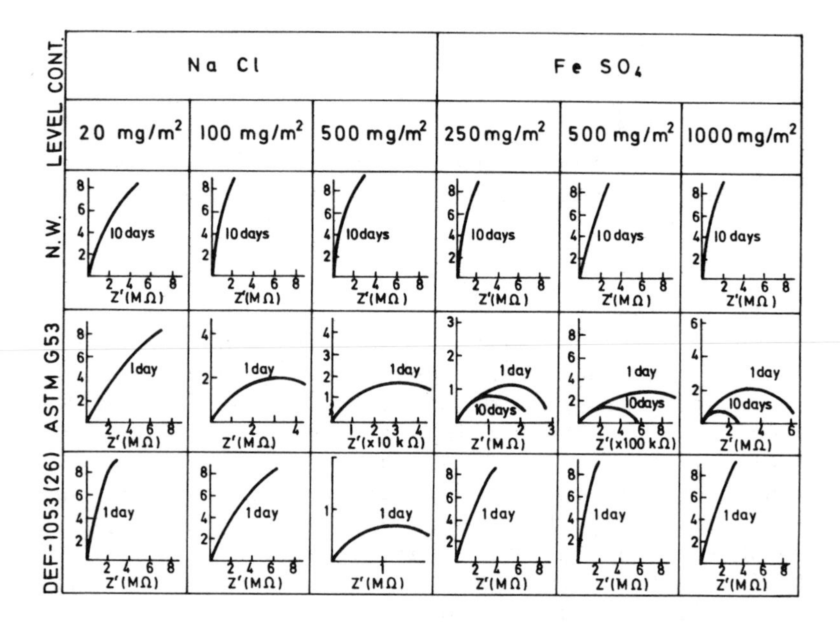

Figure 3. Chlorinated Rubber System. Effect of contaminants. Impedance diagrams of not weathered (N.W.) and weathered coatings in accelerated chambers.

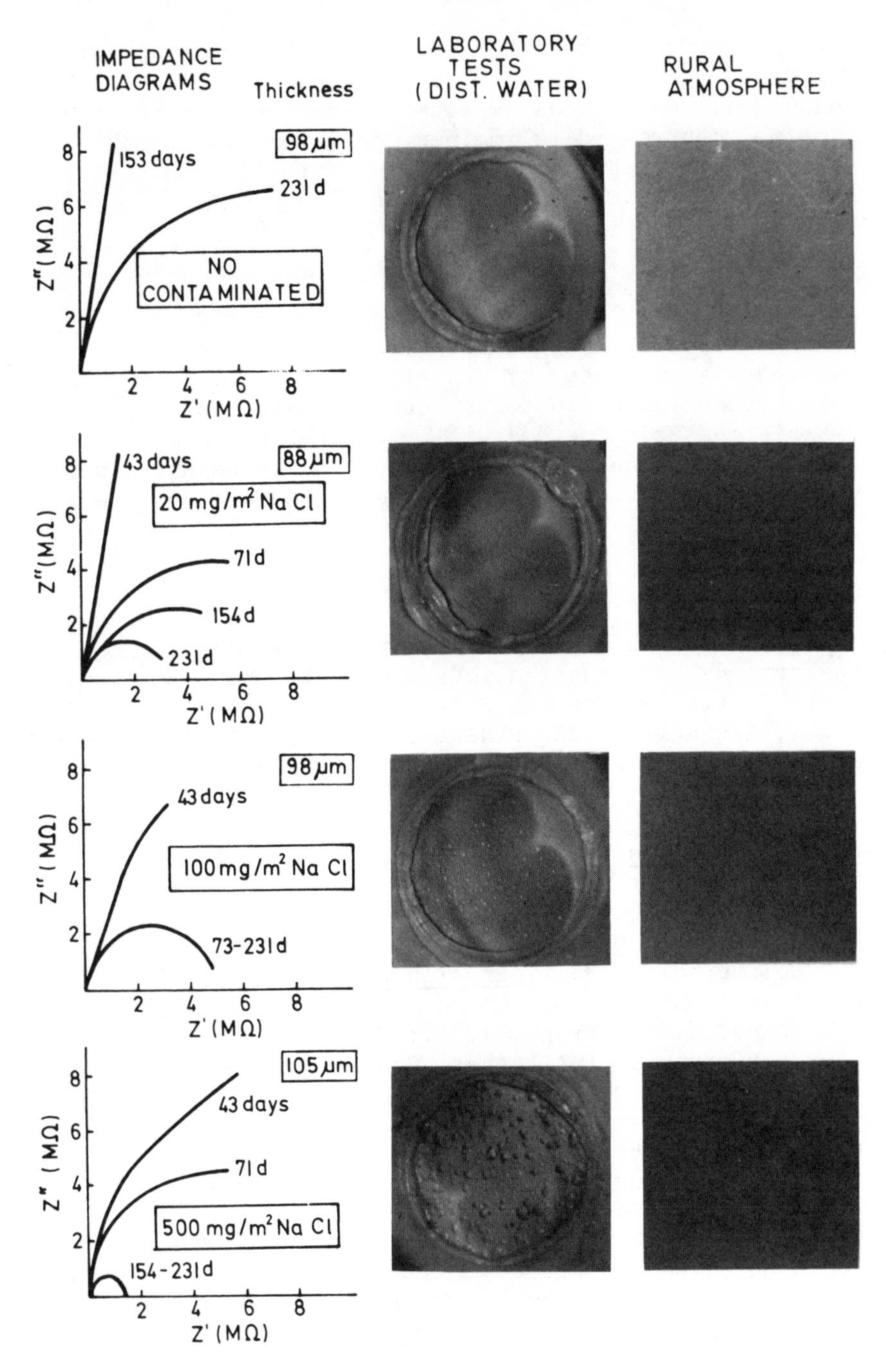

Figure 4. Effect of interface contamination by chlorides. Paint System: Chlorinated Rubber.

development stages, do not seem to be in agreement with the appearance of the painted surface, which is as yet undamaged. One may ask whether such diagrams are ahead of time with regard to imminent deterioration of the paint coating. The lapse of further investigation/test time is needed to answer this question.

Conclusions

Electrochemical impedance measurements provide a quite satisfactory correlation with the behaviour of paint coatings as evaluated by visual examination. There are well founded hopes that, in certain cases, data obtained by this measuring method will even enable the metallic corrosion underneath the paint coating to be anticipated, when no changes in the appearance of the coating can as yet be externally observed.

Acknowledgments

The authors wish to express their acknowledgments to all the persons and organizations who collaborate in this project, particularly to the following spanish firms: AESA, ANQUE, ASEFAPI, BECKER-PINER, BUFI y PLANAS, CROS PINTURAS, EMP, ERTISA, FECSA, GLASURIT, HEMPEL, PROCOLOR, RESINOR AND VALENTINE.

Literature Cited

1. Papenroth, W. Defacet 1974, 6, 282.
2. Funke, W. JOCCA 1984, 3, 71.
3. Wolstenholme, J. Corrosion Science 1973, 13, 521.
4. Scantlebury, J. D.; Ho, N. JOCCA 1979, 62, 89-92.
5. Keane, J. D.; Bruno, J. A.; Weaver, R. E. F. In "Performance of Alternate Coatings in the Environment"; Steel Structures Painting Council: Pittsburgh, 1979; P. V-8.
6. Von Fraunnofer, J. A; Boxall, J. In "Protective Paint Coatings for Metals"; Ed.; Portcullis: England, 1976.
7. Carter, V. E. In "Corrosion Testing for Metal Finishing"; Ed.; Butterworths Scientific: London, 1982.
8. Morcillo, M.; Feliu, S; Simancas, J; Bastidas, J. M. Proc. 5th Int. Cong. on Marine Corrosion and Fouling, 1984, p. 401.

RECEIVED January 21, 1986

9

Degradation of Organic Protective Coatings on Steel

Tinh Nguyen and W. Eric Byrd

Building Materials Division, National Bureau of Standards, Gaithersburg, MD 20899

The application of reflection/absorption Fourier transform infrared spectroscopy (FTIR-RA) for studying the degradation of two types of coatings on steel after exposure to an 40° C/80% RH environment is presented in this paper. FTIR-RA results indicate the occurrence of bond weakening, dehydration and bond scissions of amine-cured epoxy after exposure. On the other hand, the polybutadiene coating shows not only bond weakening but also extensive degradation which results in the formation of various oxidized products and losses in unsaturation. The characterization of complex molecules formed during the oxidation and degradation by FTIR-RA offers a powerful means for studying the degradation processes of protective coatings on steel.

Metallic corrosion is estimated to cost the U.S. more than $100 billion annually (1). Polymeric coatings are widely used to prolong the service life of corrosion-prone substrates. However, these coatings can undergo physical and chemical changes under service conditions that reduce their effectiveness. Chemical changes which occur at the metal/coating interface are particularly important in the degradation processes. Part of our overall task in predicting the service life of protective coatings is to gain a better understanding of the degradation mechanisms leading to adhesion failure and to corrosion of the protected metal beneath the coating.

Fourier transform infrared spectroscopy has been shown to be an excellent tool for surface and interface studies (2). In this paper, the application of reflection/absorption Fourier transform infrared spectroscopy (FTIR-RA) for studying the degradation of amine-cured epoxy and polybutadiene coatings on cold-rolled steel after exposure to a warm, humid environment is reported.

EXPERIMENTAL

Materials

The amine-cured epoxy used was from a formulation containing: 100 parts of epoxy resin (Epon 1001, Shell Chemical Co.), 6 parts of diethylenetriamine, and 3 parts of butylated urea formaldehyde (flow-controlled agent). All components were at 12.5% in glycol ether/xylene (50/50, w/w) solutions. The epoxy and flow-controlled agent were premixed together as the base component, which was mixed with the curing agent just before coating. The polybutadiene was Budium RKY662 (Dupont) in petroleum distillate solvent. The cold-rolled steel samples were low-carbon SAE 1010, of the size 25 x 25 x 0.8 mm. (Trade names are given solely to indicate experimental materials and equipment used and not to recommend a particular product).

Specimen Preparation

Steel substrates were mechanically polished (final polishing was with a 1μm diamond paste), rinsed repeatedly with water and methanol, blown dry with hot air, and stored in a desiccator before use (but not more than 3 days). The dried substrate specimens were immersed in methanol for 2 hours, then washed with acetone immediately prior to applying the coatings. Both methanol and acetone were ACS grade reagents. The coatings were applied by flooding the substrate with resin solutions and spinning them horizontally at 3500 RPM for 30 s using a photoresist spinner. Initial film thicknesses of 1.8μm for the epoxy and of 2.0μm for the polybutadiene were measured using a scanning spectroscopic reflected light microscope. The unaged FTIR-RA spectra were collected after curing for 3 weeks at room conditions. All specimens were then exposed in a 40°C and 80% RH chamber. Aged samples were removed from the chamber at various time intervals for IR analysis. Film thicknesses of the aged samples were not measured.

FTIR-RA Spectra

Spectra were obtained using a 60 SX Nicolet FTIR spectrometer and a Barnes variable angle specular reflection accessory. The spectrometer was equipped with a nitrogen-cooled mercury cadmium telluride detector and was constantly purged with dry air to minimize the effect of moisture. The instrument was also equipped with a laser interferometer to insure wave length accuracy. All spectra were the result of 1000 co-additions and taken at 4 cm^{-1} resolution using single reflection at an incident angle of 45 degrees. All spectra are shown in the absorbance mode and have been ratioed against the spectrum of a silver reference mirror.

RESULTS AND DISCUSSIONS

Degradation of Amine-Cured Epoxy Coating on Cold-Rolled Steel

FTIR-RA spectra of the amine-cured epoxy coating on cold-rolled steel before and after exposure at 40°C and 80% RH are shown in

Figure 1. The assignments of FTIR-RA spectra of unaged epoxy free and coated films on steel have been given previously (3). It should be noted that this aged specimen exhibited extensive filiform corrosion after 7 months exposure. Figures 2 show the spectral changes at different exposure periods. These spectra have been normalized for the baseline shift resulting from the reflection change of the steel substrate due to exposure and corrosion. Figure 2a shows that aging of amine-cured epoxy in the test environment resulted in a shift to higher frequency of the OH maximum at 3401 cm^{-1} and an increase in intensities of the band at 3560 cm^{-1}. Chen et al., (4) also noted a shift to higher frequency due to thermal degradation of the OH band in trimethoxyboroxine-cured epoxy coating on aluminum. On the other hand, Harrod (5) observed progressive shifts to higher frequency of the OH band maxima of amine-cured epoxy with increasing temperature and attributed the shift to the change from long range to short range H-bonding of both the O-H---N and O-H---O types. FTIR-RA studies of thin films on steel and free films (6) indicated that there is no strong interaction between the amine-cured epoxy and the steel substrate. Thus, the shift to higher frequency of the OH maximum at 3401 cm^{-1} of amine-cured epoxy on steel exposed to a corrosive environment is attributed to the weakening of the H-bonds within the epoxy coating itself. The intensity increase at the 3560 cm^{-1} band indicated that nonassociated OH groups were formed during exposure.

The bands at 1041 and 1085 cm^{-1} (C-O stretching) broadened and increased in intensity (Figure 2c) indicating the formation of more

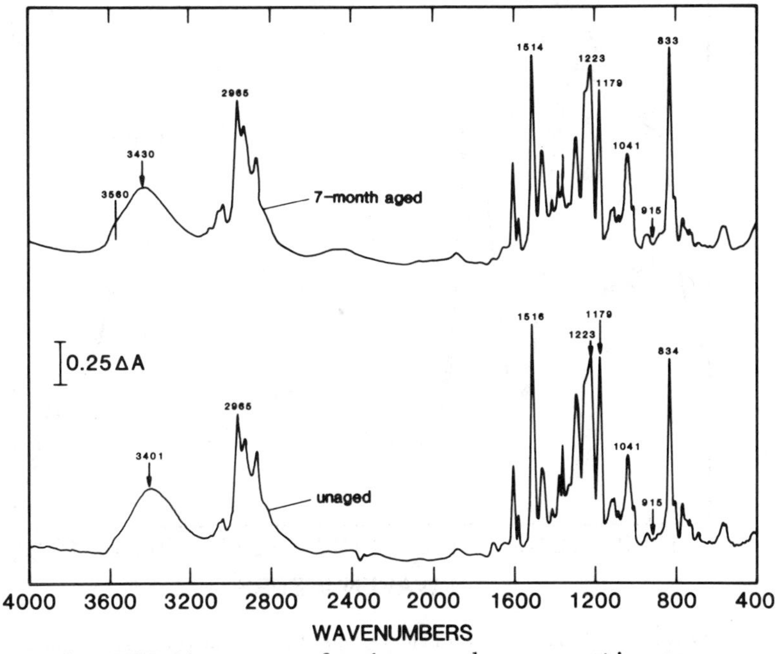

Figure 1. FTIR-RA spectra of amine-cured epoxy coating on cold-rolled steel before and after exposure to 40° C and 80% RH

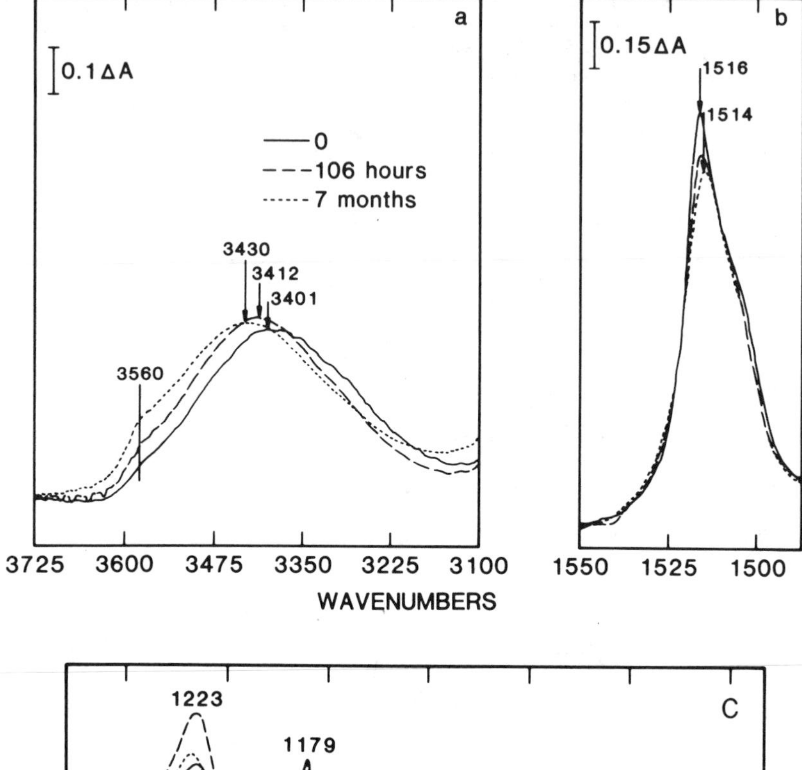

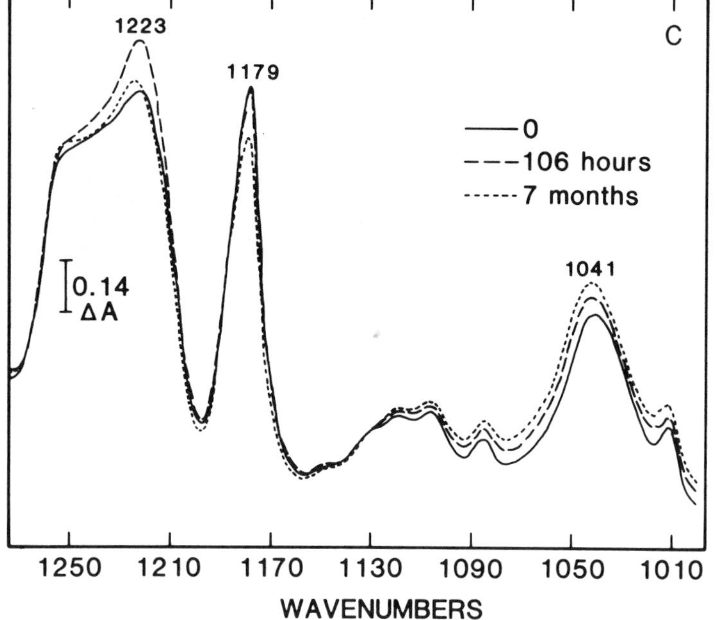

Figure 2. Normalized FTIR-RA spectra of amine-cured epoxy coating on cold-rolled steel exposed to 40°C and 80% RH for different times. Reproduced, with permission, from Ref. 3

OH groups during exposure. Despite the gradual increase in the C-O region and the reduction of the epoxide groups (915 cm^{-1} band) to form OH, the intensity in the OH region remains unchanged. This is probably due to dehydration, either through chemical reaction or loss of sorbed water. Dehydration has been identified as the major reaction and water has been identified as a major product of the degradation at low temperatures of amine-cured epoxy (7,8). The decreases in the intensities of the bands at 1516 cm^{-1} (C=C of the benzene ring) and 1179 cm^{-1} (C-C stretching of isopropylidene) suggest that the bisphenol-A structure of the coated epoxy has degraded as well. The presence of N in the cured epoxy makes it very susceptible to nucleophilic chain breaking and allyl-N bond scissions; both of these reactions result in loss of the aromatic rings (7,10). On the other hand, the loss of isopropylidene group may be explained by its low bond energy (8). Although this group is more stable than the cure linkage in amine-cured epoxy (9), it is the least stable group in anhydride-cured epoxy, and undergoes degradation during early decomposition (8).

There is also a noticeable increase in intensity of the aryl-O band at 1223 cm^{-1}, especially in the early aging stages. This indicates some association, probably through H-bonding, between the N or OH group and the phenyl ether. It is known that interaction not only shifts the frequency but also increases the intensities of an IR band (11). The association between N and O was evidenced by IR spectroscopy (12) and the intensity increase of the 1240cm^{-1}

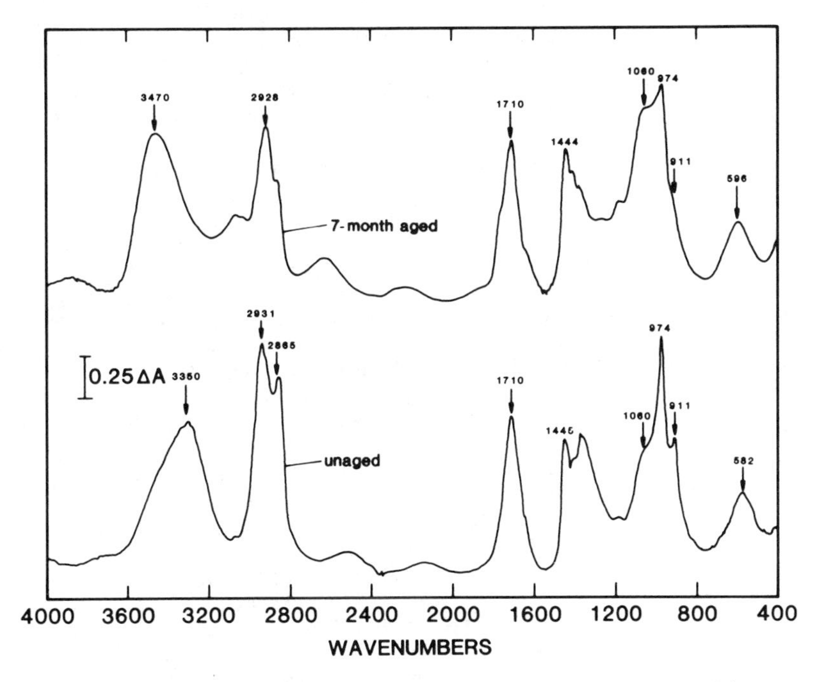

Figure 3. FTIR-RA spectra of polybutadiene coating on cold-rolled steel before and after exposure to 40°C and 80% RH

band of anhydride-cured epoxy exposed to moisture was attributed to the interaction of water molecules with the phenyl ether (13).

Degradation of Polybutadiene Coating on Cold-Rolled Steel

FTIR-RA spectra of air-cured polybuladiene coating on cold-rolled steel before and after exposure to a 40°C and 80% RH environment are presented in Figure 3. The composition of this polybutadiene and the assignments of its FTIR-RA spectra were presented elsewhere (3). Extensive corrosion was observed in the 7-month exposed specimen. The cured but unaged specimen oxidized extensively, as evidenced by the presence of OH groups in the 3150-3650 cm^{-1} region and carbonyl groups in the 1650-1800 cm^{-1} region. The normalized spectra (Figure 4) show considerable changes in the polybutadiene coating resulting from the exposure. The maximum of the OH stretching (3150-3650 cm^{-1}) shifted to higher frequency as a result of exposure (Figure 4a), and this was attributed to the weakening of the H-bonds of the coating system (3). Shifts of smaller magnitude were observed for the C-O bending band at 582 cm^{-1} (Figure 3). The intensities of the bands at 1182 cm^{-1} (ester C-O) and 1060 cm^{-1} (alcohol C-O) increased and the latter also broadened considerably as exposure time increased (Figure 4d). This indicates the formation of more and varied C-O containing products during exposure. In constrast, the OH stretching intensities increase in the early exposure stage, then level off. These results suggest that OH-containing compounds, such as alcohols, were formed in the early stages of exposure. These products continue to form in the later exposure periods but some of them are converted to highly oxidized products, such as esters and lactones as evidenced by the bands at 1737 and 1775 cm^{-1}, respectively. These compounds are known as common products found in the oxidative degradations of PBD (14,15).

The most noticeable changes of the polybutadiene coating on steel subjected to the warm, humid environment are observed in the complex 1600-1800 cm^{-1} region (Figure 4b). Numerous new peaks appeared and the intensities of existing peaks changed as a result of exposure. The assignments of these peaks were already given (3). The bands at 1573 and 1581 cm^{-1} are probably due to carboxylate ions. Various peaks in the regions between 1610 and 1660 cm^{-1} are probably associated with the C=C bonds. The bands in the region between 1680 and 1780 cm^{-1} are associated with the formation of unsaturated and saturated C=O groups (3,14).

Considerable losses of the unsaturation were also observed, as evidenced by the decreases in intensities of the bands at 974 cm^{-1} (CH out-of-plane bending of trans -CH=CH-), 911 cm^{-1} (CH_2 out-of-plane bending of $-CH=CH_2$), 2931 and 2865 cm^{-1} (CH and CH_2 stretching), and 1360 cm^{-1} (CH_2 wag). Additional evidence from IR analysis at other exposure periods indicate that most of the losses of the unsaturation in the early stages of degradation occur at the end vinyl group. This is in agreement of with an early observation (14) that under oxidative degradation at 130°C for 20 minutes, polybutadiene loses its unsaturation in the cis -CH=CH- and $-CH=CH_2$ but not trans -CH=CH- group. The loss of unsaturation of PBD is also due to crosslinking as evidenced in this study by the extensive crazing on the surface of the coated

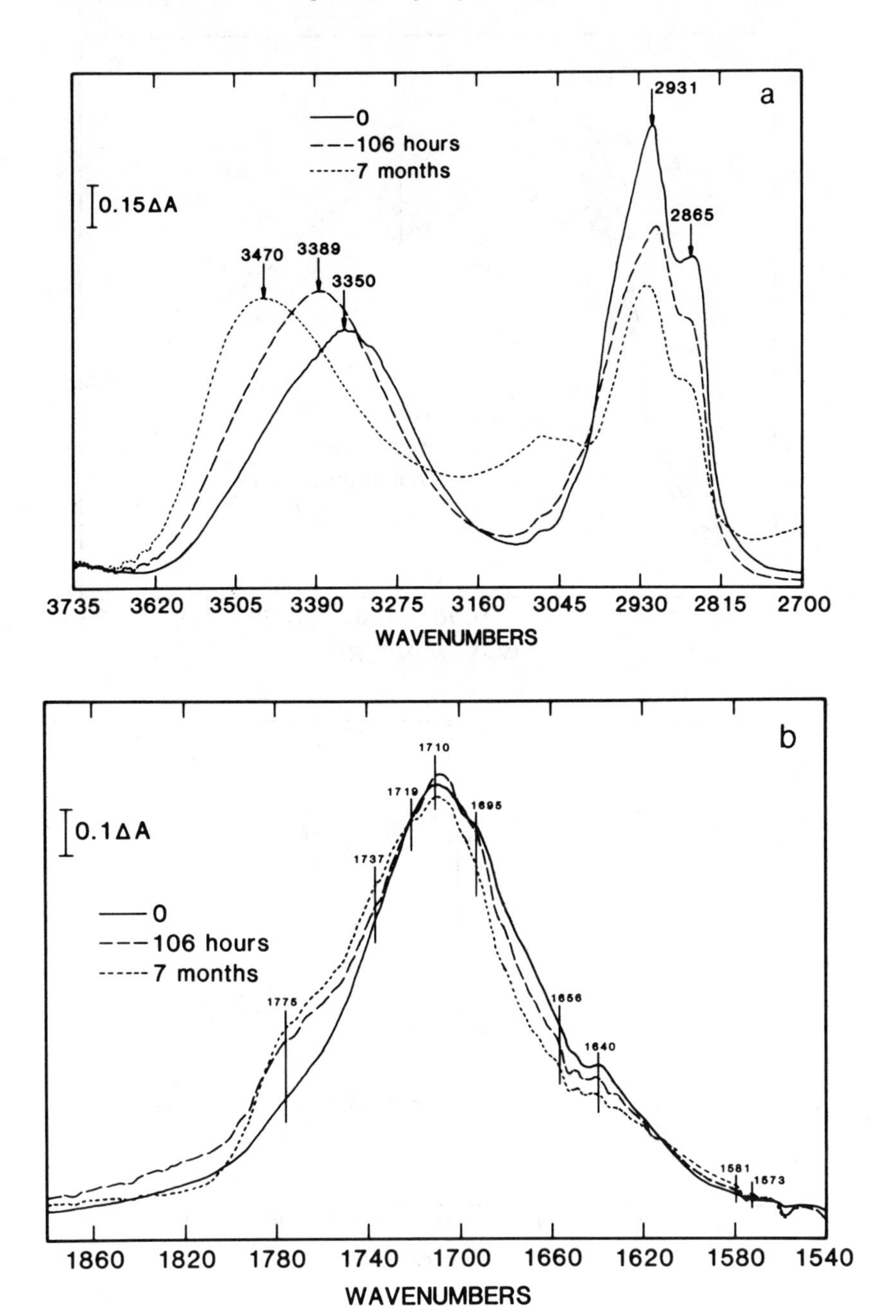

Figure 4 a and b. Normalized FTIR-RA spectra of polybutadiene coating on cold-rolled steel exposed to 40 °C and 80% RH for different times. Reproduced with permission from reference 3.

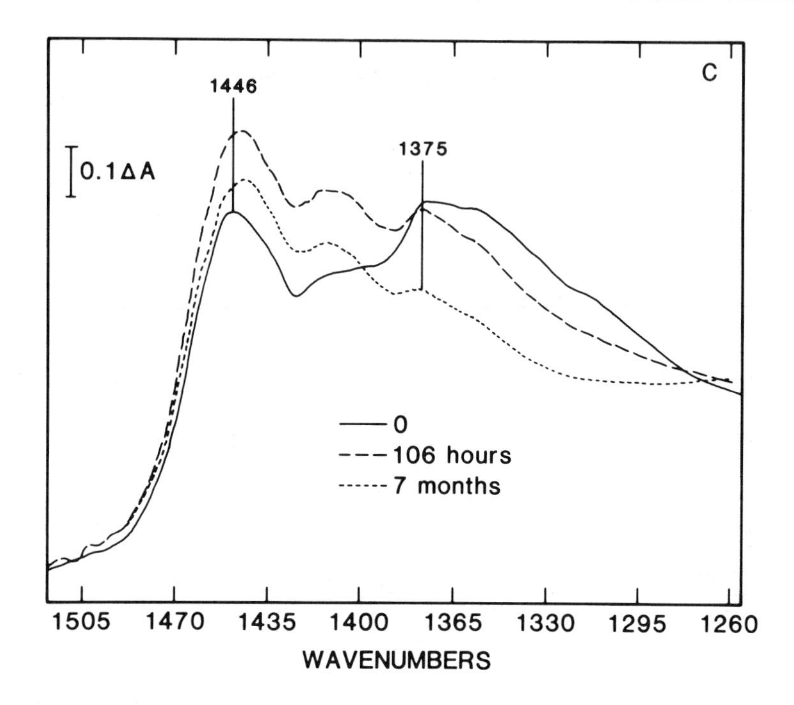

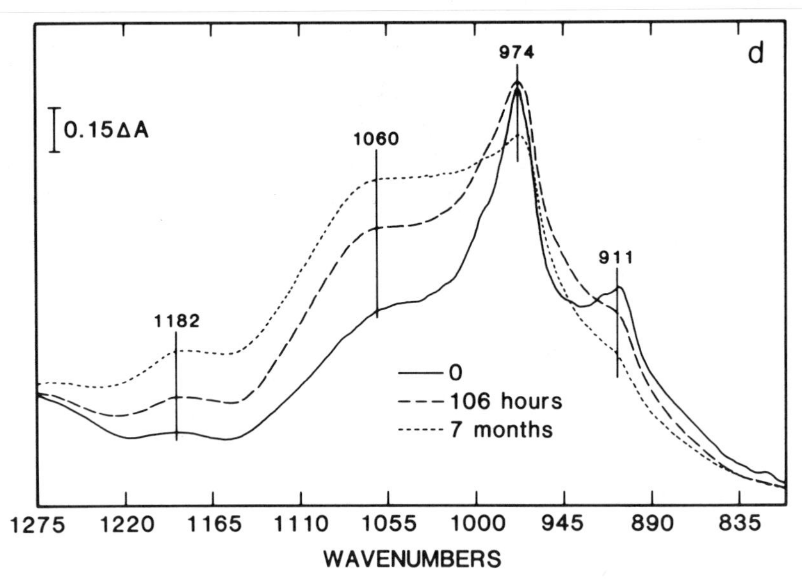

Figure 4 c and d. Normalized FTIR-RA spectra of polybutadiene coating on cold-rolled steen exposed to 40 °C and 80% RH for different times. Reproduced with permission from reference 3.

specimens. Kagiya and Takemoto (16) correlate the loss of solubility (due to crosslinking) of air-irradiated PBD with the increase of methylene groups and ether linkages. This is in agreement with Figure 4c which shows an increase of the band centered at 1446 cm^{-1} (CH_2 bending) with increasing time for short exposure periods; the decrease at the later stage is due to chain scission.

CONCLUSIONS

FTIR-RA analyses of amine-cured epoxy and polybutadiene coatings on cold-rolled steel exposed to 40°C and 80% RH environments for 7 months indicate bond weakening, dehydration and bond scissions of amine-cured epoxy during exposure. On the other hand, the polybutadiene coating specimens show bond weakening and extensive degradation which results in the formation of various oxidized products and losses in unsaturation. The characterization of complex molecules that are formed during the oxidation and degradation by FTIR-RA offers a powerful means for studying the degradation processes of protective coatings on steel.

LITERATURE CITED

1. Bennett, L. H., et al. NBS Special Publication 1978, 511-1.
2. Nguyen, T. Prog. Org. Coat. 1985, 13, 1.
3. Nguyen, T.; Byrd, W. E. Proc. XI Int. Conf. Org. Coat. Sci. Technol., 1985, p. 235.
4. Chen, C. S.; Bulkin, B. J.; Pearce, E. M. J. Appl. Polym. Sci. 1983, 28, 1077.
5. Harrod, J. F. J. Polym. Sci. 1963, A1, 385.
6. T. Nguyen and E. Byrd, FTIR-RA spectral characteristics of epoxy coatings on cold-rolled steel; effects of film thickness and angle of incidence, Appl. Spectrosc. submit for publication, 1985.
7. Paterson-Jones, J. C. J. Appl. Polym. Sci. 1975, 19, 1539 .
8. Lin, S. L.; Bulkin, B. J.; Pearce, E. M. J. Polym. Sci. Polym. Chem. ed., 1979, 17, 3121.
9. Bishop, D. P.; Smith, D. A. Ind. Eng. Chem. 1967, 59, 33.
10. Paterson-Jones, J. C.; Percy, V. A.; Giles, R. G. F.; Stephen, A. M. J. Appl. Polym. Sci. 1973, 17, 1877.
11. Hadzi, D.; Bratos, S., in "The Hydrogen Bond-Recent Developments in Theory and Experiments"; Schuster, P., et. al., Eds; North-Holland: Amsterdam, 1976, pp. 567-611.
12. Dannenberg, H. SPE Trans. 1963, 3, 78.
13. Antoon, M. K.; Koenig, J. L. J. Polym. Sci. Polym. Phys. Ed. 1981, 19, 1567.
14. Beavan, S. W.; Phillips, D. Rubber Chem. Technol. 1975, 48, 692.
15. Pecsok, R. L.; Painter, P. C.; Shelton, J. R.; Koenig, J. L. Rubber Chem. Technol. 1976, 49, 1010.
16. Kagiya, V. T.; Takemoto, K. J. Macromol. Sci.-Chem. 1976, A10, 795.

RECEIVED February 3, 1986

10

Permeabilities of Model Coatings: Effect of Cross-link Density and Polarity

W. J. Muizebelt and W. J. M. Heuvelsland

Akzo Research, Corporate Research Department, P.O. Box 60, 6800 AB Arnhem, the Netherlands

Oxygen and water vapour permeabilities have been measured for a number of model coatings. The coatings consist of pure esterdiols (oligomeric iso/terephthalates of glycol, butanediol or neopentylglycol), crosslinked with hexamethoxymethyl melamine or polyfunctional isocyanate.
By varying the length of the esterdiol, the crosslink density was varied. Differences in chemical composition resulted in variations in polarity. Differences in permeability were largely due to differences in solubility; hence diffusion through the polymeric film was not noticeably affected by crosslink density or polarity.

Water vapour and oxygen permeability of coatings is an important parameter governing their corrosion protection (1-6). Many factors influence the permeability, such as polarity, crystallinity and the presence of functional groups (7-9). Crosslink density is also mentioned in this respect (10-13). Funke and Carfagna (10) demonstrated the effect of curing temperature on permeability but they ascribed the effect to differences in glass transition temperature. Fritzwater (12) discussed the mechanism of transport of water and oxygen through pores in crosslinked materials. Gordon and Ravve (13) studied oxygen transmission of highly crosslinked materials. They concluded that permeability decreased with increasing crosslink density and the least permeable membrane was composed of a crosslinked structure of optimum space filling character and network tightness.

We have investigated the effect of crosslink density on permeability of water vapour and oxygen of high solid coatings. For this purpose we have synthesized a number of model coatings, i.e. coatings with a well-defined chemical structure. These materials consist of pure oligomeric esters of tere- or isophthalic acid with the diols glycol, 1,4-butanediol or neopentylglycol. The oligomers were then reacted by their terminal OH groups with the

0097-6156/86/0322-0110$06.00/0

crosslinkers (hexa)methoxymethylmelamine (HMMM, Cymel 303) or polyfunctional isocyanate (Desmodur N). Crosslink density of the obtained materials will be dependent on the length of the oligomer. Because the chemical composition was changed simultaneously the coatings also showed slight differences in polarity.

By determining the water vapour and oxygen permeability of the free films as well as the water solubility in the coatings, the coefficients of diffusion of water could be established.

Experimental

Synthesis of oligomers. The oligomers were esters of tere- or isophthalic acid (T or I) with the diols glycol (G), 1,4-butanediol (B) or neopentyl glycol (N). Using these symbols the materials can be indicated simply as, for instance GTG (first oligomer of ethylene terephthalate) or $(BI)_3B$ (third oligomer of butylene isophthalate).

The oligomers BTB and $(BT)_2B$ were prepared according to Hässlin et al. (14) and isolated by means of fractional crystallization from ethanol. The isophthalate oligomers NIN and BIB were prepared similarly and purified by molecular distillation (leaving the non-volatile higher oligomers in the residue) and crystallization. $(BI)_3B$ was prepared from isophthaloyl chloride and excess BIB. GCG (diglycol ester of 1,4 cyclohexanedicarboxylic acid) was made by catalytic hydrogenation of GTG. The purity of the materials was checked by means of GPC and NMR.

Preparation of coatings as free films. The oligomeric esterdiols were mixed with the crosslinkers HMMM or polyfunctional isocyanate. The molar ratio esterdiol/HMMM was 2:1 leading to an OH/OCH_3 ratio of 4:6. The OH/NCO ratio was 1:1. Some 1 wt% diethanolamine salt of p-toluene sulphonic acid, respectively 0.2 wt% Dabco were used as catalyst. The coatings were applied to Bonder 101 plates which had been sprayed with a thin layer (1-2 µm) of teflon. Curing was effected at 135°C for 30 minutes (HMMM) and one day at room temperature (isocyanate), respectively. The coatings could easily be removed from the teflon by means of a razor blade.

Film thickness was generally in the range 30-50 µm. The extent of the crosslink reaction with HMMM was checked by infrared and ^{13}C solid state NMR. The methoxy band at 915 cm^{-1} disappeared largely relative to the 815 cm^{-1} triazine ring absorption.

However, the methoxy group is present in excess relative to the $-CH_2OH$ group of the esterdiol and it may also disappear in side reactions other than the crosslink reaction. Thus the amount of methoxy groups remaining after cure is not a measure of the extent of the crosslink reaction. Solid state ^{13}C NMR spectra of the cured films showed that the $-CH_2OH$ group had disappeared virtually completely. The crosslink reaction is therefore nearly complete and the molecular weight between the crosslinks is determined by the molecular weight of the esterdiol used.

Permeability and solubility measurements. Permeability of the free films for water vapour was measured by means of the wet cup method (15). Oxygen permeability was measured using the Polymer Permeation Analyser of Dohrmann Envirotech (16,17). Results are summarized in Table I.

Also included in the table are solubilities of water in the coatings. These solubilities were determined from the weight difference of a piece of material after drying over P_2O_5 in vacuo for a number of days and after storage in water. Before weighing, the wet coating was carefully wiped with tissue in order to remove any adhering water.

Table I. Oxygen and water vapour permeabilities (P_{O_2} and P_{H_2O}) with mean deviation (m.d.), water solubility (S_{H_2O}) and diffusion coefficient (D_{H_2O}) at 21°C.

coating	$P_{O_2} \times 10^{12}$ $\frac{cc(STP)cm}{cm^2(cm\ Hg)sec}$	$P_{H_2O} \times 10^{11}$ $\frac{g\ cm}{cm^2(cm\ Hg)sec}$	$S_{H_2O} \times 10^2$ $\frac{g}{cm^3(cm\ Hg)}$	$D_{H_2O} \times 10^9$ $(cm^2 s^{-1})$
BIB/HMMM	25	5.0 ± 0.6	1.6	3.2
$(BI)_3B$/HMMM	31	4.2 ± 0.6	1.3	3.3
BTB/HMMM	47	5.2 ± 1.0	1.0	5.2
BTB + $(BT)_2B$/HMMM	-	5.4 ± 0.8	1.2	4.5
$(BT)_2B$/HMMM	-	4.7 ± 1.1	1.4	3.4
GTG/HMMM	-	4.6 ± 0.2	1.7	2.7
GCG/HMMM	30	5.5 ± 0.2	1.8	3.1
NIN/HMMM	15	4.4 ± 0.3	2.1	2.1
BIB/isocyanate	4.2	6.7 ± 0.5	4.8	1.4
GTG/isocyanate	-	11.0 ± 1.1	-	-
NIN/isocyanate	7.1	3.5 ± 0.5	-	-
B/isocyanate	-	21 ± 3	10	2.0
N/isocyanate	3.4	17 ± 4	11	1.5

Results and Discussion

Water vapour permeability. The most notable phenomenon overlooking the data presented in Table I is that the water vapour permeabilities of the HMMM-based coatings are not widely different. The isocyanate coatings show somewhat larger differences. GTG/isocyanate and the coatings made from butanediol and neopentyl glycol are more permeable.

The experimental permeability is the product of the coefficient of diffusion and solubility ($P = D \times S$). When the measured solubilities are taken into consideration it appears that the differences in permeability observed can mainly be attributed to this factor. The calculated diffusion coefficients differ at most a factor of three. However, if it is realized that this coefficient is derived from two experimentally observed variables and that the solubility measurements were somewhat less reproducible than the permeabilities, it is questionable whether the differences in diffusion coefficients are significant.

We therefore tend to conclude that the differences in permeability observed are due to differences in solubility rather than variations in the diffusion coefficient. Differences in solubility of water may be attributed to differences in polarity of the medium. It will be clear that the isocyanate coatings made from butanediol or neopentyl glycol contain a higher concentration of polar urethane linkages than those made from the oligomers (although these contain polar ester groups). Also isocyanate coatings are more polar than those with HMMM, which contain less polar ether groups.

The main conclusion we would like to advance is that the diffusion of water in the coating is not noticeably affected by the crosslink density of the films. This conclusion is in contrast to those of Gordon and Ravve (13), who found large effects of crosslink density on oxygen permeability of acrylates. Also the permeability of natural vulcanizates for various gases was found to be strongly dependent on the amount of sulfur used (18,19). It must be concluded that although the crosslink densities of our materials are in the range of the acrylates studied by Gordon and Ravve, the differences in crosslink densities do not lead to similar effects on space filling character or network tightness. The difference with the vulcanizates (18,19) could conceivably be a matter of glass transition temperature. Our measurements were carried out below Tg whereas the observations on the vulcanizates were carried out above Tg.

Oxygen permeability. Oxygen permeability measurement required a larger piece of coating with a greater chance of leaks. Therefore it was often not possible to perform these measurements. The fewer data for oxygen permeability in Table I indicate smaller values for the isocyanate coatings than for those based on HMMM. This will be due to the difference in polarity, which influences the solubility the opposite way as in the case of water. Oxygen, as a non-polar molecule, dissolves better in media with lower polarity in contrast to water. Therefore the permeability of oxygen is also larger in media of lower polarity.

Salt spray test. The model coatings of Table I are of the high solid type used in automotive top coats. Their primary function is not corrosion protection since this is first of all a matter of phosphate layer, electrocoat and/or primer. However, the topcoats may contribute to corrosion protection by their barrier function for water, oxygen and salts. Therefore their permeability is important as one of the factors in the corrosion protection by the total coating system. We feel that a salt spray test of the model coatings directly applied to a steel surface is of little relevance for their corrosion protection performance in a real system.

Nevertheless we did a number of tests of our model coatings directly applied to Bonder 101 panels. The panels were given a standard scratch just below the metal surface after which they were exposed in the salt spray test. The corrosion protection performance was at best moderate and no significant differences between the various coatings could be seen. This is in accord with the small differences in permeability observed. On this basis we do not expect significant differences when the coatings are tested on panels provided with a proper electrocoat primer, although the corrosion protection by the complete system may be expected to be on a much higher level.

Acknowledgment

Experimental assistance was given by Rianne Willems and Mark Buurman.
Solid state ^{13}C NMR spectra were taken by Ir. H. Angad Gaur.

Literature Cited

1. W. Funke, J.O.C.C.A. 62, 63 (1979).
2. W. Funke and H. Haagen, Ind. Eng. Chem. Prod. Res. Dev. 17, 50 (1978).
3. H. Haagen and W. Funke, J.O.C.C.A. 58, 359 (1975).
4. F.L. Floyd, R.G. Groseclose and C.M. Frey, J.O.C.C.A. 66, 329 (1983).
5. M. Yaseen and K.V.S.N. Raju, J.O.C.C.A. 67, 185 (1984).
6. S. Guruviah, J.O.C.C.A. 63, 669 (1970).
7. D.Y. Perera and S. Pelier, Progr. Org. Coat. 1, 57 (1973).
8. P.W. Morgan, Ind. Eng. Chem. 2296 (1953).
9. W.L.H. Moll, Kolloid Zeitschr. 195, 43 (1964).
10. W. Funke and C. Carfagna, J.O.C.C.A. 67, 102 (1984).
11. K.A. v. Oeteren, Fette, Seife, Anstrichmittel 84, 242 (1982).
12. J.E. Fitzwater, J. Coat. Techn. 53 (683) 27 (1981).
13. G.A. Gordon and A. Ravve, Polymer Eng. and Sci. 20, 70 (1980).
14. H.W. Hässlin, M. Dröscher and G. Wegner, Makromol. Chem. 181, 301 (1980).
15. M. Yaseen and W. Funke, J.O.C.C.A. 61, 284 (1978).
16. M. Lomax, Polymer Testing 1, 105 (1980).
17. P.E. Cassidy, T.M. Aminabhari and C.M. Thompson, Rub. Chem. Techn. 56, 594 (1983).
18. R.M. Barrer and G. Skirrow, J. Pol. Sci. 3, 549 (1948).
19. A. Aitken and R.M. Barrer, Trans. Farad. Soc. 51. 116 (1955).

11

Using Acoustic Emission to Investigate Disbonding at the Polymer-Metal Interface

L. M. Callow and J. D. Scantlebury

Corrosion and Protection Centre, University of Manchester Institute of Science and Technology, Sackville Street, Manchester, M60 1QD, England

There has been some discussion as to whether acoustic signals result from the fracture that occurs when adhesion is lost at the paint-metal interface during blistering and cathodic disbonding. In the present study, very sensitive Acoustic Emission measurements have been made in systems exhibiting these types of breakdown. A lacquer was applied to a panel, which subsequently suffered gross blistering, electrochemical voltage noise and acoustic emission were monitored simultaneously. The former showed large peaks as the blisters grew, whilst the latter produced no signals that were directly attributable to blistering. In the second series of experiments, artificial holidays were introduced into a commercially available epoxy lacquer, which was potentiostatically polarised to -1000 mV (SCE). Large disbonded areas grew rapidly around each holiday, but once again, no acoustic signals were detectable.

Previous work on the application of the acoustic emission technique to corrosion processes has been carried out by Mansfeld and Stocker (1) who were able to show that the detachment of hydrogen bubbles from a polarised metal surface gave measurable signals.

The application of the technique to the detachment of a paint from a stressed substrate has been shown to be viable by Strivens(2) and others. The work presented herein is an attempt to extend the above work into areas where the substrate is unstressed and the paint becomes detached as a result of stresses in the film or stresses having an electrochemical origin.

Experimental

The acoustic emission equipment used in this series of experiments is described below. Two types of A.E. transducer were employed, the first, Bruel and Kjaer type 8313 covered the frequency range 50 to 600 kHz, whilst the second, Bruel and Kjaer type 8314 covered

0097–6156/86/0322–0115$06.00/0

the higher frequency range 0.5 to 1.1 MHz. The frequency response curve characteristics are shown in Bruel and Kjaer literature[3]. The output from these transducers were fed into a Bruel and Kjaer pre-amp (model 2637) which had a gain of 40 dB. The signal was further amplified by a Bruel and Kjaer conditioning amplifier (model 2638) capable of giving an extra 60 dB of gain and giving a flat frequency response between 0.1 Hz and 2MHz. The total amplification in the system was limited by stray noise and susceptability to Radio Frequency Interface. (R.F.I.), so suppression was introduced in the system to limit the problem.

Any A.E. transients occurring were captured by a Datalab (DL 910) transient recorder; its A/D Converter was capable of operating to 20 MHz and 8k byte long transients could be stored. Further analysis was subsequently carried out on the stored data using a BBC model B microcomputer, programmed to perform Fast Fourier Transforms (F.F.T.). Preliminary experiments were carried out in which mild steel was polarised potentiostatically (Thompson ministat 251) to -1700 mV (SCE) to cause hydrogen evolution on the specimen. Following this successful test, it was decided to extend the use of the apparatus into disbonding. A deliberately underbound zinc-rich paint was applied to a standard test panel giving a wet film thickness greater than 3 mm. As the paint dried, extreme "mud cracking" was exhibited. The progress of the cracking and subsequent detachment of the film from the steel surface was observed using a video camera in conjunction with the A.E. equipment.

Further experiments were carried out in which small flakes of a 40 µm thick grey mica pigmented lacquer were mechanically detached from mild steel, the resulting transients were captured and analysed.

As A.E. signals had been observed in the above experiments, investigation proceeded with the grey mica pigmented lacquer which was prone to blistering under immersion. Specimens with film thicknesses of 50 - 110 µm were semi-immersed in 3% NaCl solution, by suspending the electrode from elastic bands, In order to conserve any signals produced. Several attachment sites for the transducer were tried on the specimen, including both the painted and non-painted sides; no detectable differences due to transducer location were observed during the several weeks immersion time of the specimen.

In a second series of experiments, an epoxy coated specimen was employed, which had 5 small (1 mm diameter) holes drilled down to the steel. Electrical connections were made to the upper edge and the specimen was potentiostatically polarised to -1050 mV (SCE) and allowed to disbond. Again, this process was monitored using the acoustic emission transducer. All the immersed specimens were masked using a mixture of beeswax and colophony resin, as described elsewhere[4].

Results and Discussions

Many of the problems encountered with the application of the Acoustic Emission (A.E) technique to corrosion problems are due to the fact that the signal level was of a similar order of magnitude to the background noise level of the instrumentation.

Figure 1 shows the noise level obtained with the maximum usable gain of 70 dB. Figure 2 is a F.F.T. of the time domain data of the previous figure. This shows the frequency distribution of the background noise.

When the transducer was connected to a piece of mild steel that was cathodically polarised to evolve hydrogen, the time domain trace of figure 3 was obtained. It can be seen that extra peaks occurred on a regular basis that could be attributed to the formation of hydrogen bubbles, as shown arrowed in figure 3. When the grey paint film was levered from the substrate mechanically, the time domain data of figure 4 was obtained. The corresponding transform is shown in figure 5. It can be seen that with a dry paint film the major A.E. events took place in the frequency band between 10 KHz and 30 KHz. The magnitude of these events was 10 times greater than those seen in the hydrogen evolution experiments.

Taking the hydrogen evolution experiments and the mechanical detachment experiments as a guideline, experiments were carried out using these on a paint known to exhibit mud-cracking (predetermined frequency bands). When the pool of zinc-rich paint first began to dry, cracks formed in the surface leaving the paint below fluid. These cracks did not give rise to any observable acoustic emission. With time the cracks eventually penetrated to the surface of the substrate and the entire pool became touch dry. It was only when the internal stresses in the cracked polymer caused it to curl up and detach itself from the substrate that A.E. signals were observed. A typical transient is shown in figure 6, and the corresponding transform in figure 7. Comparison of this data with the results obtained from mechanical detachment shows that the amount of energy in the two spectra is similar but mud cracking occurs at lower frequencies and has a more uniform distribution through the spectrum.

The immersed grey paint blistered readily. Within a period of 24 hours small blisters were observed and after a period of a week most of the surface of the specimen was blistered. No A.E. transients were observed at any time during the course of this blistering. Sharp, short transients of the type shown in figure 8 were observed from time to time throughout the course of this experiment. Considerable efforts were made to eliminate RFI, these included the use of an electrically isolated room, transient suppressed mains supplies, screened leads and passive suppression devices. Nevertheless, these signals may be attributed to radio frequency interference, as all these transients occurred throughout the working day and none were recorded overnight. Secondly, the length of the transients did not indicate that "ringing" of the substrate took place.

When the epoxy coated panel containing artificial holidays was allowed to cathodically disbond, areas of greater than one square centimetre became detached in less than 7 days. Once again at no time were realistic A.E. transients observed.

Initially, with the grey paint, it was thought that the thinness of the coating combined with its plasticity was causing the paint to peel from the surface in a ductile manner, it being further plasticised by the uptake of water. These observations were not true for the epoxy coated panels as the film thickness

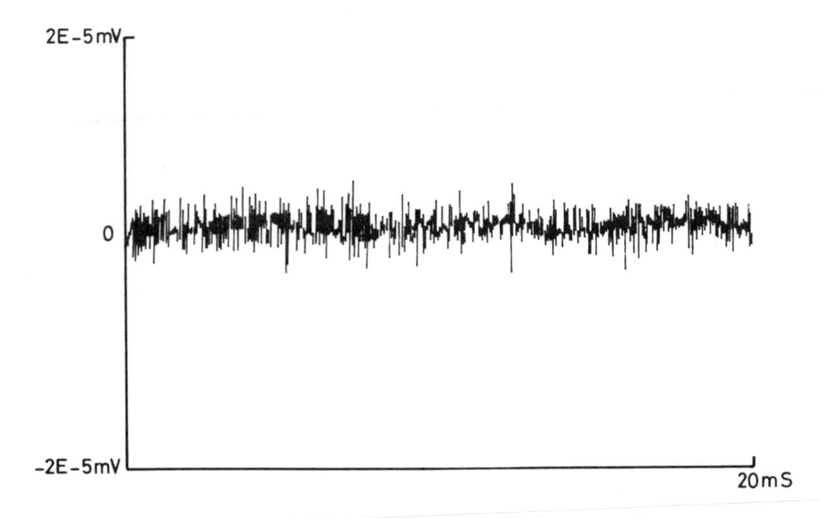

Figure 1. Background noise level from equipment.

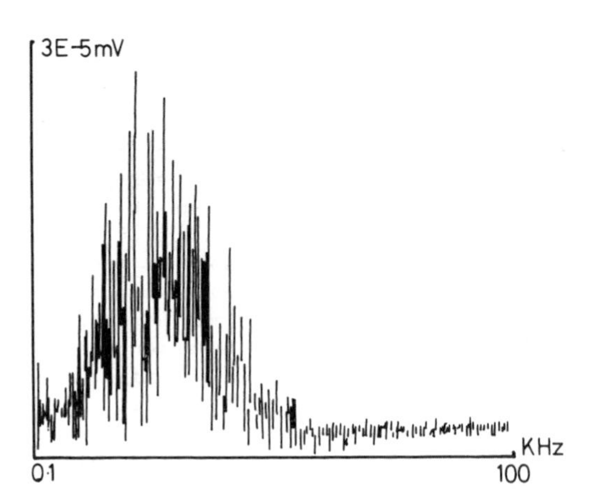

Figure 2. Fourier transform of the background noise.

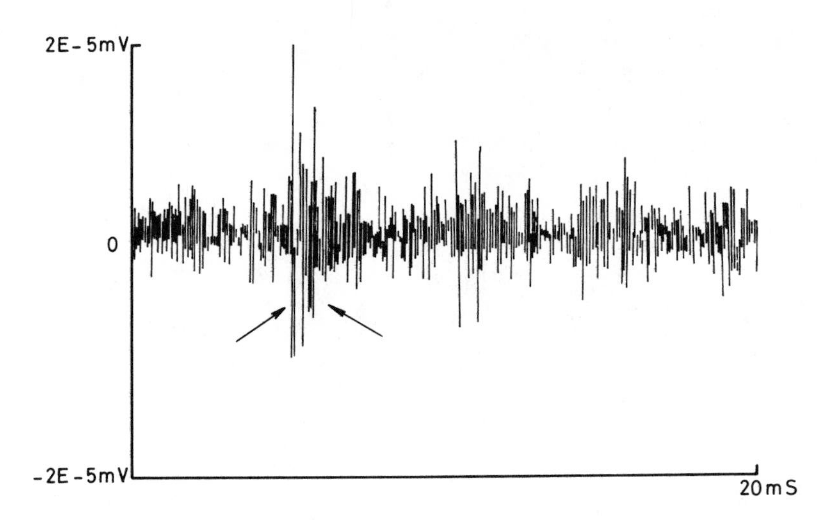

Figure 3. Time domain trace resulting from hydrogen evolution at -1700 mV SCE.

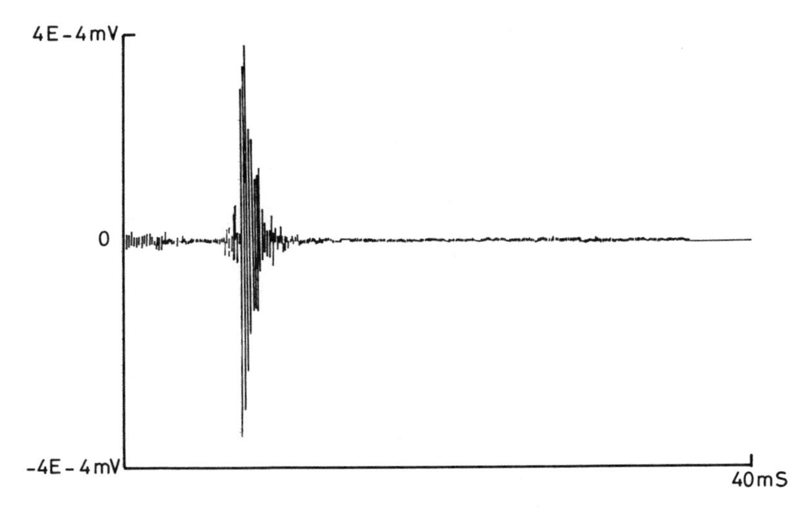

Figure 4. Time domain trace resulting from the mechanical disbonding of a chlorinated rubber substitute.

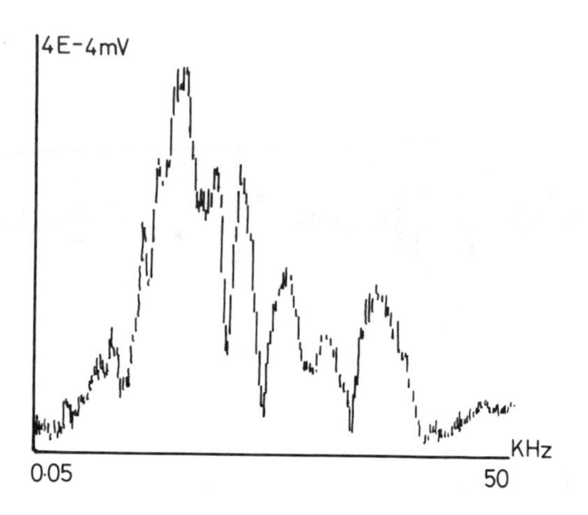

Figure 5. F.F.T. of Figure 4.

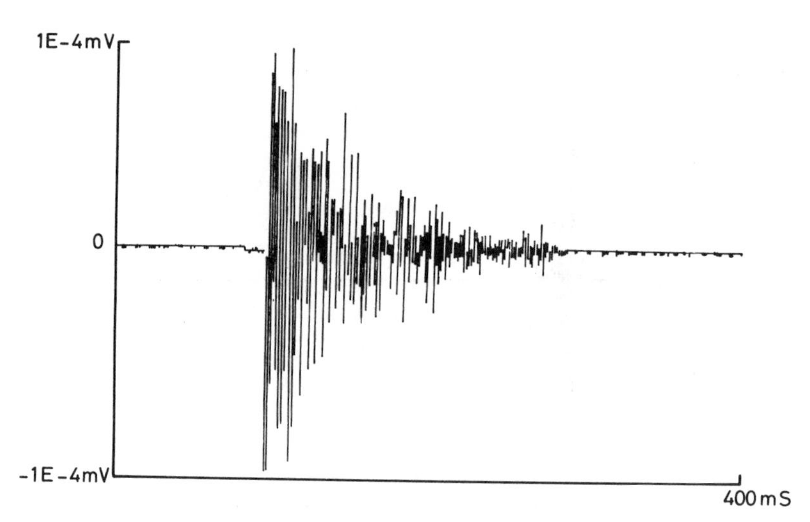

Figure 6. Long transient resulting from the later stages of "mud cracking".

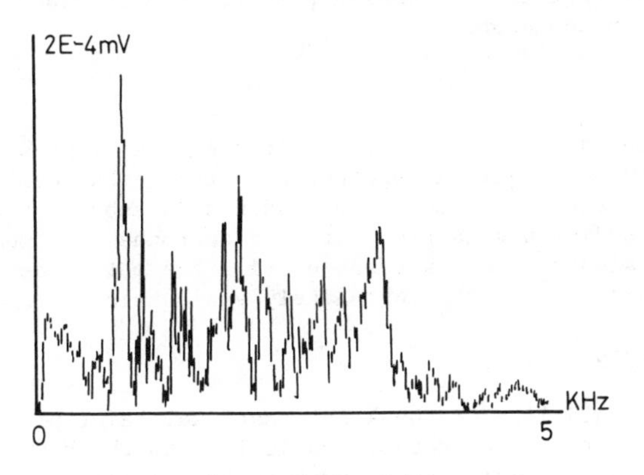

Figure 7. F.F.T. of Figure 6.

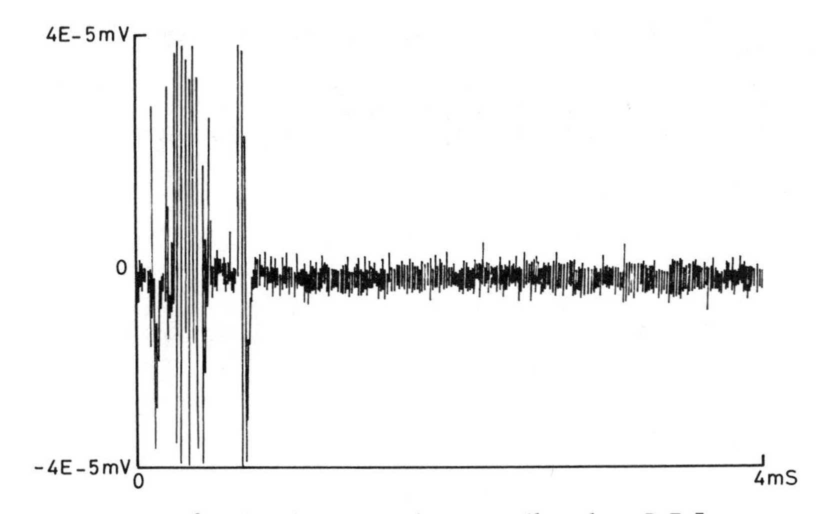

Figure 8. Spurious transient attributed to R.F.I.

and its impervious nature, together with the inherent brittleness of the film, made plasticisation unlikely. If the disbonding of the epoxy under these circumstances was giving rise to A.E. transients then these must either be below the noise level of the equipment or they were being absorbed by the water or the panel itself. However, the likely explanation for the lack of A.E. information from specimens undergoing cathodic disbonding or blistering is that the disbonding process is a chemical rather than a mechanical phenomenon.

Conclusions

The A.E. technique is capable of monitoring the progress of hydrogen evolution and mud cracking together with the mechanical disbonding of paint from the metal substrate. When this substrate was immersed and the disbonding was carried out by electrochemical means, then no A.E. transients were observeable using the method described and chemical disbonding is the most likely explanation.

Acknowledgments

L.M. Callow would like to thank International Paint plc for financial support. Both authors would like to thank Dr. R.P.M. Procter for the provision of laboratory facilities.

Literature Cited

1. Mansfeld, F. and Stocker, P.J., J. Electrochem. Soc., 77, p. 1301-2, Vol. 24. No. 8.
2. Strivens, T. and Rawlings, S. J.O.C.C.A., 63, 412-418 (1980).
3. Bruel & Kjaer, Instrumentation Handbook.
4. Callow, L.M. and Scantlebury, J.D., J.O.C.C.C.A, 64 83 (1981).

RECEIVED January 22, 1986

ADHESION AND INTERFACIAL ASPECTS OF CORROSION PROTECTION

12

Mechanisms of De-adhesion of Organic Coatings from Metal Surfaces

Henry Leidheiser, Jr.

Department of Chemistry and Center for Surface and Coatings Research, Lehigh University, Bethlehem, PA 18015

Organic coatings lose adherence to a metal substrate by many processes. The eventual consequence of this de-adhesion is corrosion of the metal beneath the coating. Among the important de-adhesion processes are: loss of adhesion when wet, cathodic delamination, cathodic blistering, swelling of the polymer, gas blistering by corrosion, osmotic blistering, thermal cycling and anodic undermining. Real-life and laboratory examples of these phenomena are given and the principles which govern the behavior are discussed. The de-adhesion processes require, with the possible exception of thermal cycling, that reactive species such as water, oxygen and ions penetrate through the coating. New studies on the migration of species through organic coatings are discussed.

Steel objects, when exposed to humid atmospheres or when immersed in electrolytes, corrode at a rapid rate. For example, abrasively polished, cold-rolled steel panels will show signs of rust within 15 minutes when immersed in dilute chloride solutions with pH in the range of 7-10. One of the methods used to control this rapid corrosion is to coat the metal with a polymeric formulation such as a paint. The role of the paint is to serve primarily as a barrier to environmental constituents such as water, oxygen, sulfur dioxide, and ions and secondarily as a reservoir for corrosion inhibitors. Some formulations contain very high concentrations of metallic zinc or metallic aluminum such that the coating provides galvanic protection as well as barrier protection, but such formulations are not discussed in this paper.

The corrosion process that occurs in de-adhered regions under paint is driven by an electrochemical process in which a portion of the area is anodic in nature and another portion is cathodic in nature. The reality of this electro-chemical process can be confirmed when pH indicators or substances sensitive to iron ions are placed beneath the coating such that the sharp distinction between

0097-6156/86/0322-0124$06.00/0

the anodic and cathodic regions is vividly illustrated. A good example is shown in the color photograph on the cover of the April 1983 issue of Materials Performance. The fact that the corrosion process is electrochemical in nature is significant in two respects. First, it is necessary that aqueous phase water be present at the interface between the coating and the paint. Second, water must migrate through the coating or through a defect in the coating and there must be a mechanism for condensation, or aqueous phase development, at the interface.

The purpose of this paper is to describe in a very general way the principles underlying the de-adhesion, when such principles are known, and to emphasize the lack of understanding when the principles are not yet recognized. Eight different types of de-adhesion processes will be discussed: loss of adhesion when wet, cathodic delamination, cathodic blistering, swelling of the polymer, gas blistering by corrosion, osmotic blistering, thermal cycling, and anodic undermining.

De-adhesion Processes

Water Aggregation. An interesting question arises at the outset as to what constitutes an aqueous phase. How many water molecules are required before an electrochemical process can be activated? Conversations with many well-known electrochemists have led us to use a 1M solution as a reference. Another basis for using 1M is the observation that the pH at the active front under a cathodically delaminating coating approaches a value of slightly under 14, i.e., approximately 1M in hydroxyl ions. A 1M solution is 55M with respect to water so that in a 1M solution of NaCl, the ratio of water to ions is 55 molecules of water for each Na^+ and Cl^- pair. We are thus using as our working guide that an aggregate of the order of 50 molecules of water represents the minimum number of water molecules that can be considered to have the properties of an aqueous phase.

This small number of water molecules thus requires a very small void at the interface between the coating and the metal in which to form a condensate. Voids may be formed where the wetting behavior of the coating is insufficient to penetrate into notches at grain boundaries, into fine scratches, into voids at boundaries between inclusions and the metal matrix, or into small recesses following abrasive blasting. Voids sufficient to contain 50 or more molecules of water are certainly present at the coating/metal interface and the important question is what are the mechanisms by which water will condense within the existing voids.

Now let us consider some of the processes which promote the development of an aqueous phase at the interface, assuming that there are voids at the interface sufficiently large to accommodate the nucleus of an aqueous phase.

Loss of Adhesion When Wet. Many coatings, particularly those applied to a roughened surface, show excellent tensile adhesion to steel but lose this adhesion after exposure to pure water at room or elevated temperatures. A thin film of water at the interface is apparently responsible for the loss of adhesion. If the coating is allowed to dry without destructively testing the adhesion, the dried coating often exhibits the original tensile adhesion. The phenomenon is reversible: the adhesion is poor when the coating is wet and is satisfactory when it is dry.

Some preliminary experiments by a student, George Rommal, indicate that a similar phenomenon can be observed at a glass/polymer interface. He studied a thin, spin-applied acrylic coating on glass and found that the water collected under the coating in discrete, approximately circular regions. The water collected under the coating in a matter of minutes and the circular regions slowly grew in size. The adherence, as measured by applying adhesive tape, was good when the coating was dry and was poor when the coating was wet. When the experiment was done repeatedly on the same sample after cyclic wetting and drying, it was observed that the water collected in the same regions. It appeared that water penetrated through channels in the coating and it was interpreted that the development of glass/water/polymer interface represented a negative free energy change relative to the glass/polymer interface.

Wet adhesion phenomena represent a potentially fruitful area of research since so little is known. Some of the important questions are: (1) How does one measure quantitatively the magnitude of the adhesion when the coating is wet? (2) What is the governing principle that determines whether or not water collects at an organic coating/metal interface? (3) What is the thickness of the water layer at the interface and what determines the thickness? A recent paper (1) correlates the wet adhesion properties of a phosphated surface with the crystalline nature of the zinc phosphate at the metal surface.

Cathodic Delamination. Most organic coatings on most metal surfaces lose their adherence when alkali is generated at a defect in the coating or at weak spots in the coating. Alkali can be generated by the cathodic half of the corrosion reaction or by driving the cathodic reaction by means of an applied potential. Previous publications (2-5) have reported extensively on the cathodic delamination phenomenon and only a brief summary will be given here.

The alkali is generated by the cathodic reaction,

$$H_2O + 1/2\ O_2 + 2\ e^- = 2\ OH^-$$

which occurs at a defect in the coating or through an electrolytic pathway at weak spots in the coating. It occurs at cathodic potentials of -0.7 to -1.5 v (vs. SCE) on all coatings, except one, that we have investigated. It has been observed on alkyd, acrylic, epoxy, epoxy powder, bitumen, vinyl ester, fluorocarbon, polyester, polybutadiene, and polyethylene coatings. The only coating in which no delamination occurred in 0.5M NaCl with an applied potential of -1.5 v vs. SCE for 60 days at room temperature is an electrostatically applied epoxy coating, 50 um thick, on a proprietary copper substrate. The reason for this lack of sensitivity to cathodic delamination is unknown, although it is suspected that the coating has a low degree of permeability to water and ions.

Some of the important facts about cathodic delamination are summarized in the following itemized statements:

(1) When the cathodic reaction occurs under the coating, the pH of the solution under the coating may approximate 14.

(2) The important cathodic reaction under the coating in most cir-

cumstances is the oxygen reduction reaction and no significant delamination occurs in the absence of oxygen when the polarization potential is insufficient to drive the hydrogen evolution reaction at a significant rate.

(3) No significant delamination is observed in the absence of metal cation. No cathodic delamination occurs in pure acid solutions.

(4) The rate of delamination is strongly a function of the catalytic activity of the surface for the oxygen reduction reaction. The activity can be decreased by surface treatment of the metal prior to the application of the coating.

(5) Reactive species reach the delamination front by migration through the coating.

(6) The area delaminated is generally linearly related to the time at constant temperature and constant potential.

(7) The rate of delamination increases with increase in the applied potential.

(8) The rate of delamination increases with increase in temperature. The activation energy in the case of polybutadiene coatings on steel is approximately 12 kcal/mole.

(9) For coatings thicker than approximately 30 um, there is an incubation period, or delay time, before the delaminated area increases linearly with time. This delay time decreases with increase in temperature or increase in applied potential.

(10) The organic coating at the metal interface is modified chemically by the strong alkaline medium that is generated under the coating.

(11) The rate of delamination is a function of the substrate metal and is very low in the case of aluminum substrates.

(12) The rate of delamination is a function of the type of coating and its thickness.

The major unknown in the cathodic delamination process is the mechanism by which the interfacial bond is broken. Alkaline attack of the polymer, surface energy considerations, and attack of the oxide at the interface have all been proposed, but none of the available evidence allows an unequivocal answer.

Cathodic Blistering. In the absence of a purposely-imposed defect in the coating, the cathodic delamination phenomenon is known as cathodic blistering. An example of cathodic blistering as a function of time is shown in Figure 1.

Swelling of the Polymer. Some polymer formulations have the property of swelling, i.e., increasing in dimension, when exposed to certain environments. An example of this effect is the swelling of some epoxy coatings when exposed to strong sulfuric acid solutions at elevated temperatures. Exposure of such a coating on steel results in the formation of multiple blisters when the substrate is sand blasted before the application of the coating and in a single large blister when the substrate is simply abraded. The process is facilitated if the coating is permeable to gaseous atmospheric constituents that may fill the void.

An interesting way to distinguish blistering by swelling of the polymer from corrosion-induced blistering is to apply the coating to a thin lead substrate and confine the area of exposure to a circular region of the order of 2-3 cm in diameter. The area may be confined

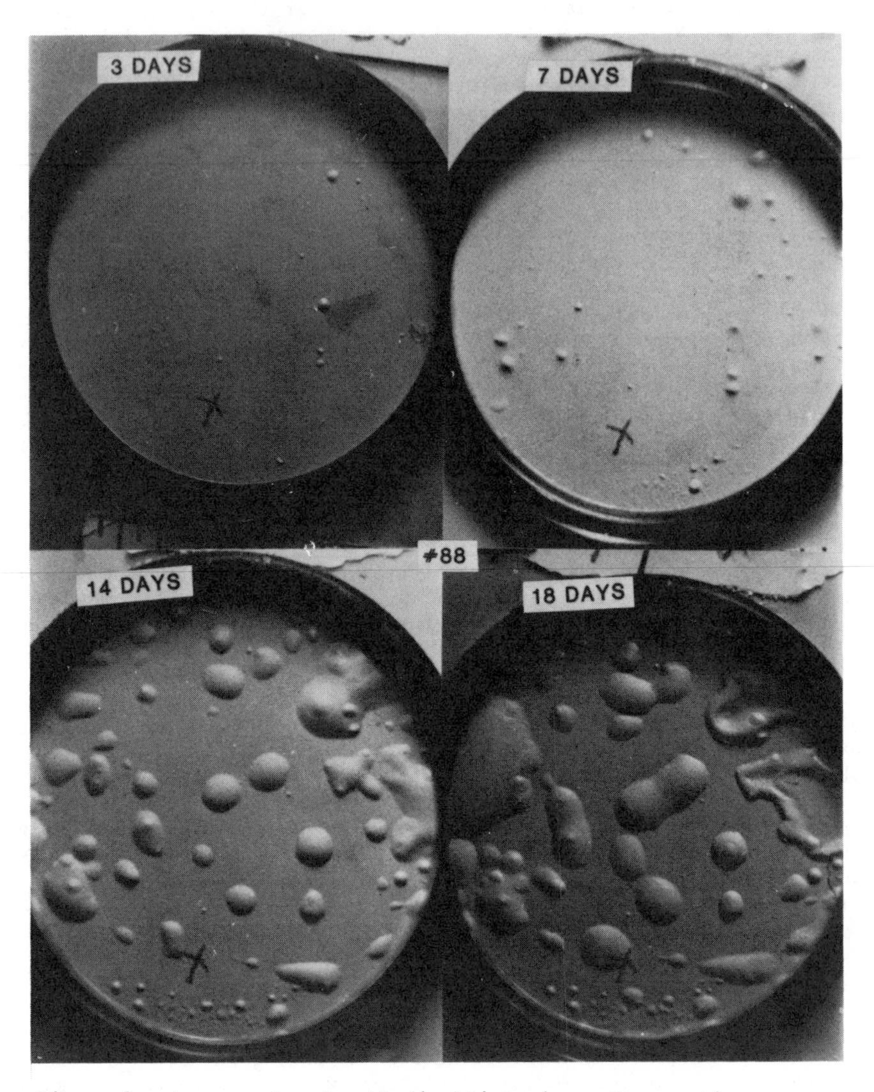

Figure 1. An example of cathodic blistering. The coating was a zinc chromate alkyd primer material. The electrolyte was 0.5M KCl and the potential of the metal was maintained at - 1.0 v vs. SCE.

by cementing a cylinder to the coating and filling the cylinder with the aggressive liquid. If the adherence to the lead is sufficiently good, the swelling of the polymer will cause the lead to deform in the same shape as the blister. An example of the deformation of a lead panel by this process is shown in Figure 2.

Gas Blistering by Corrosion. This phenomenon has been observed in a very few cases. An example is shown in Figure 3 for a coating exposed to strong sulfuric acid at 60 C. The effect was attributed to gas blistering rather than swelling of the polymer because the blister contained a large quantity of hydrogen as judged by extraction of the gas in the blister with a hypodermic needle followed by gas chromatographic analysis. The blistering must occur as a consequence of rapid penetration of the coating by hydrogen ions and slow diffusion of the hydrogen gas out through the coating. The blistering requires that the coating possess a degree of ductility since a brittle coating would be expected to fracture rather than to deform.

Osmotic Blistering. Osmotic pressures are very powerful and are a driving force for blistering. They are especially destructive under conditions where a soluble salt impurity is present beneath the coating and the coated metal is exposed to water with a low ionic content. The driving force is the attempt by the system to establish two liquids, one under the coating and the other external to the coating, with the same thermodynamic activity. The direction of water flow through the coating is inwards since dilution of the concentrated solution at the interface is the mechanism by which the two liquids strive for equal thermodynamic activity.

A quotation from a previous article (6) is worth repeating here. A discussion participant at the Corrosion 81 meeting in Toronto provided the following example of osmotic blistering. "A ship was painted in Denmark and made a voyage immediately thereafter across the Atlantic and into the Great Lakes. When it reached port, a blister pattern in the form of a handprint was observed above the water line. Apparently, the paint was applied over a handprint. No blistering occurred during exposure to sea water because of the high salt content of the water, but when the ship was exposed to fresh water, the osmotic forces became significant and the blistering occurred."

Another good example of osmotic effects is shown in Figure 4. Cathodic delamination studies were carried out on a pigmented epoxy coating at an applied potential of -0.8 v vs. SCE. Coatings of equal thickness were studied in 0.001, 0.01, 0.1, and 0.5M NaCl solution. It will be noted that the rates of delamination (slope of the curve) increased in the order 0.001, 0.01, 0.1 = 0.5M. However, the intersection point of the curves with the time axis (the so-called delay time) increased in the order 0.001 = 0.01, 0.1, 0.5M. This latter effect is attributed to the fact that the delay time is associated with the time required to form a steady-state diffusion gradient across the coating. The most important component in achieving this steady state is water since ion migration and oxygen migration probably follow aqueous pathways in the coating. The osmotic forces dominate in establishing this diffusion gradient and thus the more concentrated solutions require a longer time to establish this gradient. Once the gradient is established, the rate of delamination is determined by the rate at which cations can diffuse

Figure 2. Epoxy coating on a lead substrate. Coated metal was exposed to 1M H_2SO_4 at 60 C for 3 days. View is from the lead substrate side. Note that swelling of the coating caused a deformation of the lead.

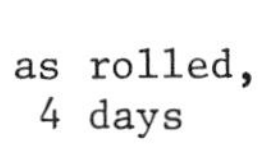

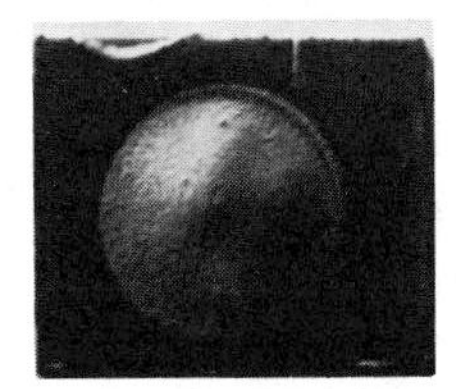

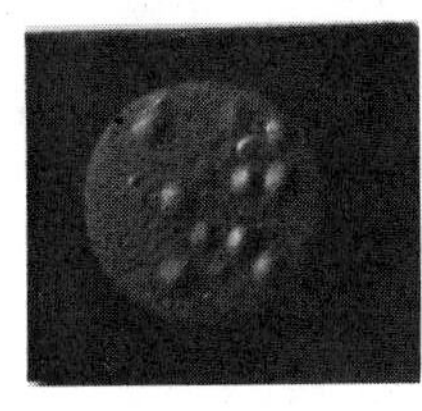

sand blasted,
3 days

Figure 3. Gas blistering of an epoxy coating on a steel substrate after exposure to 0.1M H_2SO_4 at 60 C. Blisters contained a high concentration of hydrogen.

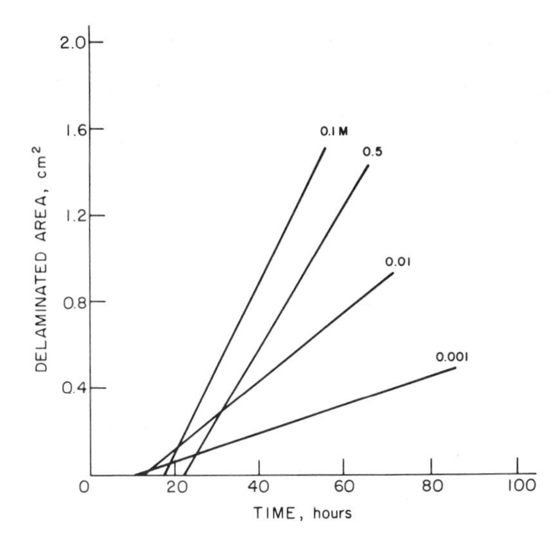

Figure 4. Cathodic delamination of pigmented epoxy coatings on steel. A defect was placed in the coating and the coated metal was maintained at a potential of - 0.8 v vs. SCE while immersed in NaCl solutions of different concentrations.

through the coating and act as charge carriers and counterions to allow the cathodic reaction to occur beneath the coating. The flux of cations across the coating increases with increase in concentration of the diffusing cation.

Thermal Cycling. Coatings that are brittle and have different coefficients of expansion than the substrate metal are very susceptible to disbonding upon thermal cycling. This disbonding may occur locally in small areas or it may occur in the most drastic cases over very large areas. A vinyl ester coating that has recently been studied in our laboratory exhibited very low rates of water transmission but the bonding had a tensile strength of the order of 70 kg/m^2, a low value compared to tensile strengths observed with many other coatings. As might be expected, this coating tends to lose adherence upon thermal cycling as shown in a recent paper by Tater (7). There is good reason to believe that one of the functions of the rough surface generated by abrasive blasting is to provide many anchor points that reduce the likelihood of large-area disbonding upon thermal cycling.

Stresses leading to disbonding of a brittle coating may also originate at welded joints or in coatings on thin substrates that suffer flexing during service.

Anodic Undermining. Anodic undermining reresents that class of corrosion reactions underneath an organic coating in which the major separation process is the anodic corrosion reaction under the coating. An outstanding example is the dissolution of the thin tin coating between the organic lacquer and the steel substrate in a food container. In such circumstances, the cathodic reaction may involve a component in the foodstuff or a defect in the tin coating may expose iron which then serves as the cathode. The tin is selectively dissolved and the coating separates from the metal and loses its protective character.

Another example is the very slight delamination that occurs when a thin copper layer is overcoated with an organic coating such as a photoresist and the system is made anodic. The rate of disbonding is a function of the applied potential and hence the rate of dissolution of the copper beneath the coating. Anodic delamination occurs very slowly relative to cathodic delamination at equal potential differences from the corrosion potential.

Anodic undermining has not been studied as extensively as cathodic delamination because there do not appear to be any mysteries. Galvanic effects and principles which apply to crevice corrosion provide a suitable explanation for observed cases of anodic undermining.

Migration of Species Through Coatings

Corrosion beneath an organic coating requires that there be an aqueous phase, that there be anions and cations to provide conductivity in the aqueous phase and that there be oxygen for the cathodic reaction. These species must all find migration pathways through the coating. Some recent experiments that provide some interesting facts about the migration of species through organic coatings will be described.

Migration of Water. Water uptake by a coating may be followed by impedance measurements (8) and by dielectric spectroscopy in the 10^9 Hz region (9). An important concern is what types of pathways do the water molecules follow in the migration through the coating. Are these pre-existing pathways that remain as the solvent is removed from the coating? Or does the water follow a random walk through the organic matrix? Immersion of some coatings in radioactive solutions followed by exposure of the coating to high resolution photographic film suggests that there are preferred pathways in the coating through which water may move relatively rapidly. It is the working hypothesis in our laboratory that the major means by which water may move through an organic coating is by pre-existing pathways where on a submicroscopic scale the density of the coating is low.

A graduate student, Hyacinth Vedage, is currently studying the pH of the liquid beneath an organic coating using an oxidized iridium wire, implanted through the steel substrate so as to be flush with the plane of the substrate/coating interface. The iridium wire is insulated from the steel by an organic coating on the wire. The pH of the liquid was determined from a calibration curve by measuring the potential of the wire relative to a reference electrode in the solution. In the case of a vinyl ester coating on steel immersed in 0.1M sulfuric acid at 60 C, it required approximately 60 days before the electrode yielded a stable potential. The potential indicated that the pH was approximately 6. It required a day or two before the potential achieved a value corresponding to a pH of 2. These measurements, which were reproducible, suggested that in this case, the water migrated through the coating first and the ionic components diffused to the interface after the water pathway was established. This work is continuing with other coating systems. It should be pointed out that this technique is useful not only for determining the pH under the coating but the time to obtain a steady-state corrosion potential indicates the length of time before an aqueous phase develops at the interface in the vicinity of the sensing electrode.

Another interesting feature about water migration is that an applied cathodic potential increases the rate of uptake of water by the coating (10). Data leading to this conclusion are summarized in Table I for three different coating systems. In all cases the water uptake as estimated from impedance measurements was more than one order of magnitude greater at an applied potential of -0.8 v vs. Ag/AgCl compared to open circuit conditions where the corrosion potential was -0.62 v. No explanation for the increased rate of water penetration with the application of a mild applied potential is apparent at the present time. Companion measurements using radioactive ^{22}Na indicated that the applied potential increased the rate of migration of sodium the same order of magnitude as the increase in the rate of migration of water. The effect of the applied potential on water uptake may be a direct consequence of the development of more effective diffusion pathways through the coating. No discrimination among these possibilities or others can be made at present.

Migration of Cations. Data are given in Table II for the rate of uptake of Na^+ and Cs^+ with and without an applied cathodic potential of -0.8 v vs. Ag/AgCl. In all cases it will be noted that the uptake was increased approximately one order of magnitude with an applied potential. This result is just what one might expect because

Table I. The Effect of an Applied Cathodic Potential on the Rate of Uptake of Water by Organic Coatings

Type of Coating	Thickness, um	Without Applied Potential	With Applied Potential
Alkyd Topcoat	37 - 40	5.9×10^{-7}	6.7×10^{-6}
Two layers of primer plus alkyd topcoat	62 - 67	3.9×10^{-8}	1.3×10^{-5}
Polybutadiene	10 - 12	1.6×10^{-7}	3.8×10^{-6}

Conditions: 0.5M NaCl, room temperature, cathode potential = -0.8 v vs. Ag/AgCl, several days exposure.

Table II. The Effect of an Applied Cathodic Potential on the Rate of Uptake of Cations by Organic Coatings (10)

Type of Coating	Thickness, um	Cation	Cation Uptake, mol/h: Without Applied Potential	Cation Uptake, mol/h: With Applied Potential
Alkyd Topcoat	37 - 40	Na^+	9.1×10^{-9}	6.9×10^{-8}
		Cs^+	1.8×10^{-8}	1.1×10^{-7}
Two layers of primer plus alkyd topcoat	62 - 67	Na^+	6.3×10^{-10}	9.3×10^{-9}
		Cs^+	3.9×10^{-10}	6.3×10^{-9}
Polybutadiene	10 - 12	Na^+	3.5×10^{-10}	7.3×10^{-9}
		Cs^+	3.3×10^{-9}	3.9×10^{-9}

Conditions: 0.5M alkali metal chloride; room temperature; applied potential = - 0.8 v vs. Ag/AgCl, 10 - 25 day exposure.

the potential of the metal substrate is such as to attract positively charged ions. It is striking that such an increase occurs when the magnitude of the applied potential is so small, i.e., 180 mv difference between the steady state potential and the applied potential.

In all cases studied to date, the rate of cathodic delamination is greater in CsCl solutions than in NaCl solutions of the same molarity. The increased rate has been attributed to the greater rate of diffusion of the hydrated cesium ion through the coating than that of the hydrated sodium ion. Unfortunately, the data in Table II are not precise enough to make a comparison between the results for Na^+ and Cs^+. The uncertainty is large because the total uptake of the cation is so small. The data also do not discriminate between the radiotracer ions that are present in the coating or in the aqueous phase at the interface between the coating and the substrate.

These data show that increased rates of migration of cations occur with small applied potentials. One may also extrapolate these data and infer that cation migration, and hence charge flow, is increased by differences in potential at local anodes and cathodes existing at the metal surface in the absence of an applied potential.

Migration of Oxygen. Our research on the migration of oxygen through organic coatings has had a very limited objective and some background is in order. The rate of cathodic delamination of many different types of coatings in alkali metal halide solutions is strongly a function of the alkali metal. In all cases studied, the rate of delamination under equivalent experimental conditions increases in the order: $Li^+ < Na^+ < K^+ < Cs^+$. The most likely explanation for this cation effect is the relative rates of diffusion of the hydrated cation through the coating or within the thin liquid in the delaminated region between the coating and the metal. Other explanations for this effect have also been considered and one that was amenable to test was an explanation based on the rates of diffusion of oxygen through the coating as a function of the type of cation under conditions where there were a simultaneous concentration gradient and potential gradient, both of which would be in the same direction through the coating as the oxygen concentration gradient.

The idea has been tested with free films of polyethylene and an acrylic spray coating. These films were mounted between two chambers in which the left hand chamber contained an oxygen probe, an electrode, and a 0.005M solution of the alkali metal chloride. The right hand chamber contained an electrode, an air bubbler and a 0.5M solution of the alkali metal chloride. The electrode in the left hand chamber was maintained at a potential of -1.2 v vs. a Ag/AgCl electrode so that the gradient between the two chambers was approximately 600 mv. The left hand chamber was deaerated before the experiment began and the oxygen concentration in the chamber was then monitored continuously with the probe as a function of time.

The rate of oxygen diffusion through the coatings in both cases was in the order $K^+ > Na^+ > Li^+$ in the presence of the potential and concentration gradient but the differences between the lowest and highest rates were of the order of 35%. In the absence of an applied potential, the rates were approximately the same with a maximum spread of 15%. These results are suggestive that the alkali metal cations do affect the migration of oxygen through a coating when there exists both a concentration gradient and a potential gradient. However, many more experiments must be performed before a conclusive statement can be made.

Acknowledgment

Much of the work reported herein was obtained in a research program supported by the Office of Naval Research. We are indeed grateful for this support. Colleagues and students who have contributed importantly to the work described herein include Dr. Richard Granata, Dr. Malcolm White, Dr. Douglas Eadline, Dr. Jeffrey Parks, Wayne Bilder, Hyacinth Vedage, Mark Atkinson, Philip Deck, George Rommal, and Valmore Rodriguez.

Literature Cited

1. Miyoshi, Y.; Kitayama, M.; Nishimura, K.; Naito, S. "Cosmetic Corrosion Mechanism of Zinc and Zinc Alloy Coated Steel Sheet for Automobiles"; paper presented at Society of Automotive Engineers, March, 1985.
2. Leidheiser, H., Jr.; Wang, W. _J. Coatings Technol._ 1981, 53 (672), 77.
3. Leidheiser, H., Jr.; Wang, W.; Igetoft, L. _Prog. Org. Coatings_ 1983, 11, 19.
4. Leidheiser, H., Jr.; Igetoft, L.; Wang, W.; Weber, K. In "Organic Coatings Science and Technology"; Parfitt, G. D.; Patsis, A. V., Eds.; Dekker: New York, 1984; Vol. 7, p. 327.
5. Wang, W.; Leidheiser, H., Jr. In "Equilibrium Diagrams and Localized Corrosion"; Frankenthal, R. P.; Kruger, J., Eds.; Electrochemical Society: Pennington, N. J., 1984; p. 255.
6. Leidheiser, H., Jr. _Corrosion_ 1982, 38, 374.
7. Tater, K. B. _Am. Painting Contractor_ 1982, 11.
8. Leidheiser, H., Jr.; Kendig, M. W. _Corrosion_ 1976, 32, 69.
9. Eadline, D. J.; Leidheiser, H., Jr. _Rev. Sci. Instrum._ 1985, 56, 1432.
10. Parks, J.; Leidheiser, H., Jr. _Ind. Eng. Chem. Prod. Res. Dev._ in press.

RECEIVED March 7, 1986

13

Chemical Studies of the Organic Coating–Steel Interface After Exposure to Aggressive Environments

Ray A. Dickie

Ford Motor Company, Dearborn, MI 48121

The chemical composition and morphology of organic coating/steel interfacial surfaces have been examined following adhesion failure in various aggressive environments. The analytical techniques employed have included X-ray photoelectron spectroscopy, dynamic secondary ion mass spectrometry, and scanning electron microscopy. Examples of cohesive and adhesive coating failure have been observed in each of several test modes in studies of model thermoset coatings. Typically, analyses of interfacial surfaces generated by simple mechanical removal of coatings from their substrates, and of those formed as a result of humidity-induced adhesion failure, indicate that there is little or no chemical change associated with the loss of adhesion. Exposure to corrosive environments can result in substantial changes in interfacial surface composition and morphology. In some instances, chemical degradation of organic coatings has been observed in the interfacial region. Chemical degradation of inorganic conversion coatings has also been observed, and appears to dominate the corrosion-induced paint adhesion loss process in some cases.

The corrosion protection afforded to steel by organic coatings is well known to be dependent on substrate composition and surface preparation, organic coating composition, and test or exposure conditions, among other variables. Organic coatings provide protection to metal substrates both by acting as barriers between the substrate and the environment and by preventing the spread of corrosion from an initial or incipient corrosion site. In general, good corrosion protection requires the establishment of good coating adhesion. For continued protection, adhesion must be maintained in the presence of water, electrolyte, and the various products of the corrosion reactions. Once corrosion starts, there is often a progressive disruption of coating adhesion; the mechanism and rate of the corrosion induced adhesion loss process has long been the subject of research (see Ref. 1 for a recent review). The nature of the chemical processes responsible for the apparent loss of adhesion has been the subject of a number of investigations using modern surface analytical techniques (see, e.g., 2-6). This paper discusses recent chemical studies of the organic coating/steel interface, with particular reference to the effect of changes in the molecular structure of the organic coating on the rate and mechanism of humidity- and corrosion-induced adhesion loss.

Surface Studies of Interfacial Composition

The locus and chemistry of adhesion loss have been studied using a wide range of analytical techniques. Among the most useful have been surface sensitive spectroscopic methods, including X-ray photoelectron spectroscopy (XPS or ESCA), Auger electron spectroscopy (AES), and secondary ion mass spectrometry (SIMS). Conventional microscopic tools (especially scanning electron microscopy) have also been widely used. Applications of various surface analytical methods to adhesion and corrosion problems have been extensively reviewed (e.g., 2, 7-12). XPS allows a quantitative elemental analysis of the topmost molecular layers and can also give useful, if somewhat limited, molecular information. A further advantage of XPS is that beam damage and charging effects are relatively minor, allowing straightforward analysis of organic materials. For polymers typically used in organic coatings, for example, high resolution carbon spectra can yield information on the presence and relative abundance of a number of common functional groups, including ether, ester, carboxylate, and carbonate moieties (cf. Figure 1). The major disadvantage of XPS as applied in most published studies is its poor lateral resolution (ca. 5 mm), although recent advances in equipment have resulted in a substantial reduction in analysis area. AES can also provide an elemental analysis of the topmost layers of a sample, and in addition can provide images with a lateral resolution on the order of 0.1 μm. Application of AES to organic materials has been limited in part due

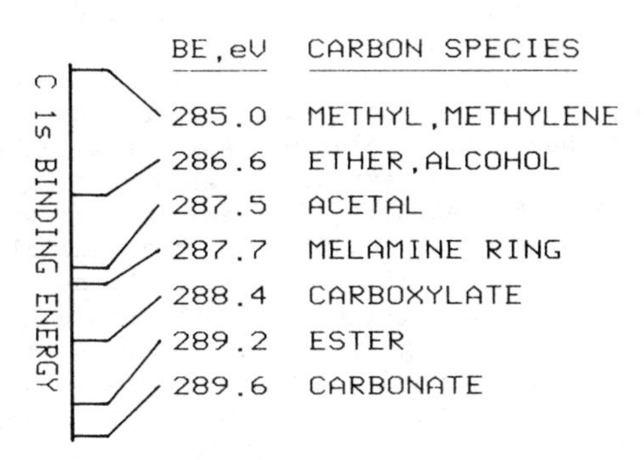

Figure 1. Experimental C 1s binding energies for selected carbon species.

to the paucity of molecular information that can be obtained, and in part due to beam damage effects during measurement. SIMS, and more recently, dynamic imaging SIMS, which provide greater surface sensitivity and substantially better lateral resolution than XPS have been applied to corrosion and adhesion problems (10, 11, 13, 14). Although only a qualitative analysis can be obtained using SIMS techniques, the ion images obtained can provide new information on the distribution and relationship between surface species. A serious problem with the application of most surface sensitive spectroscopic techniques to the study of adhesion and corrosion phenomena is that *in situ* measurements are not possible: typically, the coating film must be removed from the substrate prior to analysis. This restriction is avoided by optical methods, such as optical microscopy and ellipsometry (e.g., 15, 16), but these methods lack chemical analysis capability and, for *in situ* measurements, are limited to transparent coatings.

Humidity-induced Adhesion Loss

Good initial or dry adhesion of a coating to a substrate does not ensure good performance upon exposure to humid or corrosive environments. Exposure to high humidity is well known to reduce the adhesion of organic coatings to steel (17). Studies of moisture absorption kinetics suggest specific involvement of the interfacial region in humidity induced adhesion loss (18). In studies of humidity induced adhesion failure of organic coatings on clean, bare steel, examples of essentially adhesive failure, with little or no coating residue remaining on the substrate, and of cohesive failure of the coating film have been found (19). Typically, little or no chemical change associated with humidity induced adhesion loss is detectable in the organic coating. Of course, the presence of water-soluble inorganic salts as surface contaminants profoundly alters the interfacial chemistry and can lead to osmotic blistering and various corrosion-related blistering and adhesion-loss phenomena (20).

Figure 2 is representative of the high resolution C 1s spectra obtained in XPS analyses of the interfacial surfaces generated by humidity-induced adhesion failure; also included in the figure are spectra obtained from a reference (untested) coating surface and from interfacial surfaces generated by mechanical- and corrosion-induced adhesion failure. These spectra were obtained in a study of a thermoset coating based on a melamine-resin-crosslinked oligourethane resin (details of the resin structure and coating formulation are given in Ref. 19). Essentially identical spectra were obtained from the reference coating surface and from the interfacial surfaces after adhesion failure. Corresponding spectra of the substrate interfacial surfaces suggest that mechanical removal and humidity-induced failure leave little or no coating residue. There is little evidence for chemical change during humidity induced adhesion loss. Similar results, which are summarized in Table I, have been presented for coatings based on various other resin systems (19). For the coatings studied, there was a striking dependence of adhesion, or more correctly, of resistance to humidity-induced adhesion loss, on the ratio of resin hydroxyl to crosslinker alkoxy groups; only coatings for which this ratio was greater than about one were able to withstand condensing humidity exposure (19).

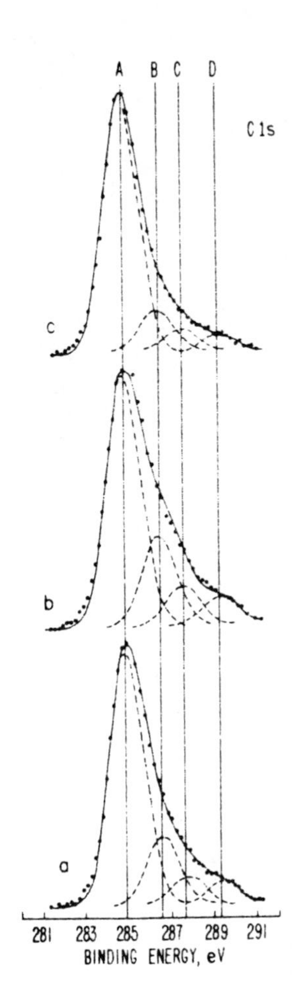

Figure 2. XPS C 1s spectra for oligourethane based coatings: (a) untested oligomer coating surface; (b) interfacial coating surface after mechanically induced adhesion loss; (c) interfacial coating surface after humidity induced adhesion loss. Spectral components A, B, C, and D attributed to methyl/methylene, ether, melamine, and urethane carbonyl carbons, respectively. Reproduced from Ref. 19, copyright 1984, American Chemical Society.

TABLE I. Summary of Failure Modes of Selected Thermoset Coatings in Humidity and Cathodic Polarization Testing.

Resin (1)	Time and Mode of Adhesion Loss			Ref.
	Humidity (2)	Cathodic Polariz'n (3)	Mode of Adhesion Loss	
Oligourethane	<0.3h (G)	<0.3h (G)	Adhesive; rapid overall loss of adhesion.	19
Oligoester	1-2h (G)	1-2h (G)	Primarily adhesive, but with some resin degradation in cathodic polarization.	19
Epoxy-diol adduct	24h (P)	4h (S)	Partly cohesive in humidity; adhesive displacement in cathodic polarization.	19
Epoxy-fatty acid adduct	>240h (NF)	2-3h (S)	Degradation and cohesive failure in cathodic polarization.	3,26
Epoxy-alkanol amine adduct	>240h (NF)	240h (S)	Slow degradation of cross-linker in cathodic polarization.	28,31
Epoxy ester-amine adduct	>48h (NF)	8h (S)	Resin degradation in cathodic polarization.	31

Polybutadiene	>240h (NF)	2-3h (S)	Principally resin degradation, cohesive failure in cathodic polarization; depends on cure, test mode (see text).	6,19 21,32

(1) Oligourethane: Reaction product of isophorone diisocyanate with 2-ethyl-1,3-hexanediol; MW = 514; f = (hydroxyl)/(molecule) = 2; r = (resin hydroxyls)/(cross-linker alkoxy) = 0.61.
Oligoester: Reaction product of 1,4-butanediol diglycidyl ether, 2-ethyl-1,3-hexanediol, and 4-methyl-1,2-cyclohexane dicarboxylic acid anhydride; MW = 830; f = 4; r = 0.75.
Epoxy-diol adduct: Reaction product of 2 moles 1,4-butanediol with DGEBA; MW = ca. 520; f = 4; r = 1.33.
Epoxy-fatty acid adduct: Reaction product of fatty acid with DGEBA; MW = ca. 880; f > 4; r > 1.7.
Epoxy-alkanol amine adduct: Reaction product of diethanol amine with DGEBA; MW = ca. 580; f = 6; r = 1.7.
Epoxy ester-alkanol amine adduct: Reaction product of ester-linked cycloaliphatic epoxy with diethanol amine.
Polybutadiene: Air oxidized films based on butadiene oligomers.
All resins except polybutadiene formulated with conventional melamine cross-linkers, applied to SAE 1010 bare steel paint test panels, and baked to yield cross-linked, solvent resistant films.

(2) Time required for general (G) or patchy (P) loss of adhesion. NF signifies that no failure occurred in test time listed.

(3) Time required for general (G) or >4 mm scribe associated (S) adhesion loss.

In studies of mechanically-induced adhesion loss of polybutadiene coatings, various results have been obtained (19, 21). Polybutadiene coatings cure by an oxidative mechanism; although the initial resin is essentially a pure hydrocarbon, the final cured coating contains approximately 10% oxygen. Interpretation of the results of studies of the interfacial chemistry of polybutadiene coatings is complicated by the dependence of film composition on position with the film and on the nature of the substrate (6, 21, 22). From XPS and infrared spectroscopic measurements, it is evident that the extent of oxidation of polybutadiene films is markedly greater near the coating/metal interface than in the bulk of the coating film; the interfacial regions typically comprise two or more times the oxygen level incorporated in the bulk. In addition, a reduction of ferric to ferrous species in the surface metal oxide has been reported (6). Mechanically-induced adhesion failure has been observed both in the bulk of the coating film (21) and near the coating substrate interface (19). Humidity-induced adhesion loss appears to occur in the more highly oxidized interfacial region of the polybutadiene film. Figure 3 reproduces the high resolution C 1s spectra for the coating interfacial surfaces of the polybutadiene coating. There appears to be a small change in coating composition in the interfacial region upon exposure to water; this change has been tentatively attributed to the hydrolysis of an acetal moiety produced during cure (19). The acetal hydrolysis does not appear to affect the humidity resistance of the polybutadiene coating significantly.

Corrosion-induced Adhesion Loss

It is commonly accepted that the loss of paint adhesion that accompanies corrosion is attributable in most cases to the increased pH associated with the cathodic corrosion sites. At initial sites of corrosion on painted steel, there is a progressive localization and separation of anodic and cathodic sites. The role of the cathodic reaction in adhesion loss has been demonstrated repeatedly. By subjecting scribed panels to cathodic polarization conditions, the anodic dissolution of iron can be suppressed, and the effects of cathodic reactions studied independent of the anodic reaction (23, 24). The extent of undercutting under cathodic polarization conditions is not simply a function of total corrosion current passed, but depends on a variety of other factors including electrolyte type, substrate and substrate preparation, and the magnitude of the cathodic potential, as reviewed in detail by Leidheiser et al. (1, 26). Under anodic test conditions there is rarely significant paint adhesion loss (24). Experimental results such as those summarized in Table II have led to the simple qualitative model of corrosion-induced adhesion loss shown in Figure 4 (25). The mechanism of delamination has variously been claimed to be dissolution of substrate surface oxide or conversion coating, displacement of polymer from the substrate oxide, and saponification or other chemical degradation of the paint resin in the interfacial region. The evidence presently available suggests that each of these mechanisms may be active under certain conditions.

The rate of adhesion loss in tests of paint performance can be greatly influenced by coating formulation and resin type (24); this is illustrated qualitatively by the results shown in Table II. To a first approximation, the anodic inhibitor pigments commonly used in

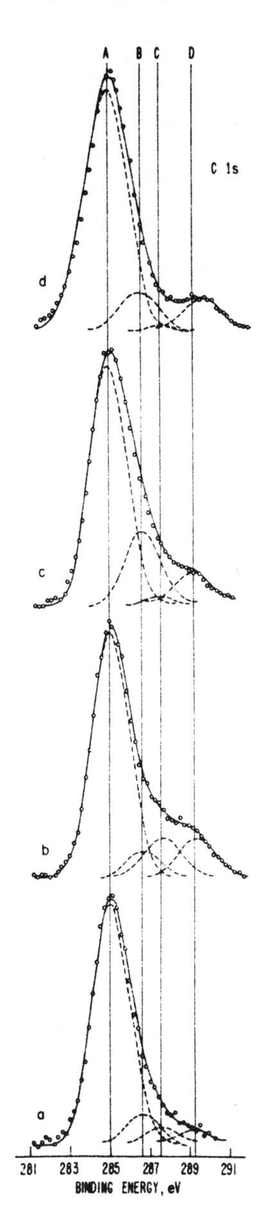

Figure 3. XPS C 1s spectra for polybutadiene coating surfaces; spectra a, b, c, and d represent spectra of the untested coating surface and interfacial surfaces generated by mechanically induced adhesion loss, by humidity induced adhesion loss, and by cathodic polarization induced adhesion loss, respectively. Components A, B, C, and D attributed to methyl/methylene carbon, ether carbon, acetal carbon, and ester carbonyl carbon, respectively. Reproduced from Ref. 19, copyright 1984, American Chemical Society.

TABLE II. Dependence of Paint Adhesion Loss on Test Conditions for Selected Formulations.

Resin Type	Delamination from Scribe, mm		
	Anodic Polarization (1)	Cathodic Polarization (2)	Salt Spray (3)
Maleinized Oil	<1.5	10	7
Epoxy-ester	<1.5	16	19
Polybutadiene/phenolic	2	9	3
Polybutadiene A	<1.5	6	3
Polybutadiene B	<1.5	6	<1.5
Epoxy-amine	<1.5	1	<0.5

(1) Anodic polarization: nitrogen-saturated 5% aq. NaCl; 200μA anodic current imposed for 7 days.

(2) Cathodic polarization: oxygen-saturated 5% aq. NaCl; maintained at -1050 mV vs. SCE for 8 hours.

(3) Salt spray: 24 hour exposure under ASTM B117 conditions.

All paints were electrodeposited on SAE 1010 bare steel paint test panels to yield 20μm dry film thickness, and were baked to yield fully crosslinked, solvent resistant films.

paint formulations are not active under either the anodic or cathodic test conditions imposed to obtain the results of Table II; thus, while the salt spray exposure can be expected to reflect the corrosion protection potentially afforded by the total paint formulation, the cathodic polarization results should reflect primarily the contribution of the binder (and possibly of inert pigments) to corrosion performance (24). A comparison of the cathodic polarization and salt spray results of Table II suggests that much of the variation in performance of the three polybutadiene based coatings listed is due to variations in effectiveness of the inhibitive pigments used, while the performance of the epoxy-amine coating may be due to the superior resistance of the resin to displacement or degradation by cathodically-produced alkali (27).

A detailed comparison of the interfacial chemistry observed for several thermoset coatings subjected to cathodic polarization conditions has been given elsewhere (28, 29). Figures 5 and 6, reproduced from the earlier work, illustrate the XPS spectra obtained for two of

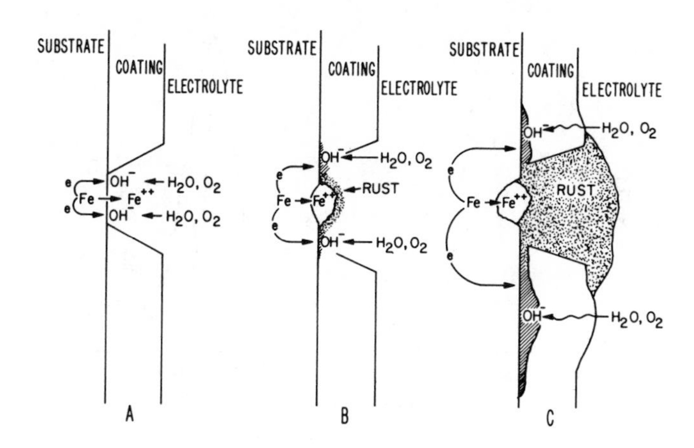

Figure 4. Schematic illustration of corrosion induced paint adhesion loss. In the initial stage (A), iron is oxidized and oxygen reduced at a site of electrolyte penetration; as corrosion proceeds (B and C), a rust deposit forms and the anodic and cathodic reaction sites become separated. Cathodically produced hydroxide progressively disrupts paint adhesion. Reproduced from Ref. 25, copyright 1980, American Chemical Society.

these materials. The two coatings in question were both based on hydroxy functional epoxy based oligomers; one was an epoxy ester (EE), the other, an epoxy alkanol amine adduct (EA). Both were formulated with conventional etherified melamine and urea crosslinking resins. When applied to cold rolled steel and baked, these coatings produced crosslinked, solvent- and water-resistant films. Neither of the coatings showed significant sensitivity to water in conventional humidity exposure tests. Despite the basic similarity of the coatings, a substantial difference in corrosion performance was observed: on bare steel, EE-based coatings failed within 24 h salt spray exposure, while the EA-based coatings survived in excess of 240 h under similar test conditions. A comparable difference in performance was observed in cathodic polarization testing. On the basis of the observed interfacial composition following adhesion loss, it was concluded that the difference in performance was attributable to the difference in resistance to attack by cathodically-produced alkali.

Figure 5 shows the XPS spectra for the EE-based coating interfacial surfaces; from the C 1s spectra, it is clear that carboxylate species are formed in the interfacial region during the corrosion process, most likely by saponification of esters. For the EA-based coating, the only significant change observed is the formation of carbonate in the interfacial region (Figure 6). The carbonate residue probably results from alkali degradation of the urea crosslinker (29). Although carbonate deposits found in similar studies of an amine-cured epoxy coating have been attributed to absorption of atmospheric carbon dioxide in the alkaline interfacial medium at the interface (30), these deposits were observed only after an induction period of several weeks. In studies of other resin systems involving relatively short exposure times (from several hours to a few days), little or no carbonate has been observed (3, 19).

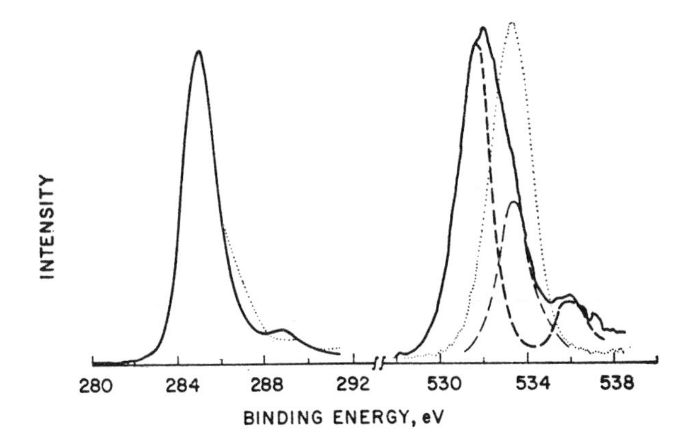

Figure 5. XPS C 1s (left and O 1s (right) spectra of the interfacial coating surface for epoxy-ester coating on bare steel following cathodic polarization testing (solid curves). Also shown for comparison are corresponding spectra for the untested surface of the coating (dots) and a component resolution of the O 1s spectrum. From left, the principal components of the O 1 s resolution are identified as carboxylate, combined ether and ester characteristic of the undegraded polymer, and adsorbed water (29) or sodium Auger (32). Reproduced with permission from Ref. 29. Copyright 1981, Pergamon Press.

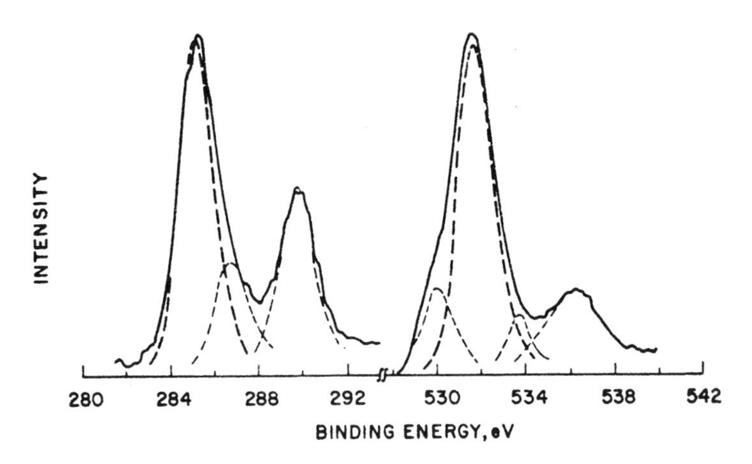

Figure 6. XPS C 1s (left) and O 1s (right) spectra of the interfacial substrate surface of epoxy-amine coating on bare steel following cathodic polarization testing (solid curves). Also shown is an approximate curve resolution for each spectrum; components of the C 1s spectrum are, from left, alkane, ether, and carbonate; principal components of the O 1s spectrum are, from left, iron oxide, carbonate, ether, ad adsorbed water (29) or sodium Auger (32). Reproduced with permission from Ref. 29. Copyright 1981, Pergamon Press.

To determine the effect of structural differences other than the presence of ester groups between the EE-based and EA-based coatings, EA-type polymers incorporating a limited amount of ester functionality were synthesized (31). In these experiments, the starting epoxy resins were chain-extended using adipic acid to provide main chain ester groups. The amount of ester was varied by using starting epoxy resins of different initial molecular weights. Although there is relatively large uncertainty associated with quantitation of adhesion loss in salt spray testing, the extent of adhesion failure observed for coatings based on these resins appears to increase monotonically with ester content (Figure 7). In an attempt to assess the importance of differences in transport properties of EA-based and EE-based resins, bilayer coatings were prepared and tested. It was found that the differences in performance of EE and EA based coatings could not be ascribed to differences in transport characteristics across the films: in each case, the observed corrosion performance was characteristic of the coating adjacent to the substrate (31).

The interfacial chemistry of corrosion-induced failure has also been studied for coatings with relatively poor resistance to water (poor wet adhesion); for these materials, the corrosion-induced failure typically involved little chemical change, but appeared to involve the same displacement mechanism observed for humidity induced adhesion loss (19). XPS spectra showing this mode of failure are reproduced in Figure 1; for this coating, essentially identical spectra were obtained from all surfaces analyzed, independent of test conditions.

Polybutadiene-based coatings have been extensively used in model studies of corrosion, and are therefore of special interest. As in the case of humidity induced adhesion loss, the corrosion induced

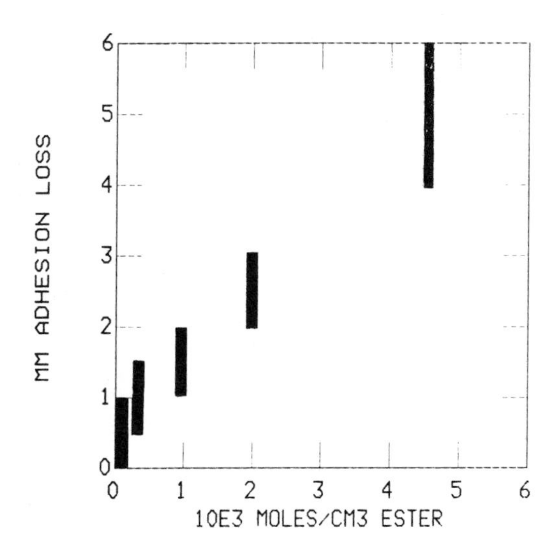

Figure 7. Paint adhesion loss in salt spray exposure (ASTM B117) as a function of ester content for chain-extended epoxy-amine and epoxy-ester resin based coatings. All coatings applied at 20-25 μm film thickness to SAE 1010 steel test panels, baked, scribed and exposed for 24 hours to salt spray conditions.

adhesion loss generally occurs within the more highly oxidized region of the coating immediately adjacent to the metal substrate (4, 6, 19, 21, 32). There appears to be an initial weakening of the resin, evidently involving alkali attack on the resin. At this stage, the coating can be readily removed with adhesive tape and appears to fail cohesively. At longer exposure times, especially under cathodic polarization conditions, there is a spontaneous loss of adhesion; little or no organic residue attributable to the coating remains on the substrate surface (32).

The interfacial chemistry of corrosion-induced failure on galvanized steel has been investigated (2); adhesion of a polyurethane coating was not found to involve chemical transformations detectable by XPS, but exposure to Kesternich aging caused zinc diffusion into the coating. Similar results were obtained with an alkyd coating. Adhesion loss was proposed to be due to formation of a weak boundary layer of zinc soaps or water-soluble zinc corrosion products at the paint metal interface.

The rate of failure of organic coatings is substantially slower on zinc phosphate conversion coated steel than on bare steel (for a recent review of conversion coating technology, see Ref. 33). The role of the conversion coating has been stated to be reduction of the rate of the cathodic reaction (34), but the effectiveness of the conversion coating depends on details of steel surface chemistry and conversion coating composition (33-37). The interaction of organic coatings with inorganic conversion coatings, and the effects of electrodeposition processes on conversion coating performance, have been widely studied (33, 38). The dissolution of conversion coating during anodic electrodeposition is well known (33). Cathodic electrodeposition of paint can also influence conversion coating performance; for example, the wet adhesion of paint films on cathodically primed conversion coated steel has been reported to be substantially affected by weakening of the phosphate by the alkali produced during electrodeposition (38).

Surface analytical studies of the interfacial chemistry of coating adhesion failure on zinc phosphated steel have been reported (28, 39). For the EE-based coating discussed previously, the locus of adhesion failure from zinc phosphated steel has been postulated to be the organic/inorganic interface, but there is evidence in the XPS studies of chemical attack on both the coating and the zinc phosphate conversion coating (28, 39). For example, Figure 8 presents XPS depth profile information on conversion coated steel before application and after corrosion-induced delamination of the organic coating. In addition to the presence of a substantial organic residue after testing (the structure of which suggests the presence of degraded EE resin), the conversion coating stoichiometry is greatly altered. SEM studies (Figure 9) reveal a corresponding change in crystal morphology, suggesting dissolution of at least the surface layers of the conversion coating. The conversion coating appears essentially intact near the leading edge of the delaminated zone (39). In recent SIMS studies, further details of the interfacial chemistry of corrosion-induced coating adhesion failure from conversion coated steel have been explored (14). These results are summarized in Figures 10 and 11. Both the SIMS step scan across the delaminated zone (Figure 10) and the ion images at the edge of the delaminated zone (Figure 11) confirm that the conversion coating is solubilized in the highly

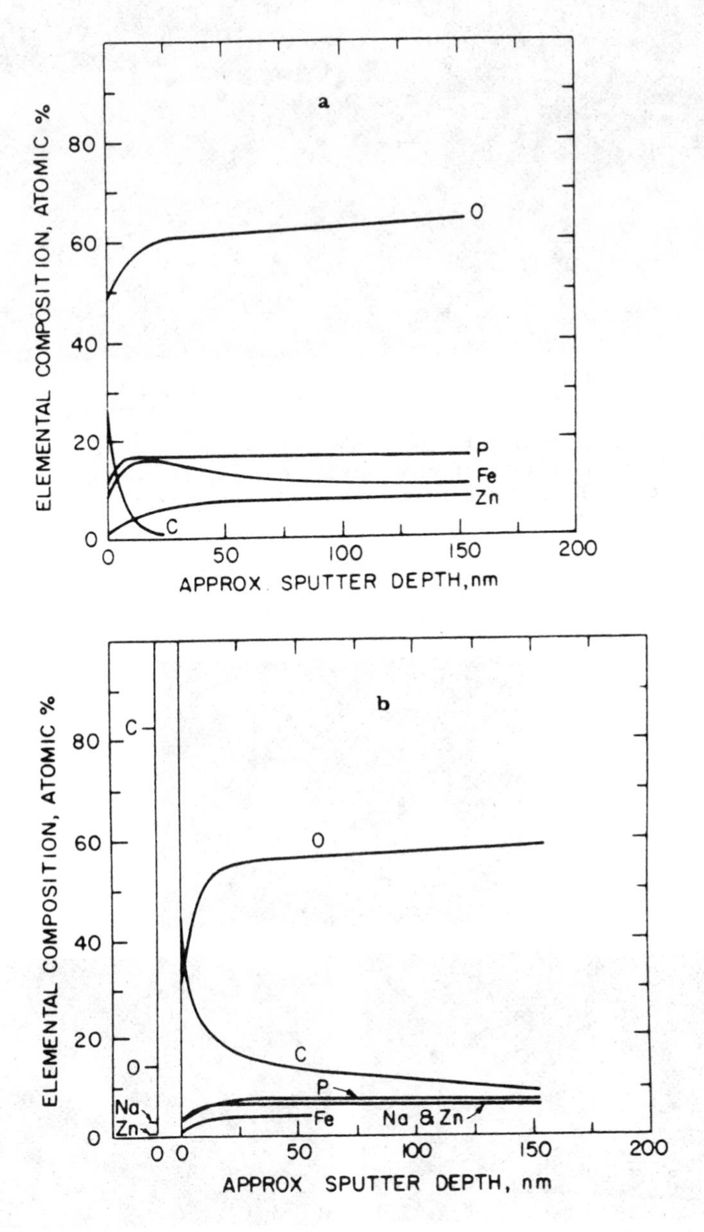

Figure 8. XPS analysis of elemental composition as a function of sputter depth: (a) zinc phosphated steel; (b) interfacial substrate surface of zinc phosphated steel after adhesion failure of epoxy-ester coating in cathodic polarization testing. Reproduced from Ref. 39, copyright 1983, American Chemical Society.

Figure 9. SEM study of the conversion-coated substrate surface after adhesion failure of epoxy-ester coating in cathodic polarization testing. Reproduced from Ref. 39, copyright 1983, American Chemical Society.

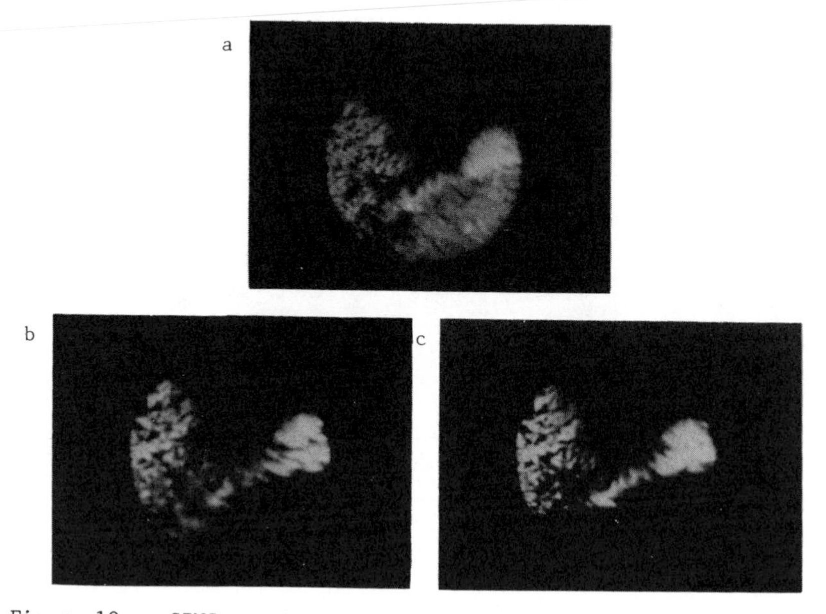

Figure 10. SIMS positive secondary ion images of the interfacial substrate surface of zinc phosphated steel after adhesion failure of epoxy-ester coating in cathodic polarization testing. Images obtained at the intact coating edge of the delaminated zone. Image diameter: 400 µm. Image a: mass 56 Fe+; image b: mass 64 Zn+; image c: mass 31 P+.

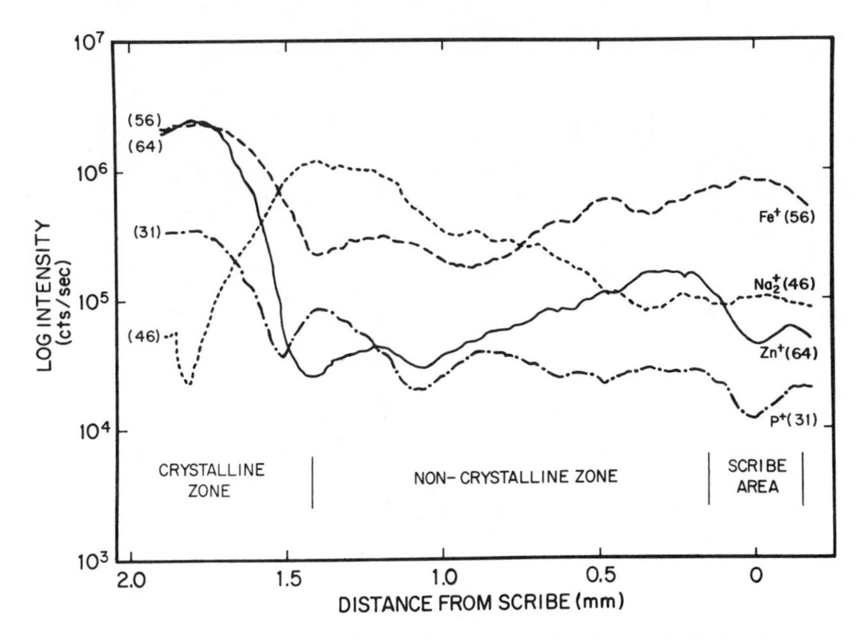

Figure 11. SIMS step scan analysis of the interfacial substrate surface between the initial scribe and the edge of the delaminated zone; coating and test conditions as in Figure 10.

alkaline medium that comprises the leading edge of the cathodic area. Based on the presence of organic coating residues on the substrate interfacial surface; the absence of conversion coating residue on the coating interfacial surface; the retention of conversion coating structural features and composition near the leading edge of the delaminated zone; and the continued strong dependence of total system performance on alkali resistance of the organic coating, it is speculated that the locus of initial adhesion loss for the EE-based coating applied over zinc phosphate conversion coating is the organic coating/inorganic conversion coating interface.

Highly alkali resistant coatings, such as EA-resin based materials, are very slow to lose adhesion in standard salt spray and cathodic polarization tests when applied over zinc phosphate conversion coated steel (28). With long term exposure, a patchy adhesion loss is observed which appears to involve primarily dissolution of the conversion coating. Various cyclic exposure and alternate immersion testing schemes have been applied to such systems in order to accelerate failure (35). One of the important mechanisms of failure appears to be a dissolution of the conversion coating. In the example schematically illustrated in Figure 12, undercutting of the film occurred from sites along the edge of the test panel; the interfacial region was characterized by more or less evenly spaced white deposits on the substrate which proved to be sodium phosphate (40). SIMS results are presented in Figure 13. In this study, as in other results on organic coatings applied over zinc phosphate (28, 39), little or no trace of the conversion coating was found on the interfacial surface of the organic coating.

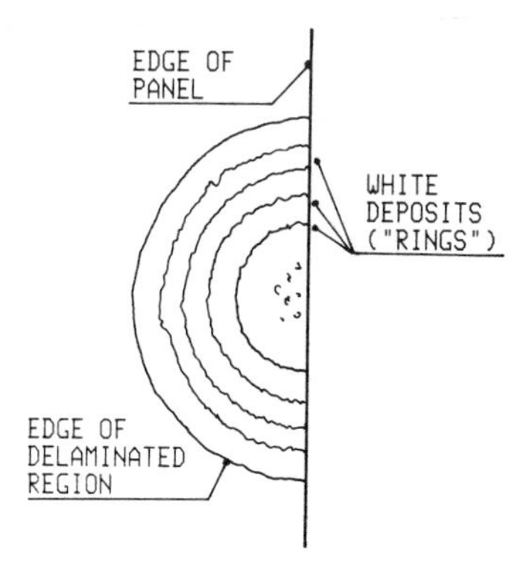

Figure 12. Schematic illustration of zone of adhesion loss in cyclic exposure test.

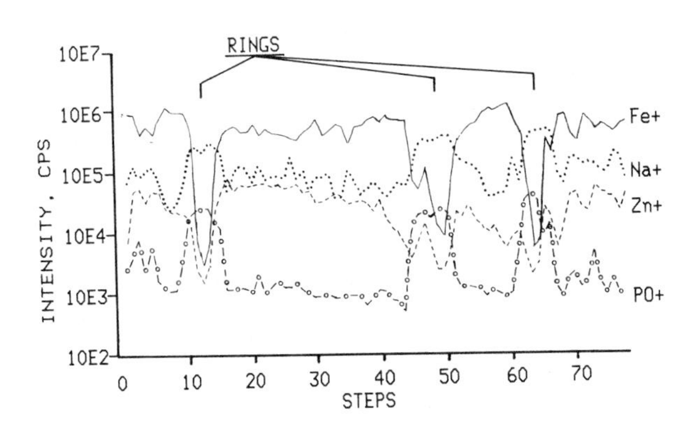

Figure 13. SIMS step scan analysis of a portion of the zone of adhesion loss illustrated in Figure 12. Step size approximately 150 μm.

Acknowledgment

This paper was presented in March, 1985, at the U.S.-Japan Seminar on Critical Issues in Reducing the Corrosion of Steels held in Nikko, Japan, and first appeared in the Proceedings of the Seminar. The complete Proceedings are available from National Association of Corrosion Engineers, PO Box 218340, Houston, TX 77218.

Literature Cited

1. Leidheiser, H. L., Jr.; Wang, W.; Igetoft, L. Prog. Org. Coatings 1983, 11, 19.
2. Van Ooij, W. J.; Kleinhesselink, A.; Leyenaar, S. R. Surf. Sci. 1979, 89, 165; Van Ooij, W. J. In "Organic Coatings Science and Technology"; Parfitt, G. D.; Patsis, A. V., Eds.; Dekker: New York, 1984; Vol. 6, p. 277.
3. Hammond, J. S.; Holubka, J. W.; Dickie, R. A. J. Coatings Technol. 1979, 51 (655), 45.
4. Castle, J. E.; Watts, J. F. In "Corrosion Control by Organic Coatings"; Leidheiser, H. L., Jr., Ed.; NACE: Houston, 1981; p. 78.
5. Dickie, R. A. In "Adhesion Aspects of Polymeric Coatings"; Mittal, K. L., Ed.; Plenum: New York, 1983; p. 255.
6. Watts, J. F. Anal. Proc. 1984, 21, 255.
7. Castle, J. E. Surf. Sci. 1977, 68, 583.
8. Holm, R.; Storp, S. Surf. Interf. Anal. 1980, 2, 96.
9. Van Ooij, W. J.; Kleinhesselink, A. Appl. Surf. Sci. 1980, 4, 324.
10. Baun, W. L. Appl. Surf. Sci. 1980, 4, 291.
11. Baun, W. L. In "Characterization of Metal and Polymer Surfaces"; Lee, L.-H., Ed.; Academic: New York, 1977; Vol. 1, p. 375.
12. Van Ooij, W. J. In "Physicochemical Aspects of Polymer Surfaces"; Mittal, K. L., Ed.; Plenum: New York, 1983; p. 1035.
13. Schuetzle, D.; Riley, T. L.; deVries, J. E.; Prater, T. J. Mass Spectrom. Rev. 1984, 3, 527.
14. deVries, J. E.; Riley, T. L.; Holubka, J. W.; Dickie, R. A. Surf. Interf. Anal. 1985, 7, 111.
15. Ritter, J. J.; Kruger, J. Surf. Sci. 1980, 96, 364.
16. Ritter, J. J. J. Coatings Technol. 1982, 54 (695), 57.
17. Walker, P. Off. Dig. 1965, 37, 1561.
18. Funke, W.; Haagen, H. Ind. Eng. Chem. Prod. Res. Dev. 1978, 17, 50.
19. Holubka, J. W.; deVries, J. E.; Dickie, R. A. Ind. Eng. Prod. Res. Dev. 1984, 23, 63.
20. Funke, W. Prog. Org. Coatings 1981, 9, 29.
21. Dickie, R. A.; Hammond, J. S.; Holubka, J. W. Ind. Eng. Chem. Prod. Res. Dev. 1981, 20, 339.
22. Dickie, R. A.; Carter, R. O., III; Hammond, J. S.; Parsons, J. L.; Holubka, J. W. Ind. Eng. Chem. Prod. Res. Dev. 1984, 23, 297.
23. Wiggle, R. R.; Smith, A. G.; Petrocelli, J. V. J. Paint Technol. 1968, 40 (519), 174.
24. Smith, A. G.; Dickie, R. A. Ind. Eng. Chem. Prod. Res. Dev. 1978, 17, 42.
25. Dickie, R. A.; Smith, A. G. Chemtech 1980, 10, 31.
26. Leidheiser, H. L., Jr.; Wang, W. J. Coatings Technol. 1981, 53 (672), 77.

27. Smith, A. G. Unpublished.
28. Holubka, J. W.; Hammond, J. S.; deVries, J. E.; Dickie, R. A. J. Coatings Technol. 1980, 52 (670), 63.
29. Hammond, J. S.; Holubka, J. W.; deVries, J. W.; Dickie, R. A. Corrosion Sci. 1981, 21, 239.
30. Watts, J. F.; Castle, J. E. J. Materials Sci. 1984, 19, 2259.
31. Holubka, J. W.; Dickie, R. A. J. Coatings Technol. 1984, 56 (714), 43.
32. Watts, J. F.; Castle, J. E. J. Materials Sci. 1983, 18, 2987.
33. Bender, H. S.; Cheever, G. D.; Wojtkowiak, J. J. Prog. Org. Coatings 1980, 8, 241.
34. Zurilla, R. W.; Hospadaruk, V. Soc. Automotive Engrs. Trans. 1978, 87, 762.
35. Hospadaruk, V.; Huff, J.; Zurilla, R. W.; Greenwood, H. T. Soc. Automotive Engrs. Trans. 1978, 87, 755.
36. Cooke, B. A. In "Organic Coatings Science and Technology"; Parfitt, G. D .; Patsis, A. V., Eds.; Dekker: New York, 1984; Vol. 7, p. 197.
37. Maeda, S. Prog. Org. Coatings 1983, 11, 1.
38. Maeda, S.; Asai, T.; Yamamoto, M.; Asano, H.; Okada, H. In "Organic Coatings Science and Technology"; Parfitt, G. D.; Patsis, A. V., Eds.; Dekker: New York, 1984; Vol. 7, p. 223.
39. deVries, J. E.; Holubka, J. W.; Dickie, R. A. Ind. Eng. Chem. Prod. Res. Dev. 1983, 22, 256.
40. deVries, J. E.; Prater, T. J.; Dickie, R. A. Unpublished.

RECEIVED June 16, 1986

Adhesion Loss of Ultraviolet-Cured Lacquer on Nickel-Plated Steel Sheets

S. Maeda[1], T. Asai[1], and M. Kakimoto[2]

[1]R & D Laboratories-I, Nippon Steel Corporation, 1618 Ida, Nakahara-ku, Kawasaki, Japan 211
[2]Production Technology Department, Daiwa Can Company, Ltd., 800 Nakanogo, Shimizu, Japan 424

Nickel plated steel sheet with a cathodic dichromate treatment is increasingly being used as a welded can material. It has been found that ultraviolet curing (UVC) of the exterior lacquer coatings is unsuitable for oiled (dioctyl sebacate, DOS) Ni-plated sheet since it shows poor adhesion of UVC lacquer on the surface contacted with a wicket during prebaking for the interior lacquer coatings. An x-ray photoelectron spectroscopy study of this problem showed that the adhesion loss can be ascribed to the dehydration of the hydrated chromium oxide, less evaporation, and poor oxidation of surface oil (DOS) in the wicket-contact areas during prebaking. Acetyl tributyl citrate (ATBC)-oiled surfaces perform well on the adhesion of UVC lacquer on all surfaces including wicket-contact areas because of the higher polar group concentration of ATBC.

Nickel plated steel sheet with cathodic dichromate (CDC) treatment has been developed as welded can stock. Recently, ultraviolet curing (UVC) of organic coatings (or inks) has been introduced for the external lacquering of welded cans made from this stock. Since the external coating is usually applied after the internal coatings, the sheet surface is first prebaked before applying UVC lacquer. It has been found that the UVC lacquer shows poor adhesion on oiled (dioctyl sebacate, DOS) nickel plated sheet, particularly on the surface where the sheet comes into contact with the wicket (supporting instrument) during prebaking.

Generally, prebaking is known to improve the adhesion of UVC lacquer to passivated tinplate because it increases surface energy by evaporating the surface oil and altering the oxide film composition (1), but the details of this phenomenon are not yet clear. During prebaking, the wicket-contact surface of the sheet is considered to change in a manner different from the other surfaces of the sheet. In order to clarify this problem, X-ray photoelectron spectroscopy (XPS) measurements were made on the CDC-nickel plated sheets with various surface pretreatments.

0097-6156/86/0322-0155$06.00/0

Experimental Method

Samples. CDC-treated nickel-plated steel sheets (nickel coating weight of 600 mg/m^2 and chromium coating weight of 6 mg/m^2) were oiled with about 5 mg/m^2 of dioctyl sebacate (DOS) or acetyl tributyl citrate (ATBC) and were employed as original samples.

Pretreatment. Ten types of samples were prepared, including as-oiled, solvent-washed, open-prebaked and tight-prebaked samples (to simulate wicket-contact portions), and tight-prebaked samples followed by flame-heating, UV-preradiation and solvent-washing. Prebaking was conducted at 205°C for 10 min. using a conventional oven.

XPS Measurement(2). XPS spectra were obtained using a VG-ESCA III electron spectrometer. An aluminum anode X-ray source (1486.6 eV) was used. The instrument was operated with a resolution of 1.36 eV FWHM (full width at half maximum peak height) on the Au 4f 7/2 line. The binding energies for the spectra were referenced to the hydrocarbon line at 285.0 eV. Analysis of high resolution spectra was accomplished using computer curve-fitting and graphics routines available in our own data system.

To obtain information in the direction of depth of the surface layer, the angle of the sample with respect to the incident X-ray beam was changed and the resultant photoelectron count rate was measured. The principle is shown in Fig. 1 (3).

When the peak intensity ratio I_i/I_j is measured for each of all the elements observed (carbon, oxygen, chromium and nickel), the surface atomic concentration ratio n_i/n_j of the elements can be given by the following equation, if surface contamination is negligible:

$$\frac{n_i}{n_j} = \frac{I_i}{I_j} \cdot \frac{\sigma_j \cdot \lambda_j \cdot S_j}{\sigma_i \cdot \lambda_i \cdot S_i} \qquad (1)$$

where σ_i and σ_j are the photoionization cross sections of elements i and j, λ_i and λ_j the electron escape depth and S_i and S_j are apparatus functions (detection efficiency dependent on the kinetic energy of electrons).

For calculating atomic concentration ratios of the elements, photoionization cross-section by Scofield (4) and apparatus function by Vulli (5) were adopted. Electron escape depth (λ) is determined by an experimental equation $\lambda = E^{0.7}$ (where E is kinetic energy of the electron) proposed by Hirokawa, et . al. (6).

Spectral intensities were measured as integrated peak area of each element and relative error for ratios of elemental intensity is about 10 % for all elements except carbon. The adventitious surface carbon (contamination) was estimated at approximately 10% of the total carbon measured. The concentration of carbon is, therefore, supposed to be in relative error by ca. 20 %.

UVC Lacquer and Test. A white epoxy-acrylate lacquer was applied on all nickel plated sheets with different pretreatments to a thickness of 16 μm and was cured by irradiation with four 10-kW ultraviolet lamps for about 1.2 sec.

The adhesion of the UVC lacquer was tested immediately after radiation with Scotch tape in a "peel back" pull-adhesion test and was evaluated by a "pass/fail" test. No delamination is rated as "good" and others are rated as "bad". "Bad" samples usually showed a large delaminated area (>50 % of the test surface) and the reproducibility of the test results was very good.

Critical Surface Tension of Wetting. According to Zisman's method (7), the critical surface tension (γ_c) on oiled CDC-Ni plated sheets was measured using a homologous series of liquid (alcohol/water). The advancing contact angle of a drop of the liquid was determined on the test surfaces of the Ni-plated sheets with Erma Contact Anglemeter, Goniometer Type, Model G-1. The critical surface tension (γ_c) was determined by the value of γ_l (surface tension of the liquid) at the intercept of the plot of cos θ vs. γ_l with the horizontal line, cos θ= 1.

Results and Discussion

XPS Spectrum of Surface. Figure 2 shows XPS spectra of CDC nickel-plated sheets with various pretreatment. Carbon, oxygen, chromium and nickel are observed in the surface of CDC nickel-plated sheets oiled with DOS and ATBC. The carbon spectra show ester carbonyl carbon and hydrocarbon species on both samples. The ester carbonyl carbon reflects the ester bond of DOS ($C_8H_{17}COOC_8H_{16}OCOC_8H_{17}$) and ATBC ($(C_3H_7COO)_3C(CH_2)OCOCH_3$).

The chromium and nickel are identified as chromium hydroxide and nickel (III) oxide, respectively.

Prebaking usually decreases the hydrocarbon and increases the ester carbonyl carbon (polar group) for DOS-oiled sheets, indicating the oxidation and evaporation of surface oil. Prebaking also caused the oxidation of surface nickel. Although not shown, the samples that were first tight prebaked and then flame heated underwent the same changes as open-prebaked samples. UV preradiation and hexane washing after tight prebaking hardly changed the surface carbon concentration and the polar group concentration.

The samples oiled with ATBC which is originally higher in polar group ratio than DOS showed a higher polar group concentration on the surfaces, and both surface carbon and polar group concentration changed little under either open or tight prebaking conditions as shown in Fig. 2. The high resolution C1s spectra for typical DOS and ATBC-oiled sheets are shown in Fig. 3 and the surface composition (atomic %) of the oiled CDC Ni-plated sheets with various pretreatments is summarized in "Table 1".

According to the high resolution spectra (Fig. 3), the apparent percentage of the ester species (286. 5 eV) appears smaller than that of the carbonyl species (288. 8 eV) on both DOS-and ATBC-oiled sheets, indicating the contribution of the contamination by adventitious hydrocarbon species. The measured percentage of the carbonyl carbon of as-received sheets is 10.9 % for DOS and 18.0 % for ATBC, those of which roughly correspond to the calculated ratios of DOS (9.2 %) and ATBC (21.0 %), respectively. But the effect of contamination makes it difficult to have further insight into the deviation from the calculated values.

Identification of Locus of Adhesion Failure. To clarify whether the disruption is cohesive failure of the lacquer or interfacial failure between the substrate and lacquer, the lacquer and metal sides of the fracture surface were both measured by XPS. The results are shown in Fig. 4. For the purpose of comparison, UVC lacquer coated and DOS oiled nickel-plated sheets are shown in the top and bottom of the diagram, respectively. The contribution at ca. 286.5 eV of the lacquer surface is attributed to carbon singly bonded to oxygen

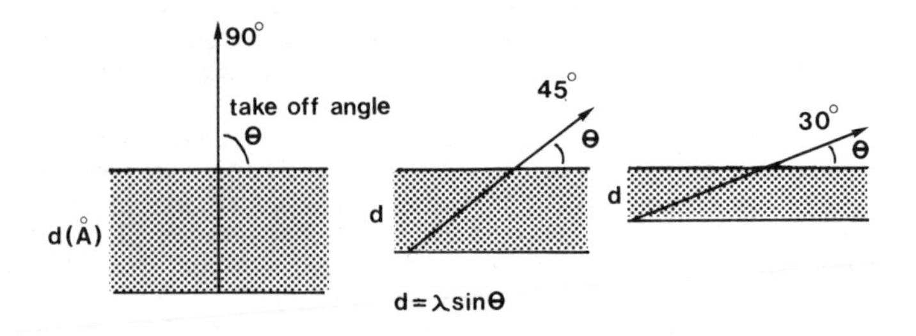

Figure 1. Relative sampling depth for angular dependent XPS.

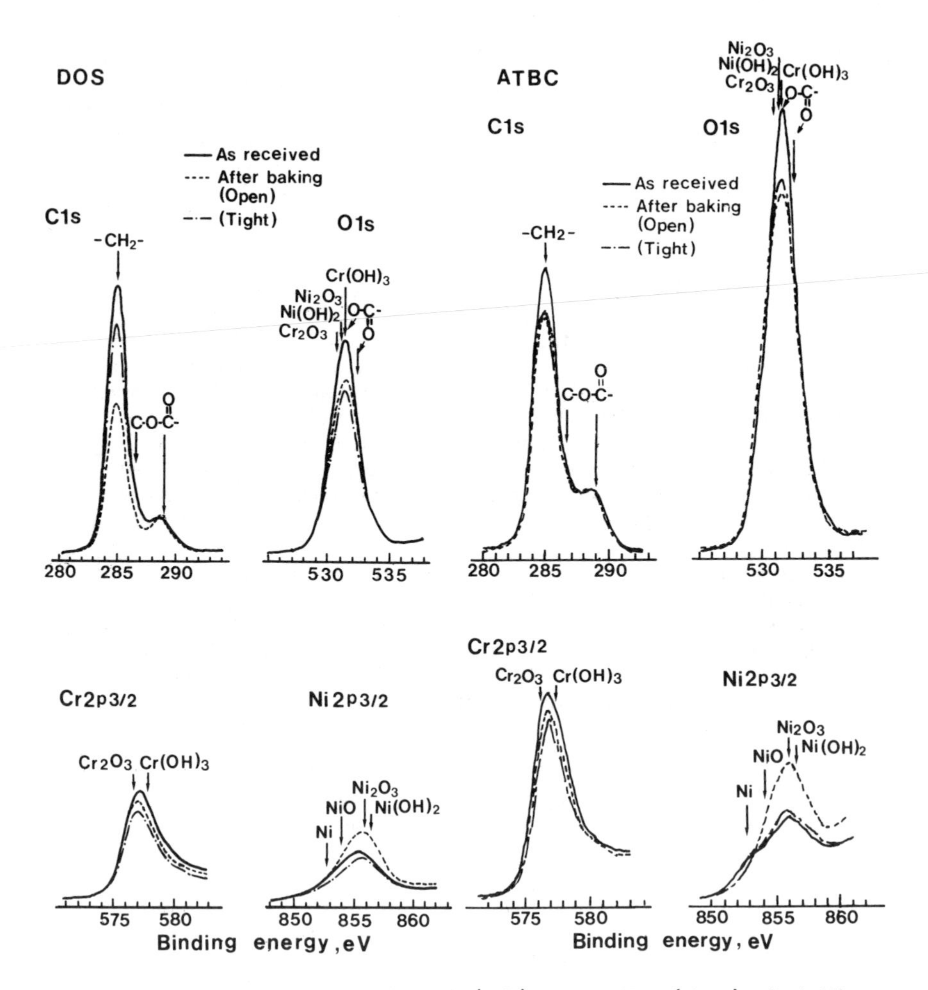

Figure 2. XPS spectra of DOS (left) and ATBC (right) oiled Ni-plated sheets before and after baking.

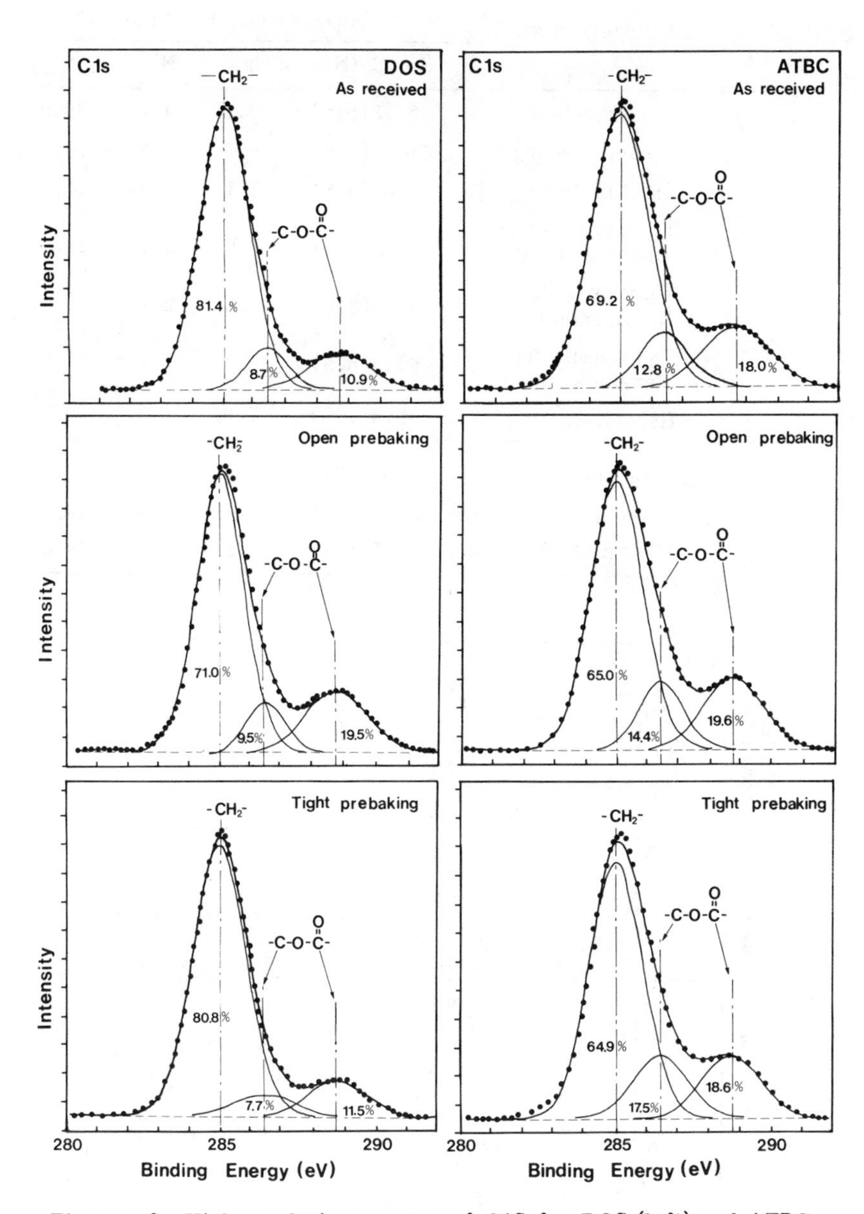

Figure 3. High resolution spectra of C1S for DOS (left) and ATBC (right) oiled surfaces.

Table I. Surface composition of the oiled CDC-Ni plated sheets

Oil	Pretreatments	Surface composition (at %)			
		C (>C=O)	Cr	Ni	O
DOS	As-received	51.8(5.6)	7.5	1.8	38.9
	Open prebaking	43.7(7.9)	9.4	3.5	43.5
	Tight prebaking	56.3(6.8)	7.1	1.8	34.7
	Tight prebaking + Flame heating	39.3(7.9)	9.9	3.1	47.6
	Tight prebaking + UV preradiation	51.8(6.8)	7.8	2.2	37.3
	Tight prebaking + Hexane washing	57.4(6.9)	6.7	1.7	34.2
	Hexane washing	49.4(5.9)	7.9	2.8	39.9
ATBC	As-received	37.9(6.8)	9.0	1.7	51.3
	Open prebaking	38.5(7.8)	9.0	3.9	48.6
	Tight prebaking	38.7(7.2)	9.2	2.4	49.6

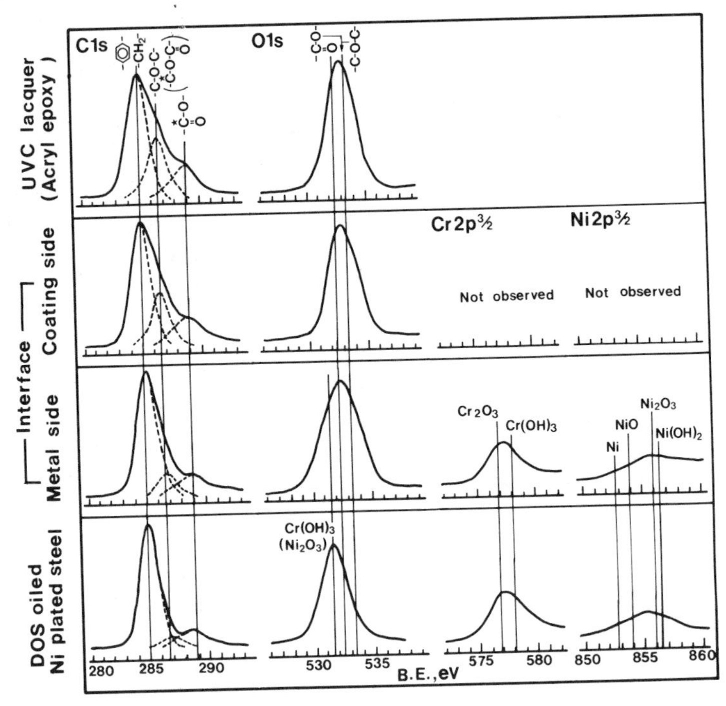

Figure 4. XPS spectra showing the interfacial coating surface and the interfacial metal surface (Top and bottom are UVC lacquer and DOS oiled surface)

(characteristics of ether and ester species), but the observed greater intensity of this species relative to ester carbonyl carbon (at 288.8eV) clearly shows the contribution of ether carbon of lacquer component. Detection of no chromium and nickel in the peeled lacquer side indicates that the fracture occurred at and near the substrate/coating interface. Identification of the ether bond (-C-O-C-), a component of the lacquer, on the substrate, however, shows that the lacquer component remained on the substrate to some degree.

Figure 5 shows the XPS angular dependence of the fracture surface of a sample. The ratios of the ester carbonyl carbon and ether carbon to the total amount of carbon are plotted on the ordinate. The 30° angle means that the thickness measured is assumed to be about a half of that with an angle of 60°. The results show that the concentration of the ester bond observed on the interfacial substrate surface decreases toward the fracture interface, suggesting that the hydrocarbon groups are oriented toward the outside (lacquer side). The concentration of the ether group (a component of the lacquer) on the interfacial substrate surface is about one-third that on the interfacial lacquer surface, and it is presumed that the lacquer remained like islands on the substrate surface at an area ratio of 1 : 3.

On the interfacial lacquer surface, the concentration of the ester group decreased toward the outside (the fracture interface), suggesting the diffusion of DOS into the lacquer film. SEM observation of the fracture surface of metal side is shown in Fig. 6. It can be seen that some lacquer remains as an island state. The apparent disagreement with XPS data seems to be due to the presence of invisible lacquer by this magnification (x 550). According to XPS results and SEM observation the locus of failure may be schematically represented as shown in Fig. 7.

Change in Composition of Surface Film with Treatment. Since trivalent chromium hydroxide can be generally expressed as $Cr_2O_3 \cdot nH_2O$, the composition of the outermost surface layer is assumed to be expressed by "Equation 2" and the hydration degree n and covering rate α of chromium hydroxide are calculated by "Equations 3 and 4" from the atomic percent of each of the elements measured by XPS. (The escape depth of nickel is smaller than that of chromium and oxygen. Although it may pose a problem, strictly speaking, the effect of this condition is neglected because the atomic concentration of nickel is low.)

$$\alpha Cr_2O_3 \cdot nH_2O + (1-\alpha)Ni_2O_3 \qquad (2)$$

When $[O]/[Cr] = X$ and $[Ni]/[Cr] = Y$,

$$\alpha = 1/(Y + 1) \qquad (3)$$

$$n = 2X - 3Y - 3 \qquad (4)$$

where [Cr], [Ni] and [O] are the atomic concentration of each of the element, and $[O] = [O]_{total} - 2\,[C]_{coo}$. The oxygen bonded with carbon is assumed to be approximately twice as much as the atomic percent of carbonyl carbon (288.8 eV) which seems to be more reliable than that of ester carbon (286.5 eV).

According to the binding energy of Ni 2P (8), nickel oxide was identified as Ni_2O_3 but the possibility of the state as $Ni(OH)_2$ is not always denied. If $Ni(OH)_2$ is adopted for analysis, "Equations 3 and 4" take a different formula (9), leading to slightly smaller value for both α and n , but the relative values among the individual sheets are invaliable.

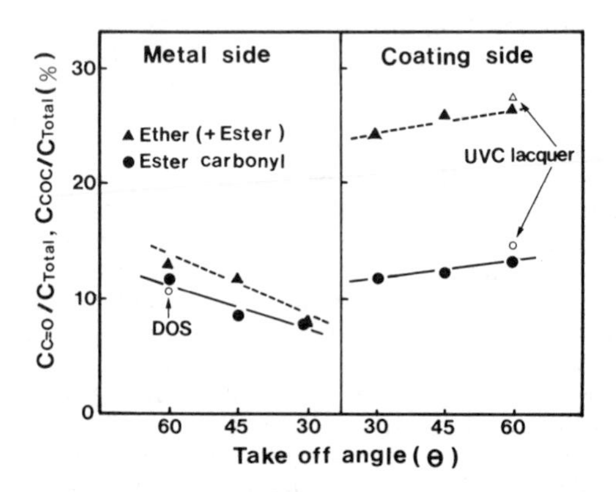

Figure 5. Variation of the polar groups (ester carbonyl and ether) with photoelectron take off angle.

Figure 6. SEM observation of the fracture surface of metal side (x 550)

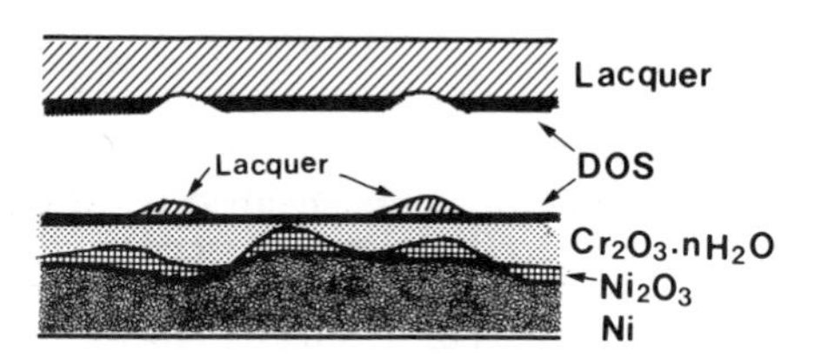

Figure 7. Schematic model showing a locus of failure

The covering rate α and hydration degree n of chromium hydroxide and the polar group ratio $[C]_{co}/[C]_{total}$ of oil are shown in Fig. 8. It is found that n = 3.7 and α = 0.81 for as-received DOS-oiled sheet and n = 4.8 and α = 0.85 for as-received ATBC-oiled sheet, respectively. (The n-values of 3 and 5 correspond to the composition of $Cr(OH)_3$ and $Cr(OH)_3H_2O$, respectively.) Different n values on two as-received sheets seem to be due to the different manufacturing chances. The hydration degree n thus determined approximately agrees with the hydration degree n (3.6 to 4.2) of passivation films on tinplate measured by Auger electron spectroscopy (AES) in our previous paper (9).

Prebaking is found to lower the hydration degree n and the covering rate of chromium hydroxide. The decrease in the value of α suggests the volume shrinkage of chromium hydroxide by dehydration. When open and tight prebaking are compared, DOS oxidizes under open prebaking but little under tight prebaking and ATBC, naturally high in polar group concentration, changes little depending on the prebaking condition (tight or open). Although the sheet is first tight prebaked, the oxidation of oil occurs to a considerable degree but the dehydration of hydroxide does not proceed further by the following flame heating. UV preradiation and solvent washing after tight prebaking do not appreciably alter the composition of the surface film.

Composition of Surface Film and Adhesion of UVC Lacquer. As the de-adhesion of UVC lacquer has occurred at the interface between the substrate and lacquer (or in the oil film), both chromium oxide and oil film affect the adhesion performance. From this standpoint, the adhesion of the UVC lacquer is summarized in Fig. 9. When chromium oxide is hydrated to a high degree or surface oil is oxidized to a greater level, the adhesion of the UVC lacquer does not deteriorate.

The chromium hydroxide with a high OH group concentration is generally said to provide the good primary adhesion of organic coatings (10). Figure 9 shows that when chromium hydroxide is dehydrated by prebaking, the adhesion loss due to low OH concentration in the hydroxide is considered to be compensated for by the oxidation of the oil. Since the dehydration of chromium hydroxide occurs but the oxidation and vaporization of oil can not easily occur on the tight prebaked sample, the de-adhesion of UVC lacquer is thought to have occurred in the wicket-contact areas. Therefore, ATBC with a high polar group concentration is considered to improve the adhesion loss of the UVC lacquer. In fact, it was confirmed that the UVC lacquer did not peel in wicket-contact sheets on the practical test using a commercial UVC coating line.

The Effect of Surface Composition on Wettability. Critical surface tension (γ_c) of the samples is shown in Fig. 10. γ_c of as-received oiled surfaces is greater on ATBC oiled sheet than on DOS oiled sheet, and it remarkably increases on the open-prebaked surfaces but little on the tight-prebaked surfaces of both DOS and ATBC oiled sheets. Flame heating improves the wettability of tight-prebaked samples but UV preradiation and hexane washing hardly change the wettability. The improvement of wettability by open prebaking and flame heating suggests the vaporization and oxidation of the oil films.

A single-variable correlation was run between γ_c and the surface composition to see what is the most significant factor for determining the wettability ("Table II").

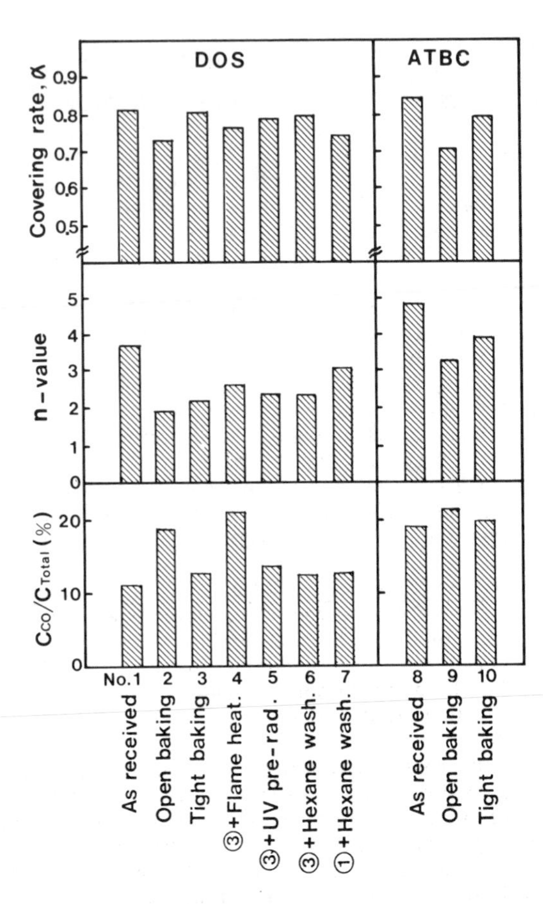

Figure 8. Change of covering rate (α), n-value and the polar group ratio of surface oil with pretreatment

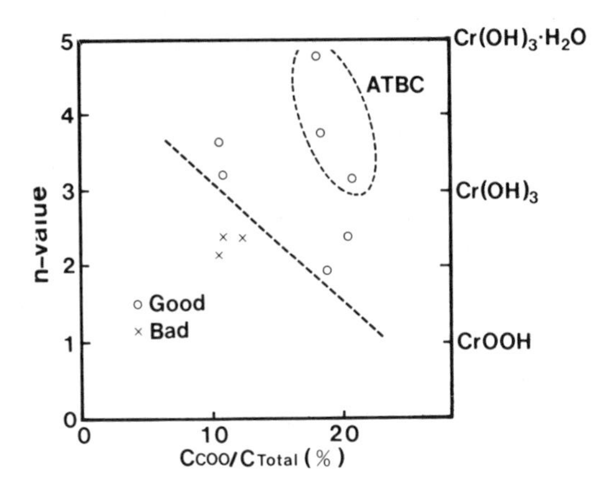

Figure 9. Effect of the polar group ratio of oil and the hydration degree of chromium oxide on lacquer adhesion.

Table II Single variable correlation between γ_c and surface composition

	C	>C=O	Cr	Ni	O	α	N	γ_c
C	1.000	-0.523	-0.933#	-0.546	-0.989#	0.262	-0.491	-0.569
>C = O		1.000	0.646	0.648	0.444	-0.498	-0.371	0.903#
Cr			1.000	0.652	0.883#	-0.366	0.212	0.730*
Ni				1.000	0.433	-0.940#	-0.314	0.830#
O					1.000	-0.143	0.603	0.463
α						1.000	0.480	-0.692*
N							1.000	-0.399
γ_c								1.000#

99% confidence level

* 95% confidence level

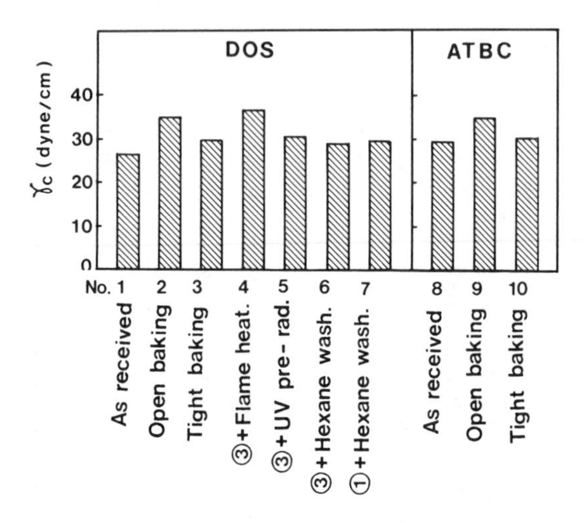

Figure 10. Critical surface tension (γ_c) of oiled Ni-plated sheets

A high correlation coefficient (r = 0.903, 99 percent confidence level) was found between γ_c and polar group concentration (>COO) and relatively high correlation coefficients were obtained between γ_c and the metallic elements (Cr and Ni), but carbon (main component of oil), oxygen, hydration degree (n) and covering rate (α) showed low correlation coefficient. In Fig. 11, γ_c versus the polar group concentration is presented.

The multiple regression analysis by four independent variables is shown in Fig. 12 (here, [O]= 1- [C] - [Cr] - [Ni], α and n are all dependent variables). The multiple regression equation is expressed as follows:

$$\gamma_c = 0.149\ [C] + 2.274\ [C]_{coo} + 1.418\ [Cr] + 1.469\ [Ni] - 7.053 \qquad (5)$$

In Fig. 12, the ordinate indicates the calculated values from the equation (5) and the abcissa is the experimental values of γ_c.

A high correlation coefficient (r = 0.969) and a relatively low intercept value (-7.053 dyne) indicate that γ_c on the oiled CDC - Ni plated sheet is substantially determined by the polar group of oil and the metallic elements.

It is of interest to note that hydration degree (n) does not directly contribute to the wettability, suggesting that it affects the bond strength formed during the cure of UVC lacquer.

Conclusions

The surface of oiled-CDC nickel-plated steel sheets was analyzed by XPS and was investigated for its effect on the adhesion of UVC lacquer. The following findings were obtained:

(1) When the composition of the surface film is indicated as $\alpha Cr_2O_3 \cdot nH_2O$ + $(1 - \alpha)\ Ni_2O_3$ (where α is covering rate and n is hydration degree), n = 3.7 to 4.8 and α = 0.81 to 0.85.

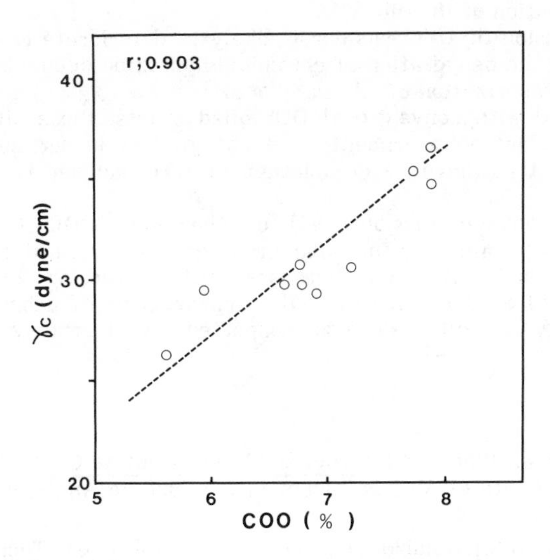

Figure 11. Correlation of γ_c and the polar group of carbon of surface oil.

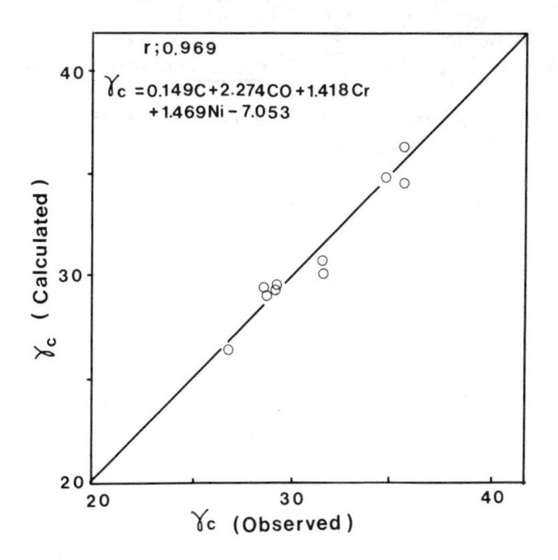

Figure 12. Multiple regression analysis of γ_c.

(2) During prebaking for the internal coating, dehydration of chromium hydroxide (decrease in n-value) has occurred. If surface oil is oxidized at the same time, the adhesion loss of UVC lacquer due to the dehydration is made up for by the oxidation of the oil.
(3) The adhesion of UVC lacquer is likely to deteriorate in wicket-contact portions where the dehydration of chromium hydroxide occurs but only limited oxidation and vaporization of oil take place.
(4) Compared with conventional DOS-oiled sheets, those oiled with ATBC with a high polar group concentration change less in surface condition by prebaking and thus, have no de-adhesion of UVC lacquer in wicket-contact portions.
(5) Critical surface tension (γ_c) of the oiled CDC Ni-plated sheets apparently is determined by the polar group concentration of oil and metallic elements (Cr and Ni). γ_c increases with increasing the polar group concentration of oil and, therefore, the improvement of adhesion due to the higher polarity of oil was also supported by thermodynamical aspect (wettability).

Literature Cited

1. Helwig, E.J.; Black, M.L. Proc. 2nd Inter. Tinplate Conf., 1980, p.407.
2. Japan Institute of Metals "Quantitative Surface Analysis" (in Japanese), Tokyo, 1978; p.64.
3. Hammond, J.S.; Holubka, J.W.; Dickie, R.A. J. Coat. Technol., 1979, 51, 45.
4. Scofield, J.M. J. Electron Spectroscopy, 1976, 8, 129.
5. Vulli, M.; Starbe, K. J. Phys. E., 1977, 10, 158.
6. Hirokawa, K.; Oku, M.; Honda, F. Z. Anal. Chem., 1977, 286, 41.
7. Zisman, W.A., Ind. Eng. Chem., 1963, 55, 18.
8. Kim, K.S.; Winogard, N. Surf. Sci., 1974, 43, 625.
9. Maeda, S.; Asai, T.; Sawairi, T. Proc. 2nd Inter. Tinplate Conf., 1980, p.407
10. Bolger J.C., In "Adhesion Aspect of Polymeric Coating"; Mittal K.L. Ed.: Plenum: New York, 1983; p.3

RECEIVED January 27, 1986

15

Cathodic Delamination of Protective Coatings: Cause and Control

J. S. Thornton, J. F. Cartier, and R. W. Thomas

Texas Research Institute, Inc., 9063 Bee Caves Road, Austin, TX 78733

The results of some recent tests, directed at understanding the role of the hydroxide ion in the cathodic delamination of thick rubber adherends, are discussed. In addition to contributing to the breaking of adhesive bonds, hydroxide ions appear to cause some components of commercially available adhesive systems to become swollen. Thus, at the debond interface, where the cathodic reaction is producing a strongly basic solution, the adhesive bond may be additionally strained by volume changes. Evidence is presented which suggests that the selection of an adhesive for marine applications which includes exposure of the metal substrate to a cathodic potential, should be preceded by an examination of the predisposition of the adhesive system to volume changes in the presence of high concentrations of hydroxide ion.

Cathodic protection is a common approach to reducing corrosion of metals in marine service. Virtually every steel ship in the U.S. Fleet is protected by the placement of zinc anodes which corrode sacrificially, thereby protecting the steel. The benefits of cathodic protection are enormous - steel hulls would be quite short lived without it.

There are complications however. The cathodic potential effectively stops corrosion on the metal substrate but it also contributes to early debonding of adherends, the development of leak paths under seals, and the blistering and peeling of coatings. Degradation of protective coatings is a basic life-limiting problem for underwater equipment exposed to a cathodic potential.

In this paper we are concerned with the adhesive systems beneath thick rubber adherends. Unlike the technology for the paint industry, where the science of qualification testing for durable thin coatings has produced paints with 10, possibly 20 years life

0097–6156/86/0322–0169$06.00/0

expectancy, the adhesives for underwater use are not so well qualified. The relationship between formulation and performance reliability certainly is not obvious for this particular harsh environment.

There is a need for an accurate assessment of the mechanism of cathodic debonding of thick adherends. This would contribute to better control over the selection of formulations resistant to attack. One of the objectives of this work was to develop an accelerated screening test. The screening test was used to evaluate the relative performance profiles of a number of commercial adhesives which were recommended for marine applications. It is hoped that a screening test such as this one could be used to isolate measurable primer properties which can be related to the long term prospectus of the primer, and the variability of this prospectus under permutation of substrate type or additives.

The Role of the Hydroxide Ion in Cathodic Delamination

The relationship between performance reliability and adhesive formulation is not simple. The key step in improving the reliability of adhesives on cathodically protected substrates is fully understanding the cathodic delamination process. Various mechanisms have been proposed in the literature. A large number of investigators have focused attention on the damage hydroxide ion does to coating adhesion.

During the cathodic delamination process there are two important reactions which can occur at the cathode and which are catalyzed on the thin layer of metal oxide which covers the cathode surface. These reactions are;

$$1/2\ O_2 + H_2O + 2e^- = 2\ OH^- \tag{1}$$

$$2H_3O^+ + 2e^- = H_2 + 2H_2O \tag{2}$$

Either reaction will result in an increase of the pH near the reaction site. The hydrogen reaction will proceed even in strongly basic solutions if the applied potential is increased sufficiently. Which reaction predominates depends upon the circumstances. The equilibrium potential for the oxygen reduction reaction is 1.24V more positive than the equilibrium potential for hydrogen reduction. On the other hand, the exchange current densities for hydrogen evolution on corrodible metal surfaces are far greater than the corresponding values for oxygen reduction. From cathode polarization curves for steel in 0.6M salt water (1) it can be seen that the oxygen reduction reaction is favored at potentials less than -0.8V (versus a standard calomel electrode) and the hydrogen reduction reaction is favored at potentials more negative than -1.0V. Thus, in neutral or basic solutions, where the H_3O^+ concentration is low, where dissolved oxygen is present and where an applied voltage less than -0.8V is present, we expect to find the oxygen reduction reaction dominating.

It is generally (2-4) agreed that the formation of the OH- is a crucial step in progression of circumstances which lead to delamination. As long as there is a growing pocket of caustic solution sequestered between the coating and substrate layers, further delamination is occuring. However, this is where the agreement ends. The actual mechanism by which OH- initiates the debond is not clear.

Various theories exist. These theories generally revolve around two central tenets: either the OH- is attacking the polymer surface and disrupting polymer to metal bonds, or it is attacking the metal oxide layer that covers the metal surface. In support of the first mechanism, using surface analysis techniques, Dickie, Hammond, and Holubka (5) have reported that carboxylated species present at the interface can be seen as a result of OH- attack of the polymer. On the other hand, Leidheiser (4) reports that Ritter has observed the attack of the metal oxide using ellipsometric techniques to study a polybutadiene coating on steel. In support of this, Ritter and Kruger (1) have measured pH values as high as 14 at the delamination site under natural corrosion conditions - this is certainly high enough to cause the dissolution of some metal oxides. Koehler (6) presented arguments for the case that the root cause of cathodic delamination is the displacement of the coating by a high pH aqueous film that grows in the interfacial region. In that description the interfacial water drastically reduces the dispersion forces between polymer and metal.

It is also true that if the metal to polymer bonds were principally of ionic character, then the water which forms at the interface would seriously degrade these bonds due to the high solvation energy released during the dissolution of ionic bonds. It has been observed (7) that the commercial adhesive system Chemlok 205/220 used for bonding rubber to metal will fail a adhesive tape peel test after submersion in seawater and exposure to cathodic potential. However, it may recover up to 80% of the original bond if it is dryed for several days before subjecting it to the peel test. This reversiblity strongly suggests that ionic bonds or dispersion forces are a more important source of bonding strength (in at least the case of this Chemlok system).

Other explanations of the nature of the polymer to metal bond include; mechanical adhesion due to microscopic physical interlocking of the two faces, chemical bonding due to acid/base reactions occuring at the interface, hydrogen bonding at the interface, and electrostatic forces built up between the metal face and the dielectric polymer. It is reasonable to assume that all of these kinds of interactions, to one degree or another, are needed to explain the failure of adhesion in the cathodic delamination process.

In addition, we have observed that the OH- appears to be responsible for a surprising degree of increased water absorption. Studies conducted on the Chemlok 205/220 bonding system showed substantially increased weight gains when conducted in 0.1 N NaOH over those observed when conducted in water or seawater.

The importance of the OH- ion is not really disputed. It is the role of the OH- which is in question. To further understand the effect of the OH- on the adhesion of metals to polymers, some exploratory tests were conducted. Delamination rates were compared to study the effect of cathodic action and the effect of hydroxide ion concentration.

Delamination Rate Studies

Much of the focus of the following work was aimed at the previously mentioned Chemlok adhesive system. This was because it had been used extensively in a familiar marine application, we were acquainted with its long term performance so it could be used as a baseline for comparison with other commercial adhesive formulations. All of the commercial adhesives used, including Chemlok 205/220, are proprietary formulations, thus the exact functional nature of each adhesive is not available.

A series of delamination rate tests were conducted on the Chemlok 205/220 adhesive system. The test specimen was developed from a modification of the standard (ASTM D-429) peel test. Each specimen had two one inch square bond areas on a monel substrate. The test specimens were suspended in five different test tanks. One tank had 1.0 N NaOH, two had 0.25N NaOH and two had 3.5% (by wt.) NaCl solution. One 0.25N NaOH and one 3.5% NaCl solution tank were purged of oxygen with a nitrogen atmosphere (oxygen concentration of less than 3%), the complimentary pair of tanks had natural air exposure (oxygen concentration 20%). A summary of the initial status of each tank is given in Table I.

Table I. Initial Conditions in the Delamination Rate Tanks

Tank	Solution	Oxygen*	Conductivity	Current	pH
1	1N NaOH	22%	$1.8x10^{6}$mohm	1.0 amp	13.1
2	0.25N NaOH	23%	0.5	1.95	12.95
3	3.5% NaCl	23%	0.6	3.6	8.8
4	0.25N NaOH	2%	0.6	1.5	12.92
5	3.5% NaCl	2%	0.6	3.5	8.75

* Temperature of the all four tanks was maintained at 35°C. The oxygen content refers to the degree of saturation of the solution for that temperation of solution.

Each tank was continuously scrubbed of CO_2 and potentiostats were employed to maintain an applied potential of -1.2 volts vs SCE (a typical value for the potential of monel or steel on a ship's hull in close proximity to a zinc anode). Having eliminated the possibility of buildup of carbonate ions and zinc ions, the test could continue for extended periods without changing the balance of ions in the electrolyte. This approximates actual application conditions where the instrumentation is immersed in the ocean which provides a relatively constant electrolyte environment. Each tank contained a set of four delamination test specimens, each sample painted everywhere except under the rubber adherend. Delamination was measured on all bondlines. This yielded then 16 measurements per tank.

The concentration of the 0.25N NaOH was selected because it has equivalent conductivity to the 3.5% NaCl solution. The initial current levels in the NaCl versus NaOH tanks were different despite the equivalent conductivities and the equal applied potential. The difference was due to the difference in pH. The more basic solution reduced the exchange current density for the hydrogen reaction. At this potential, the principal reaction is the hydrogen reaction. When the pH is increased, decreasing the concentration of hydronium ion, thereby decreasing the concentration of the reactants, the reaction is slowed down.

Figure 1 illustrates the comparitive delaminating rate in each of the five tanks. The delamination rates in the NaCl solutions were slightly greater than the NaOH solutions of equivalent conductivity. The basic electrochemical difference between the solutions was the current. Figure 2 illustrates the current due to the cathodic reaction in each of the five tanks. The current in the NaCl solutions of equivalent conductivity was much higher because of the lower pH, consequently higher hydronium ion concentration, favoring the hydrogen reaction. Thus, the cathodic reaction was proceeding more rapidly in the NaCl solution and delamination is proceeding correspondingly, more rapidly.

The rate of the cathodic reaction is not the only factor influencing the debonding however. Consider the 1N NaOH solution. The 1 N NaOH solution is clearly delaminating the fastest, despite its lower current. The lower current is due to the fact that the hydrogen reaction is hampered by the higher concentration of hydroxide ions. The other tanks have higher average current densities but slower debond rates. As a result we see that the hydroxide ion concentration is a separate and distinct accelerating factor.

Decreasing the availability of dissolved oxygen in the NaCl and equivalent NaOH tanks had the expected effect. In each case, lowering the oxygen concentration reduced the cathodic current. However, at this potential the hydrogen reaction is clearly dominating and the reduction in oxygen is not of great magnitute.

These results indicate that the hydroxide ion is an accelerating factor of its own right. In the next section we present some evidence that suggests that one of the reasons hydroxide ion is a delaminating agent is that it causes unusual swelling of some adhesives.

Swelling of Adhesives in the Presence of Hydroxide

To further understand the action of the hydroxide ions on the primer properties, weight gain tests were undertaken on various, commercially available adhesives. We attempted to prepare neat samples of ten types of adhesive agents (ie top coats and primers) which had been recommended by the manufacturers as suitable for marine applications. The neat samples were prepared with the use of a commercial adjustable wet film applicator supplied by the Paul Gardner Company. This applicator will lay films up to 0.25inches

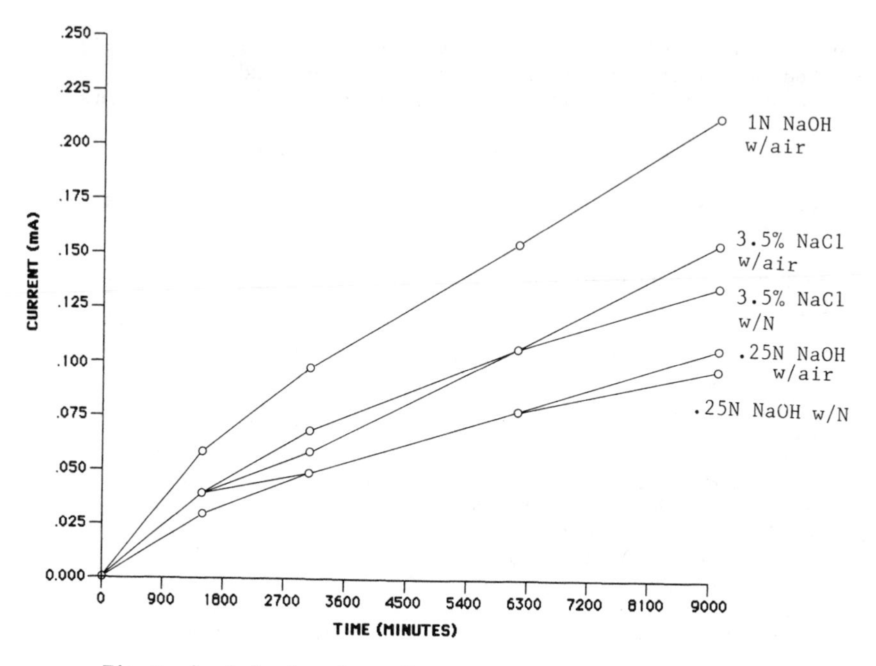

Figure 1 Delamination of peel test specimens in five tanks versus time.

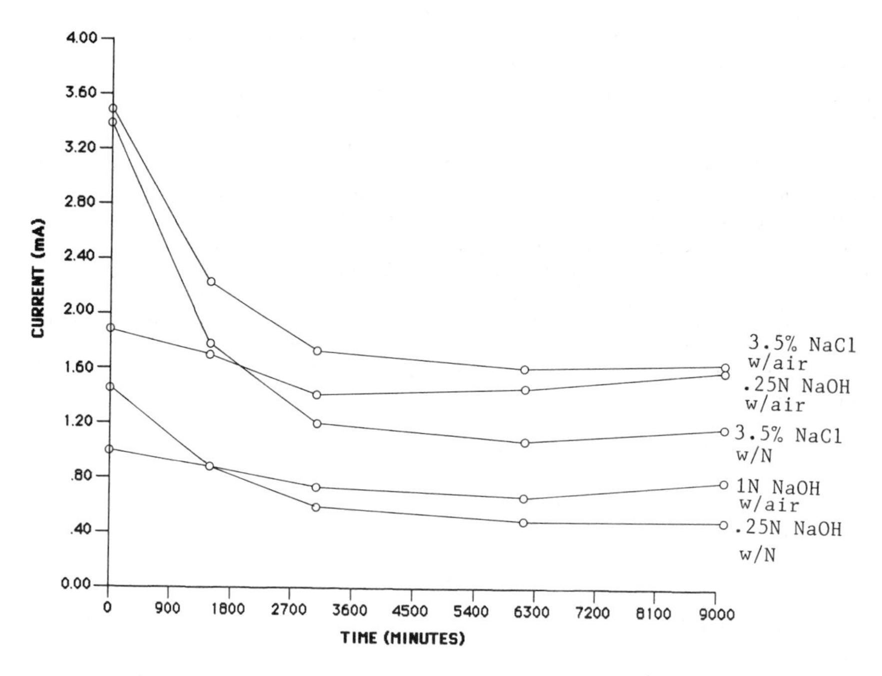

Figure 2 Current versus time in the five tanks.

thick and six inches wide. The films were cast on a glass substrate. The films were composed of two layers of 0.035 inches wet thickness. Multiple layers were laid down. After drying for about four hours, the slightly wet film was cut into one inch strips and removed from the glass substrate. It was important to remove the strips before they had completely dried or they would become brittle or stick to the glass surface. The free films were placed between teflon sheets and allowed to dry for several days. This method gave reproducible dry film thicknesses of 0.020 inches for the primer and 0.015 inches for the top coat. Cured neat samples were obtained by pressure curing the specimens at 315 F for 50 minutes under 25000 pounds of platen pressure. The average sample size was .8" X 1.00" x 0.07", and the weights ranged from 1.3 to 1.8g. Three of the adhesive systems were not prepared because they were so viscous it was not possible to draw down a thin film. Another system was deleted because its neat sample was too porous after being cured.

The remaining six adhesive agents were tested for their weight gain and volume change properties in 3.5% NaCl, 1N NaOH and deionized water. Two specimens of each type were placed in each of the solutions. The solutions were maintained at 35°C and conditions monitored included: pH, temperature, specific gravity and electrical conductivity. Density changes and volume changes were measured by first weighing each sample in air, then in water. All samples were rinsed thoroughly to prevent contamination of the water used for weighing. Before weighing in air they were dried with a thin stream of nitrogen until they were visibly dry.

Figures 3-4 are representative of the range of results on the weight gain tests. The range of results was dramatic. Table II includes a tabulation of the volume changes observed in the six adhesives. Some of the neat samples showed extreme increases in weight and volume in the OH solution over the NaCl solution. If you regard the OH^- solution to be typical of the solution conditions at the substrate surface where the cathodic reactions have made the local environment very basic, then the OH^- curve in Figure 3-4 portray what is going on at the bond line. Thus, primers, like 205, will become swollen and stretched with increased water absorption at the bond line. This swelling can contribute to the stresses that lead to debond. On the other hand a few of the neat samples showed almost no change in volume, in neither the NaCl nor the NaOH solutions. It might be expected that rubber to metal bonds employing these agents would have less strain and better bond preformance than the swollen ones.

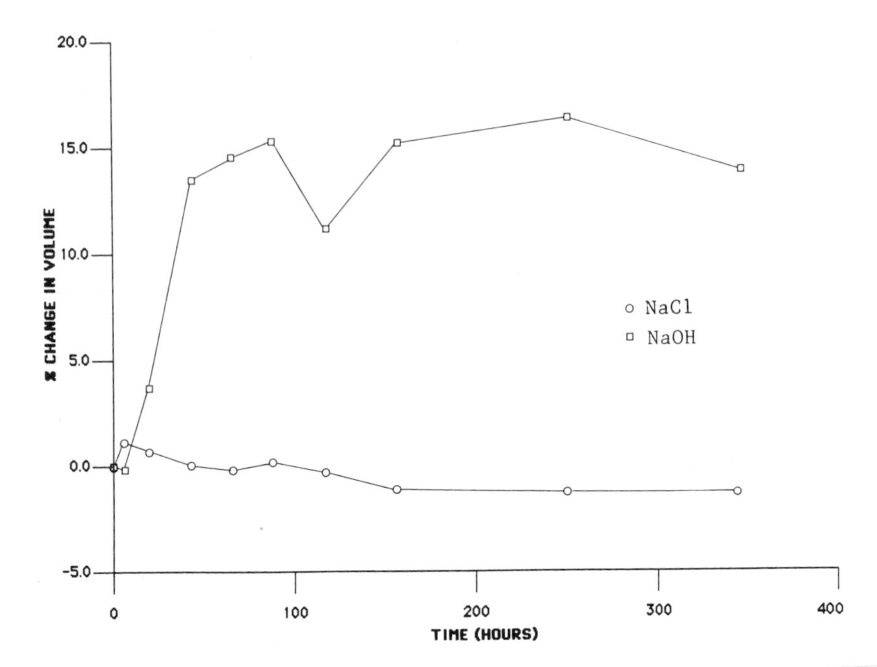

Figure 3 Change in volume of Chemlok 205 primer in 1N NaOH solution and in 3.5%(Wt.) solution of NaCl. Thus, the upper curve, showing the volume change in the presence of hydroxide ion, will be typical of the behavior of the primer at the debond during cathodic delamination.

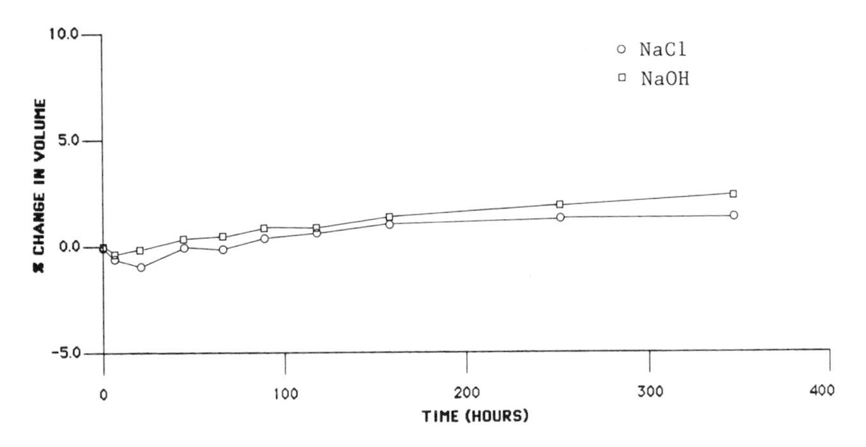

Figure 4 Change in volume of Chemlok 220 top coat in 1N NaOH solution and in 3.5%(Wt.) solution of NaCl. Volume changes were minimal, illustrating the potential range of responses of adhesive agents to exposure to hydroxide ion.

Table II. Commercial Vulcanizing Adhesives

SUPPLIER	PRODUCT	VOLUME CHANGE
Lord Corp.	Chemlok 205	15%
	Chemlok 220	3%
	Chemlok 252	16%
Dayton Chemicals	Thixon P-10	6%
	Thixon 511T	10%
	Thixon GPO	2%

All this evidence points to the conclusion that while hydroxide ion may be responsible for breaking metal oxide or polymer bonds, it also acts to increase the degreee of saturation of the polymer with water. It may, in fact, be swelling the primer, weakening bonds and pooling water at the interface.

Screening Test for Adhesives and Additives

Initially, we tried to devise a screening test for adhesives and additives which imitated naturally occuring conditions. The environment we were imitating was characterized by salt water immersion, temperatures ranging from just above freezing to 30°C, and a cathodic potential ranging anywhere from -0.8 to -1.2 volts. In light of the evidence about the importance of the hydroxide ion in swelling the adhesive, artificial enhancement of OH^- concentration was taken as an accelerating factor. Thus, instead of the artificial seawater solution, a 1N solution of NaOH was used.

The adhesives were applied according to manufacturers directions to scaled down models of an application. Failure of the vulcanized, rubber to metal bond was detected by a loss of resistance resulting from the establishment of a leak path under the rubber. The samples were connected electrically to the zinc anode. The temperature, specific gravity, electrical conductivity and pH of the solutions were monitored during the test.

After a seventy seven day exposure, failures were observed in the quantities listed in Table III. We see from this table that Thixon P-10/GPO is clearly the best performing system. It is instructive to compare the failure rates from the screening test with the volume changes observed in the primers in the water absorption test. Where the primer is resistant to volume changes in the presence of hydroxide ion, the bond generally has a low incidence of failure. There is one notable exception to this generalization. The Chemlok 205/252 system has a primer distinguished by large volume changes, yet when combined with the topcoat 252, it performs reasonably well. This is puzzling and needs to be investigated. One possible explanation is that the volume changes in the topcoat are also important. If the topcoat is very much stiffer than the primer, stress will develop and contribute to debonding when the primer starts to expand with moisture. Comparing the behavor of the topcoat and primer in the 205/252 system we see they share similar expansion tendencies in the high pH environment. Thus the topcoat may be reducing stress in the primer by accommodating the volume changes more easily. A further investigation of this phenomenon would contribute to our understanding of the correct procedure for selecting the topcoat to go along with a chosen primer.

Table III. Correlation Between the Incidence of Failure in the Screening Test and the Absorption of Water in the Presence of OH- Ions

Adhesive System	SCREENING TEST	WATER ABSORPTION TESTS	
		%volume changes in	
	#failed	Primer	Topcoat
Chemlok 205/220	83%	15%	3%
Chemlok 205/220/A1100	100%	15%	3%
Chemlok 252	66%	---	16%
Thixon P-10/511T	17%	6%	10%
Thixon P-10/GPO	0%	6%	2%
Chemlok 205/252	33%	15%	16%

These results indicate that the degree of volume change due to the uptake of water is related to the resistance of the bond to cathodic action. Tentatively we can conclude from this test, pending further information, that a primer with high resistance to swelling in the presence of OH- and a topcoat with similar volume change behavior in the presence of OH-, will perform best as an adhesive system.

Conclusions

The preceding has been a description of our efforts to develop a screening test for adhesives which will be used on steel or monel substrates which are maintained at a cathodic potential, immersed in seawater at temperatures ranging from just above freezing to 30° C. Because of the quantity of evidence that points to the hydroxide ion, not only as a potential bond breaking agent, but also as the agent which induces swelling and concentrates stress at the bondline, the hydroxide ion was taken to be a useful accelerating factor for the screening test.

At least one conclusion can be drawn from all of this. Primers that experience large increases in swelling in the presence of hydroxide ion, despite stability and considerable bond strength in salt water, would probably perform poorly if a cathodic potential was applied. Thus, the qualification of an adhesive system for a marine application that includes exposure of the metal substrate to a cathodic potential, should include examination of the predisposition of the adhesive to volume changes in the presence of hydroxide ion.

Literature Cited

1. J. S. Thornton, and R. W Thomas, "Adhesive Systems - Durability Testing", Sonar Transducer Reliability Improvement Program STRIP 4'th Qtr. Rpt., NRL Memo. Rpt., R. W. Timme, Ed., Oct., 1985.
2. E. L. Koehler, Corrosion, 40, 5 (1984).
3. E. L. Koehler, Corrosion, 33, 209 (1977).
4. H. Leidheiser, Jr., Corrosion, 38,374 (1980).
5. J. S. Hammond, J. W. Holubka, and R. A. Dickie, Org. Coatings Plastics Chem., 39,506 (1978); J. Coatings Tech., 51,655 (1979); Org. Coatings Plastics Chem., 41,499 (1979); J. S. Hammond, J. W. Holubka, J. E. DeVries, and R. A. Dickie, Corrosion Sci., 21, 239 (1981).
6. E. L. Koehler, Corrosion 40, 984, TRI 128
7. D. A. Dillard, and H. F. Brinson "Adhesive Systems - Mechanical and Thermal Properties", Sonar Transducer Reliability Improvement Program STRIP 4'th Qtr. Rpt., NRL Memo. Rpt., R. W. Timme, Ed., Oct., 1985.

RECEIVED January 22, 1986

16

Effect of Surface Preparation on the Durability of Structural Adhesive Bonds

C. A. Gosselin

Coatings Research, Research & Technology, Armco Inc., 703 Curtis Street, Middletown, OH 45043

Effects on durability due to substrate type and environmental exposure as a function of surface preparation were examined using single lap shear adhesive joints. Zincrometal, cold rolled, aluminized and galvanized steel adherends were left untreated, alkaline cleaned, lubricated, or zinc phosphated prior to bonding. Samples using chemically precleaned bare metal adherends exhibited good initial strength, but poor durability. Conversely, metal adherends onto which a phosphate coating had been deposited prior to bonding yielded greatly enhanced durability in wet environments. Samples prepared from zincrometal adherends retained the highest percentage of initial strength even in severe environments. Results indicate that in addition to chemically precleaning the surface, a moisture resistant barrier layer must be deposited between the metal/adhesive interface in order to promote durability in humid environments.

One of the most important requirements of a structural adhesive bond is durability: that is, the ability to retain a significant portion of its load bearing capability for long periods of time under the wide variety of environmental conditions which are likely to be encountered during service life.(1) Unfortunately, the poor durability of metal/adhesive bonds in wet, hostile environments has proven to be the major obstacle to widespread development and practical usage within many industries.

Adhesion between metallic/organic interfaces is facilitated by a combination of mechanical interlocking, chemical and physical bonding. Physical bonding alone cannot provide for durable, temperature resistant bonds, as van-der-Waals forces present between the metal surface and adhesive molecules are relatively weak. Rather, chemical bonding reactions between the two surfaces are at least an order of magnitude stronger than van-der-Waals forces and account for a large amount of both joint strength and durability. Depending upon the adhesive and cure temperature involved, some hydrogen bonding can also occur between the adhesive and the metal

0097–6156/86/0322–0180$06.00/0

oxide and hydroxide surfaces. Mechanical interlocking refers to the ability of the adhesive to interpenetrate a modified porous oxide surface prior to and during the early stages of cure. The success of this type of bonding depends heavily upon the viscosity of the uncured resin as well as the ability to modify the oxide without creating a weak, easily fractured micro-layer. Structural adhesive bonds generally exhibit good chemical interaction between the resin and the metallic substrate together with some degree of mechanical interlocking.

Moisture acts as a debonding agent through one of or a combination of the following mechanisms: 1) attack of the metallic surface to form a weak, hydrated oxide interface, 2) moisture assisted chemical bond breakdown, or 3) attack of the adhesive.(2) A primary drawback to good durability of metal/adhesive bonds in wet environments is the ever present substrate surface oxide. Under normal circumstances, the oxide layer can be altered, but not entirely removed. Since both metal oxides and water are relatively polar, water will preferentially adsorb onto the oxide surface, and so create a weak boundary layer at the adhesive/metal interface. For the purposes of this work, the detrimental effects of moisture upon the adhesive itself will be neglected. The nitrile rubber modified adhesive used here contains few hydrolyzable ester linkages and therefore will be considered to remain essentially stable.

Within the aerospace industry, etched or anodized aluminum surfaces provide a mechanism for microscopic mechanical interlocking at the adhesive/metal interface, which enhances and stabilizes the strength obtained through chemical and physical bonding. A comparatively featureless oxide surface, such as that found on bare steel, must rely primarily on chemical bonds which are unstable in the presence of moisture.(3) In addition, divalent metals, (such as Fe^{+2}) have a more basic (proton acceptor) surface than the higher valence oxides (Al^{+3}, Fe^{+3}, Si^{+4}), and actually promote dehydrogenation reactions which lead to hydrolysis at the interface or anion formation and chain scission within the adhesive.(4) The moisture resistance can be improved significantly by depositing as little as a monolayer of an acidic silane on the surface prior to bonding.(5) However, some of these specialized surface enhancement techniques require time frames not available to assembly line operations.(6)

In essence, the durability of metal/adhesive joints is governed primarily by the combination of substrate, surface preparation, environmental exposure and choice of adhesive. As stated earlier, the choice of the two-part nitrile rubber modified epoxy system (Hughes Chem - PPG) was a fixed variable, meeting the requirement of initial joint strength and cure cycle and was not, at this time, examined as a reason for joint failure. Durability, as influenced by substrate, surface preparation, and environmental exposure were examined in this study using results obtained from accelerated exposure of single lap shear adhesive joints.

Experimental Details

Substrates

Four types of substrates were examined in this study. Cold rolled steel (CRS), galvanized steel (ZGUS), and an aluminum-silicon alloy coated steel (Al-T1) represented bare and metallic coated steels, respectively. Zincrometal (ZM), a composite consisting of a bare steel substrate, a chrome rich primer and a zinc-rich organic top-coat was selected in order to illustrate the effects of a painted surface.

Sample Preparation

Metallic adherends were cut from .032" (.08 cm) thick sheets into 1x4-inch (2.54 cm x 10.16 cm) specimens. Cleanliness of the surface, is required in order to facilitate good adhesion. Steel, as bonded within the automotive industry, often experiences a variety of surface contaminants which are not removed prior to bonding. Since the choice of surface pretreatment prescribed for a metallic adherend has a direct effect on the performance of a joint in humid conditions, four types of commonly utilized automotive surface preparations were examined. The effects upon durability of no cleaning, alkaline cleaning, lubricating or zinc phosphating were examined. Accordingly, adherends were prepared using one of the four methods detailed below.

Untreated adherends generally referred to samples which had been bonded in the "as-received" condition. These surfaces were usually covered in a non-uniform manner with residual mill oil and/ or dirt. Sets of alkaline cleaned samples were immersed and agitated in a 66°C (150°F) aqueous solution (pH 13) of Parker 338 alkaline cleaner for approximately 10 seconds, rinsed in 150°F deionized water and wiped with a soft cloth. Oil and dirt were removed from a third set of adherends by vapor degreasing. Once clean, a thin film of Nalco 314 (a water soluble drawing compound) was roller applied to the surface. The oiled samples were secured vertically overnight in order to remove excess oil from the bond area. The final set of substrates was zinc phosphated with an 8-stage automotive phosphate system (Chemfil Chem. Co.) which concluded with a chrome rinse. Immediately prior to bonding, the phosphated adherends were treated in air at 350°F for five minutes in order to remove excess moisture.

Following the appropriate surface preparation, adherends were assembled into single lap shear adhesive joint configurations having a .59" (1.50 cm) overlap and a wire shimmed .005" (.013 cm) bondline (Figure 1). The adhesive joints were cured overnight at room temperature followed by a 45-minute high temperature cure at 171°C (345°F). Cured samples were placed in the appropriate accelerated environments for up to 60 days or retained for testing in order to obtain initial joint strength. All samples were eventually tested to failure at a crosshead speed of .14 in/min. (.005 cm/sec) using a Universal Testing Machine.

Under the best of conditions, single lap joint samples do not fail in pure shear due to the tensile and peel forces present at the ends of the overlap. These non-shear forces are exacerbated when using thin gauge adherends. Because of this, the lap joint dimensions as well as the testing rate were modified from the ASTM D-1002 standard as a result of earlier work on thin gauge steel adherends.

Aging Environments

The aging environments in which cured lap shear adhesive bonds were exposed included:

1. Room Temperature 23°C in air (control)
2. 23°C Temperature Water Bath
3. 60°C Constant Temperature Water Bath
4. Humidity Cabinet (95% rh, 58°C)
5. Salt Fog Cabinet (ASTM B-117)
6. Arizona Proving Grounds (APG) Cycle Test (laboratory equivalent)
7. Fisher Body Cycle Test

At appropriate intervals (5, 10, 20, 40, or 60 days), a minimum of five samples from each set were removed and tested to failure.

Results and Discussion

Initial Bond Strength

All failure associated with initial bond strength was found to be cohesive within the adhesive. These results are illustrated in Figure 2, where the effects of substrate type and adherend pretreatment can be examined. Al-Tl/adhesive bonds exhibited consistently higher initial bond strength under all pretreatment conditions, while joints prepared using ZGUS adherends demonstrated the lowest joint strength. Surface preparation of the adherends adversely affected initial bond strength only when the pretreatment was able to alter the mechanical properties of the adhesive itself. With the exception of samples prepared using Al-Tl adherends, lap shear joints prepared with lubricated adherends demonstrated significantly lower failure strength when compared to those otherwise prepared. The presence of the lubricant at the adhesive/adherend interface reduced the wettability of the substrate and caused plasticization of the adhesive during cure, thereby lowering the strength of the adhesive joint. The Al-Tl samples did not exhibit this problem. SEM indicated that vapor degreasing, which preceded lubrication, caused a thickening of the porous aluminum oxide on the surface which facilitated better mechanical interlocking between the adhesive and the adherend. In addition, the thin, low viscosity oil film applied to the surface may have penetrated the porous oxide and therefore not interfered with the cure of the adhesive. Within a substrate set, less strength difference was noted between joints which had been alkaline cleaned, left untreated or phosphated. The only exception to the latter observation can be found in the untreated ZM data, where a heavy, uneven coating of mill oil on the substrate may have contributed to the reduction in the average initial joint strength.

Durability

The relative aggressiveness of the environments proved to be consistent for all substrates, with the room temperature control the least hostile (virtually no loss of adhesion), and the cycle tests the most aggressive (up to 100% loss of adhesion within 60 days). Humidity cabinet exposure and 60°C water immersion yielded very similar values. As a result, for reasons of clarity, only water immersion data is actually presented here. Joint strength data obtained from either the Ford APG or Fisher Body Cycle Tests were identical, and were therefore also represented by one set of data points. The relative aggressiveness of the host environments toward

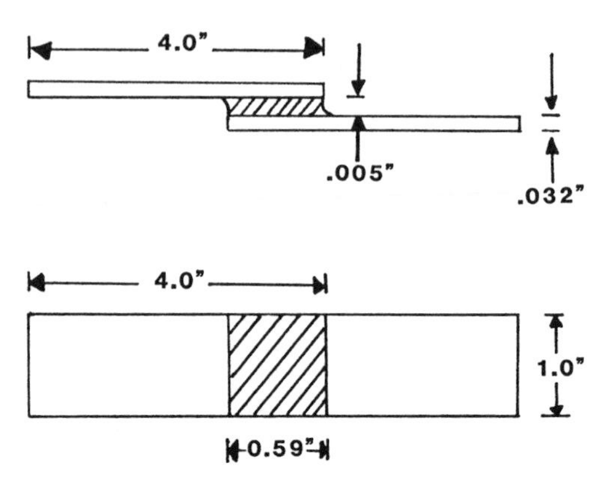

Figure 1. Details of single lap shear adhesive joints.

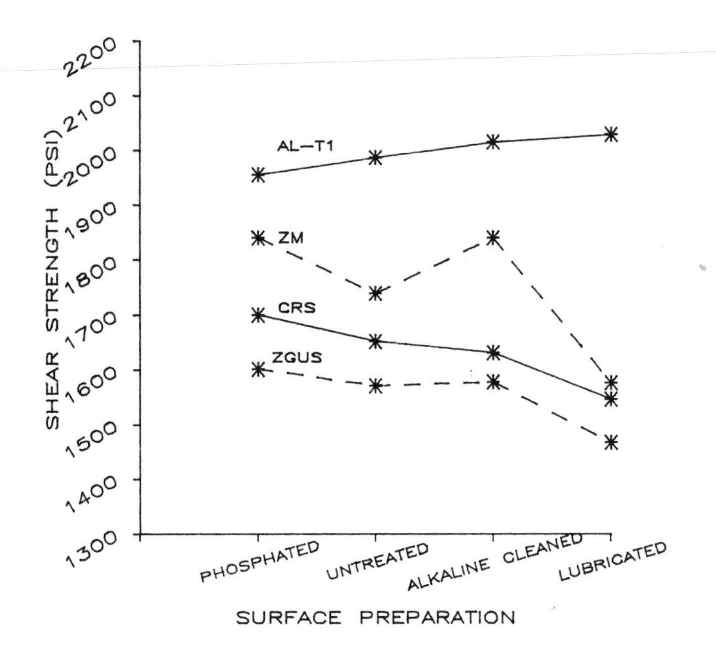

Figure 2. Graph of initial bond strength as a function of surface preparation.

retained joint strength is ranked from most to least hostile as follows: APG (Fisher Body) Cycle Tests > 60° water immersion (humidity cabinet at 95% rh, 58°C) > salt fog > 23°C water immersion > room temperature control. Note one exception in the case of aluminized steel. The 60°C water immersion test was less hostile than the salt fog exposure because of the detrimental effects of chloride environments on aluminum surfaces.

The first set of lap-shear samples to be evaluated following accelerated environmental testing were those which had been manufactured using CRS adherends. Little difference in joint strength could be observed in durability data taken at 5, 10, 20, 40, or 60 day intervals between samples constructed from adherends which had been left untreated, alkaline cleaned or lubricated prior to bonding (Figure 3). All room temperature control samples exhibited no significant loss of adhesion and failure remained cohesive within the adhesive. After 5 days, SEM analysis indicated that samples which had been exposed to salt fog, 60°C water immersion or cycle testing failed at or very near the adhesive/oxide interface. There was considerable evidence of progressive metal surface corrosion proceeding inward from the bond edge. These corrosion products eventually completely separated the adhesive from the metal substrate in 60°C water immersion and cycle test samples where virtually all of the initial strength was lost. Samples subjected to salt fog tests appeared to be less corroded, but still lost approximately 60% of initial bond strength after 60 days. This phenomenon was slower to develop on samples which had been immersed in a 23°C water bath. Failure remained cohesive within the adhesive for approximately 20 days, at which point some corrosion products became visible around the edges of the bonded area. After 60 days, failure occurred at or near the interface in all but the very center of the bonded area, and initial joint strength was reduced by only 30%.

CRS which had been phosphated prior to bonding exhibited a significant enhancement of durability and corrosion resistance under the same accelerated conditions (Figure 4). The crystalline barrier layer restricted the exposure of the metal oxide to moisture by reducing the rate of water penetration at the interface. Even samples exposed to the cycle test were able to maintain failure within the adhesive for up to 10 days, after which varying amounts of interfacial failure were noted. Again, room temperature control samples maintained initial joint strength and failure remained cohesive within the adhesive.

The second set of lap-shear samples were constructed from galvanized steel adherends. Once again, the room temperature control samples in all instances retained initial joint strength values and failure remained cohesive within the adhesive. Adherends which had been left untreated, alkaline cleaned or lubricated exhibited similar long term behavior within each accelerated aging environment. These simple surface preparations, which left no moisture resistant barrier layer at the surface, were not effective in enhancing the durability of the samples (Figure 5). Samples immersed in a 23°C water bath lost up to 35% of initial joint strength and SEM indicated that failure occurred in the oxide after only 5 days. The effect of cycle testing, 60°C water immersion and salt fog on ZGUS samples was even more drastic. All of these environments proved hostile enough to reduce joint strength by almost 100% after 40

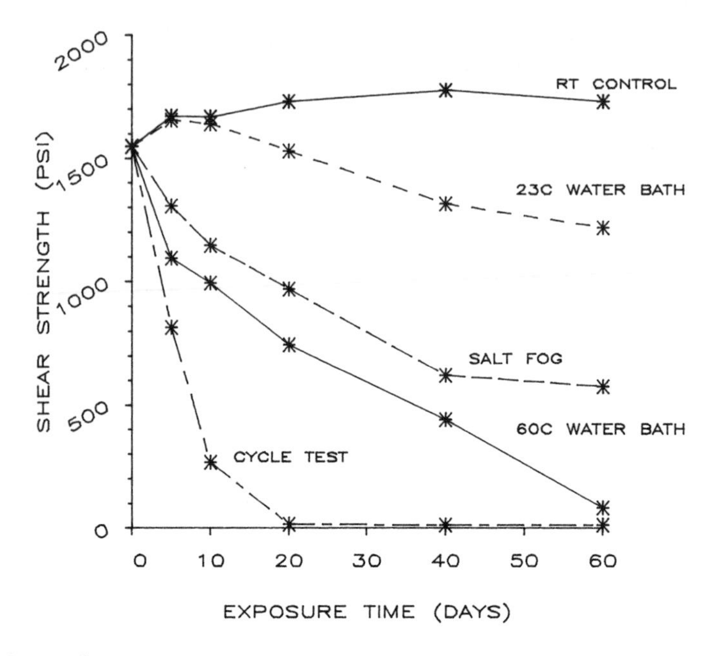

Figure 3. Graph of shear strength vs. exposure time for lap shear joints constructed from untreated, alkaline cleaned or lubricated CRS adherends.

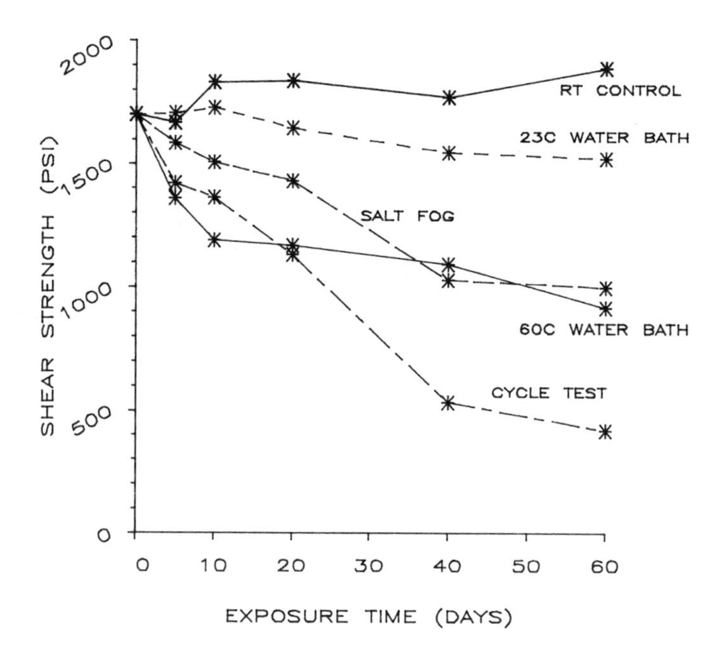

Figure 4. Graph of shear strength vs. exposure time for samples constructed from phosphated CRS adherends.

days. In the case of samples exposed to cycle testing, most of the initial strength was lost after 10 days. The large corrosion deposits present at the interface indicated that the failure of the oxide layer had caused joint strength deterioration. This was to be expected, as galvanized steel is coated with a sacrificial zinc layer in order to improve corrosion resistance of the bare metal. As a result, when bare ZGUS is used in an adhesive joint, zinc at the interface corrodes at a rapid rate in the presence of moisture, leading to severe debonding and ultimate catastrophic failure.

Phosphating ZGUS adherends prior to bonding led to a marked decrease in the rate of adhesion loss. Room temperature control and 23°C water immersion samples retained almost 100% of initial joint strength and failure remained cohesive within the adhesive. As illustrated in Figure 6, samples which had been subject to the other more hostile environments also exhibited a marked improvement in joint strength after 60 days. Placing a protective barrier layer between the metal/adhesive interface improved durability by reducing the rate of water permeation at the interface, which in turn reduced the amount of zinc corrosion product present within the bonded area.

A third set of samples was constructed from Al-T1. As observed using other metallic adherends, little difference in joint strength was observed over a period of 60 days between samples in which the substrates had been alkaline cleaned, lubricated or left untreated. (Figure 7). In addition, the most rapid deterioration in bond strength occurred in environments where high concentrations of chloride ions were present, as depicted by salt fog and cycle test data.

Joint strength deteriorated much more rapidly after immersion in a 60°C water bath as opposed to a 23°C water bath. The increased rate of hydrolysis of the surface oxide in the presence of water at elevated temperatures was the cause of joint strength deterioration, as failure in these cases occurred within the oxide layer. This observation agrees with others who have extensively analyzed failed aluminum/epoxy interfaces and found that hydration and subsequent weakening of the oxide layer was the manner of failure in aluminum alloy joints.(7) On the other hand, joint strength did not begin to fall significantly until after 40 days of immersion in a 23°C water bath. SEM analysis indicated that failure remained cohesive within the adhesive near the center of the bonded area, but occurred within the hydrated oxide layer at the outer edges.

Lap shear samples exposed only to dry room temperature environments were the most durable. In some cases, joint strength appeared to increase slightly over time, indicating some degree of residual post cure and/or stress relief at the oxide/adhesive interface. All of these failures were found to remain cohesive within the adhesive. Unlike results obtained using the other metallic adherends in this study, zinc phosphating Al-T1 prior to bonding did not improve the durability of the adhesive bonds exposed to any of the chloride containing accelerated environments. However, significant improvements in joint strength were obtained after immersion in both the 23°C and especially the 60°C water baths (Figure 8). SEM analysis indicated that, unlike that found on ZGUS, no phosphate crystals had been deposited on the aluminized substrate. Rather, the alkaline precleaner had exposed the silicon dendritic layer by etching away both the surface aluminum oxide as well as some of the bulk aluminum in the coating prior to phosphating (Figure 9).

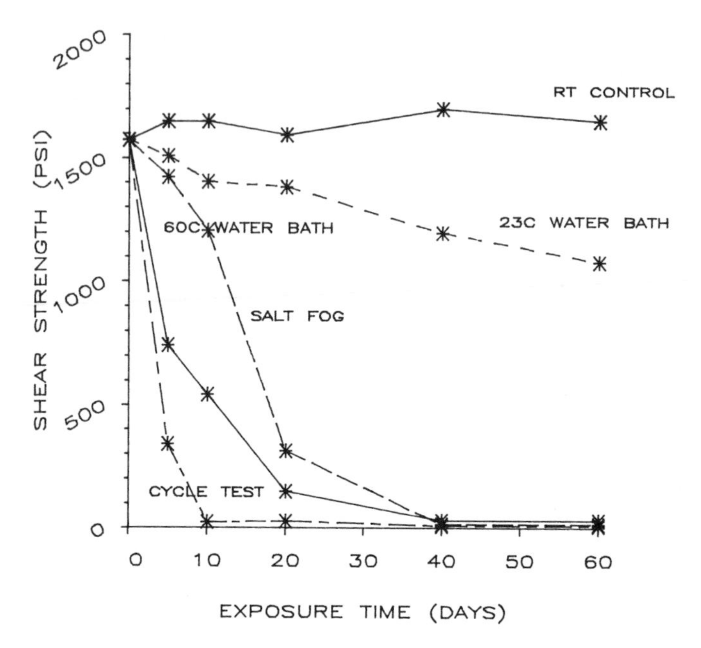

Figure 5. Graph of shear strength vs. exposure time for lap shear samples constructed from untreated, alkaline cleaned or lubricated ZGUS adherends.

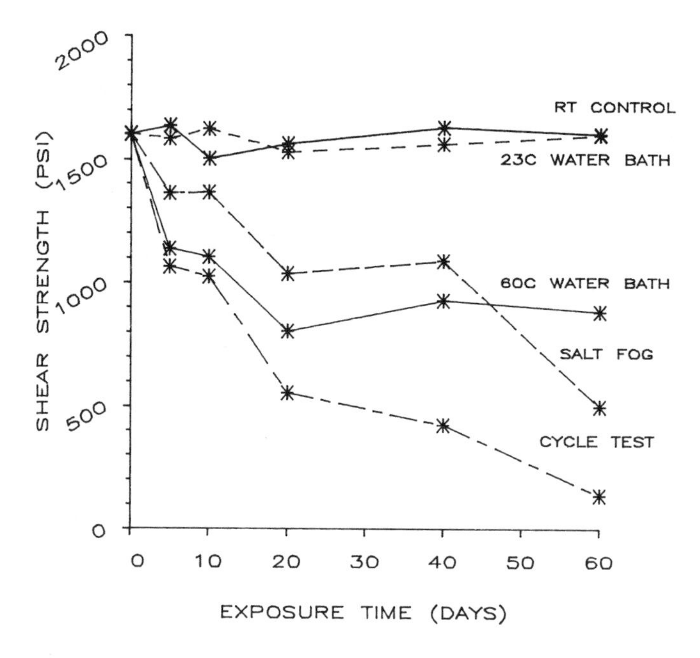

Figure 6. Graph of shear strength vs. exposure time for samples constructed using phosphated ZGUS adherends.

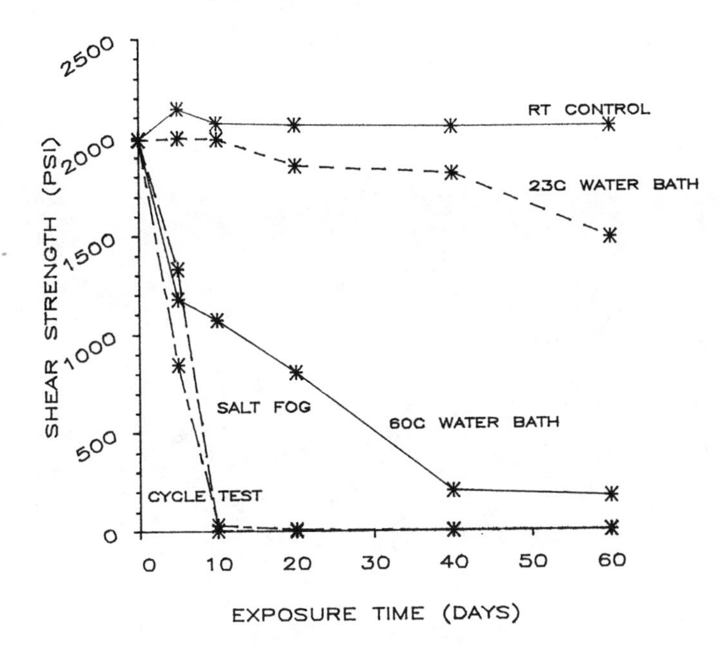

Figure 7. Graph of shear strength vs. exposure time for samples constructed from Al-Tl adherends which had been left untreated, alkaline cleaned or lubricated prior to bonding.

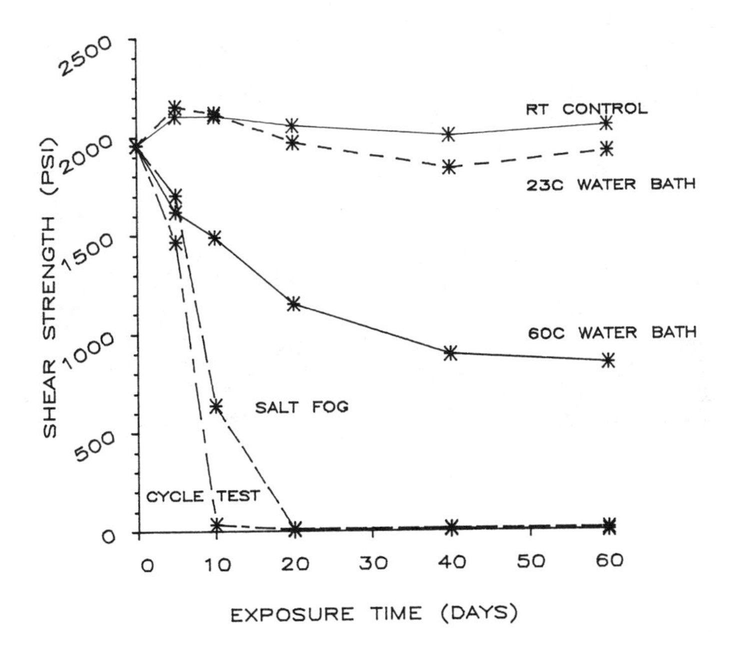

Figure 8. Graph of shear strength vs. exposure time for lap shear samples constructed from phosphated Al-Tl adherends.

a

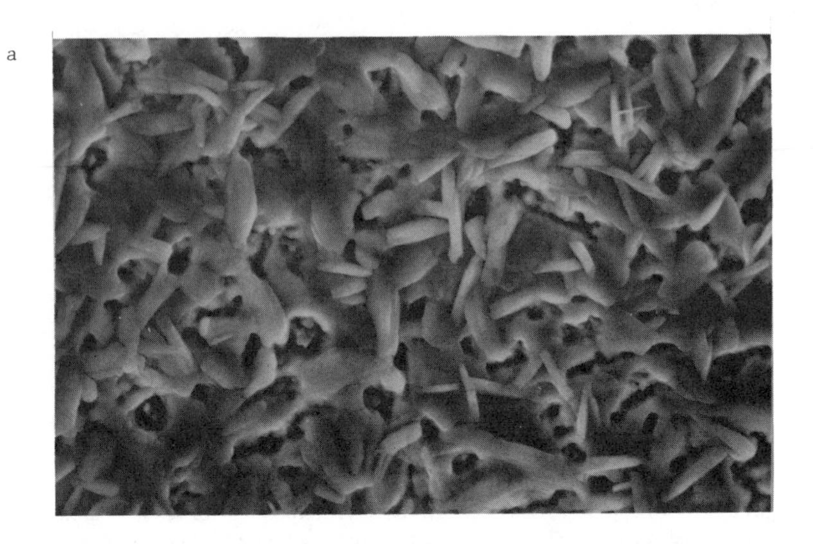

b

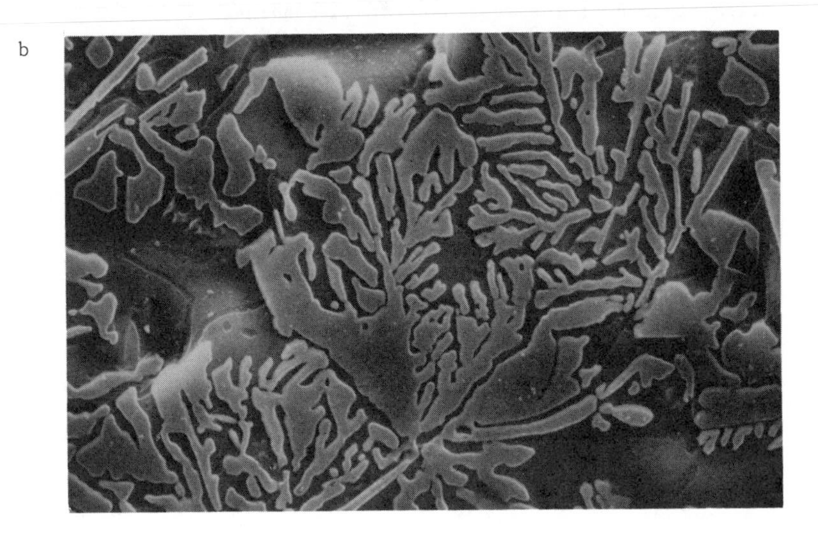

Figure 9. SEM micrographs comparing phosphated (A) ZGUS and (B) Al-T1 surfaces. (2000X)

The final step in the 8-stage automotive phosphating procedure is immersion in a chromic acid rinse. Generally, phosphate crystals (or films) provide bonding sites for the chrome ions. No evidence of chrome was detected using both ESCA and Auger. However, ESCA did indicate that a sodium phosphate (Na_2HPO_4) complex was present in some form on the Al-Tl surface. The indicated Na_2HPO_4 complex may have been incorporated into a thin, amorphous film covering the surface of the substrate. This surface modification, while not a typical crystalline phosphate configuration, may have provided a mechanism for the observed improvement of long-term durability of Al-Tl/adhesive bonds in wet environments.

The final set of samples was prepared using ZM adherends. These samples demonstrated exceptional durability regardless of surface preparation, due to the moisture resistant chemical and physical interlocking of the paint/adhesive interface. In addition, the zinc particles within the paint are encapsulated by the organic resin. Since few, if any, metal oxide sites are available for hydrolysis by moisture at the interface, more durable joints are the end result.

Samples constructed from adherends which had been alkaline cleaned, lubricated or left untreated exhibited similar joint strength values and durability trends (Figure 10). Adhesive joints placed in the room temperature control environment or the 23°C water bath retained 100% and 92% of initial joint strength, respectively. Failure remained cohesive within the adhesive for all of the control samples and for the first 20 days of exposure in the 23°C water bath. After 20 days, some failure began to initiate at both the primer/steel and primer/topcoat interfaces. The adhesive/topcoat interface proved to be more durable than those found between the substrate/primer/topcoat layers. Samples exposed to the more severe salt fog, 60°C water bath and cycle tests were able to retain 70% to 50% of their initial strength over a 60-day exposure period. Failure occurred primarily in the adhesive for up to 5 days exposure in these aging environments. After 5 days, a mixture of failure sites could be identified with more than one interface often exposed on a given sample. This indicated that during extended exposure to humid environments, any non-uniformity or interfacial weakness could be attacked and eventually become the locus of failure.

While a non-phosphated topcoat/adhesive interface provided an excellent, moisture resistant, occlusive seal even under the most severe cycle testing, phosphated ZM adherends did not prove to be as durable in comparison (Figure 11). The reason for this lies in the fact that phosphate coverage on Zincrometal is incomplete. Partially crystalline phosphates are non-uniformly interspersed on randomly exposed zinc dust spheres at the surface. Consequently, the moisture resistance normally provided at the adhesive/topcoat interface was reduced due to the incomplete sealing between the topcoat/adhesive surfaces. This became apparent as most of the failures examined after aging in these environments were concentrated at the adhesive/phosphate/paint interface. Results obtained on these samples were similar to those obtained for phosphated CRS joints, indicating that the locus of failure occurred at phosphate crystal sites. Note, however, that the durability of these joints was still considered to be very good in comparison to other metallic oxide/adhesive interfaces.

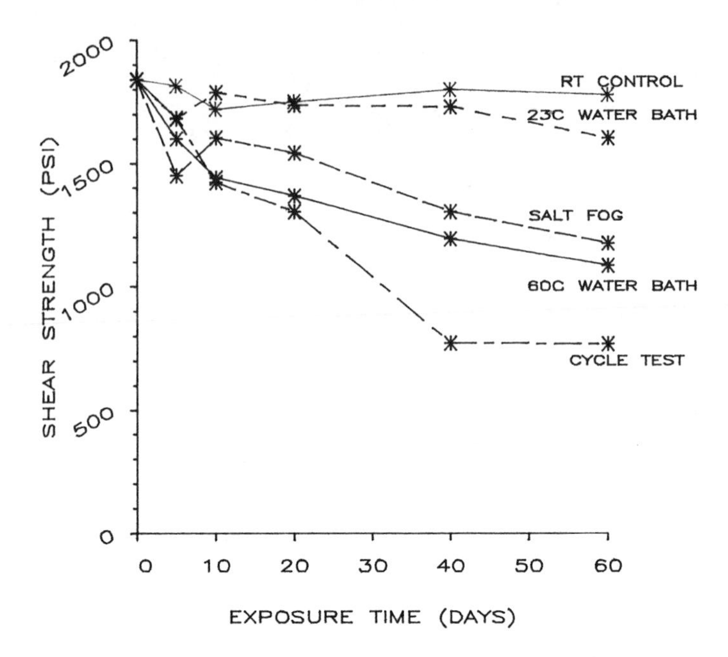

Figure 10. Graph of shear strength vs. exposure time for lap shear samples constructed from untreated, alkaline cleaned or lubricated Zincrometal adherends.

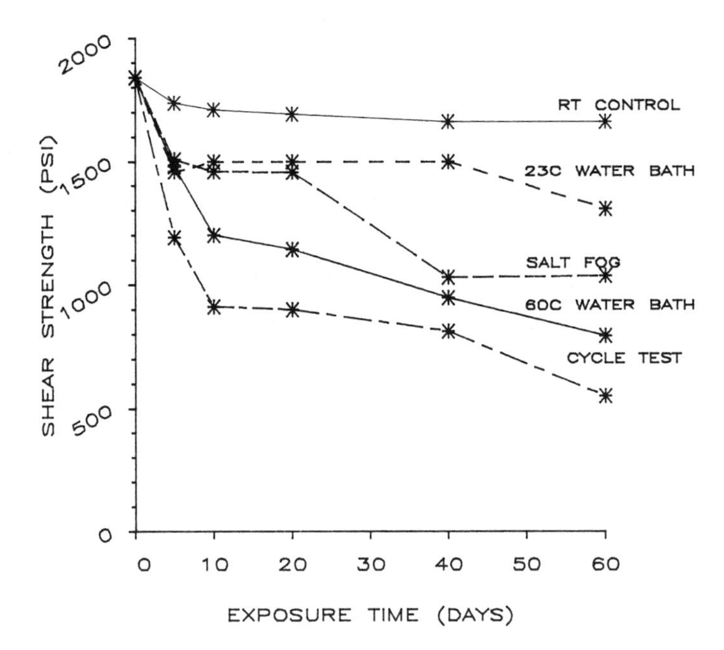

Figure 11. Graph of shear strength vs. exposure time for lap shear samples constructed from phosphated Zincrometal adherends.

Apparently, the chemical bonding present at the paint/adhesive interface is much stronger than that occurring at either the phosphate/adhesive or the phosphate/topcoat interfaces. In the case of ZM, phosphating to improve durability is not necessary, and in fact, was proven to be detrimental. The paint provides a moisture resistant barrier layer which reduces the activity of water at the interface providing for a surface receptive to the chemical and physical bonds necessary to promote good adhesion and enhance durability.

Conclusions

Initial bond strength depended heavily upon substrate type rather than surface preparation. Regardless of pretreatment, initial bond strength was highest when using Al-Tl adherends and lowest when the adherends were galvanized steel.

However, high initial bond strength did not necessarily promote good durability in humid environments. The most durable samples proved to be those constructed from non-phosphated ZM (painted) adherends or bare metal onto which a crystalline phosphate layer had been successfully deposited. This led to the conclusion that simply chemically precleaning the bonding surface was not sufficient to enhance durability. Rather, durability was improved only if a moisture resistant barrier layer was deposited between the oxide/adhesive interface. This aided both in retarding the rate of initial debonding as well as reducing the activity of water once penetration into the bond area had occurred.

Acknowledgments

Thanks are extended to R. Kelly, D. J. Robbins, and D. M. Smith of Armco Inc. for their assistance in the completion of this project. Thanks are also extended to F. J. Boerio of the University of Cincinnati for use of the ESCA equipment.

Literature Cited

1. Kinloch, A. J. ed. Durability of Structural Adhesives. Applied Science Publishers. London & New York. 1983. p. 5.
2. Travinski, D. L. et al. "Adhesive Bonding to Conversion Coated Steel Surfaces." 15th National SAMPE Technical Conference. October 4-6. 1983.
3. Paboda, E. A. et al. Applied Surface Science. 9. 1981, pp. 359-376.
4. Patrick, R. L. ed. Treatise on Adhesion and Adhesives V.3. Marcel Dekker. Inc. New York. 1973. Chapter 1. p. 6.
5. Boerio, F. J. and Gosselin, C. A.. "Structure-Property Relationships of Silane Films on Aluminum Substrates." 36th Annual Conference Reinforced Plastics/Composites Institute, SPI. February 16-20. 1981. Session 2-G.
6. Houwink, R. and Salomon, G. Adhesion and Adhesives V.2. Elsevier Publ. Co. Amsterdam. 1967. p. 91.
7. Kinloch, A. J. and Smart, N. R.. J. Adhesion. 12. 1981. pp. 23-25.

RECEIVED June 10, 1986

17

Effects of Corrosive Environments on the Locus and Mechanism of Failure of Adhesive Joints

J. W. Holubka, W. Chun, A. R. Krause, and J. Shyu

Polymer Science Department, Ford Motor Company, Dearborn, MI 48121

The effects of corrosive environments on the locus and mechanism of failure of adhesive bonds has been investigated. Adhesive chemistry resistant to degradation and swelling by water and corrosion products form the most durable adhesive bonds. The nature of the chemistry at the adherend/adhesive interface plays an important role in determining the durability of an adhesive bond in a corrosive environment. The durability of adhesive bonds to metal (steel) adherends is greatly influenced by the surface chemistry of the metal. For adhesive systems that employ adhesive primers for the metal adherend prior to bonding, the durability of the adhesive bond is largely controlled by the corrosion resistance of the primer. Surface analysis of adhesive bonds which failed after exposure to corrosive environments suggests that the degradation of interfacial layer is one of the principal modes of bond failure.

The extent of adhesive bond failure under corrosive environments is greatly accelerated when cyclic mechanical stresses are imposed on the adhesive bond during exposure. Three to four orders of magnitude reduction in fatigue life of adhesive bonds is observed for bonds exposed to environment prior to fatigue testing.

Although numerous studies (1-3) have described work aimed at establishing criteria for the durability of adhesive joints, a thorough understanding of effects of the chemical and mechanical properties, on the durability of adhesive bonds is lacking. More specifically, the effects of surface preparation and dynamic loading, especially under environmental service conditions, has not been explored in detail for automotive structures. In this paper, a description of the effects of environment on the durability of adhesive bonds is presented. Particular attention is given to

0097-6156/86/0322-0194$06.00/0

correlating the effects of adhesive chemistry with durability of the bond. The effects of environment, with and without loading of the bond are also described.

Experimental

Materials. The adhesives and primers used in this study were model and commercial materials that were cured according to conditions appropriate for the specific adhesive chemistry. Adhesives A and B were conventional epoxy/Versamid and epoxy/dicyandiamide adhesives, respectively. Adhesives C and D were commercial urethane and epoxy/polyamide adhesives, respectively. Adhesive E was a conventional two-part epoxy/amidoamine adhesive. Adhesive F was a vinyl plastisol adhesive. The adhesive primers used in this study were a urethane crosslinked epoxy electrocoat primer and spray primers based on tall oil modified epoxy ester, and polyesterpolyol/isocyanate chemistry. Dicyandiamide was obtained from Aldrich Chemical Company. Epon 828 was obtained from Shell Chemical Company. Genamid 250 and Versamid 115 were obtained from Henkel Company. The steel substrates used were cold rolled steel and Bonderite 40 phosphated steel from Parker. The composite was sheet molding compound from Rockwell.

Adhesive Bonding Technique. Standard procedures for preparing adhesive bond specimens were used. The composite was initially sanded with 240 grit emery paper and then thoroughly rinsed with methylene chloride. Steel substrates were rinsed thoroughly with methylene chloride. Bonds were prepared as one inch overlap shear specimens; bond thickness was 0.76 mm. Bond thickness was defined using 1.5 mm long wires of the appropriate thickness.

X-Ray Photoelectron Spectroscopy. Spectra were obtained using a Leybold-Heraeus LHS-10 Spectrometer using a Mg Kα anode. All binding energies for the obtained spectra were referenced to the hydrocarbon alkane line at 285.0 eV.

Fatigue Testing. Fatigue testing was conducted using an MTS closed-loop electrohydraulic test machine that was operated in load control, tension-tension mode. Loads in this experiment ranged from a minimum of 10% of the maximum load to maximum loads of 200-700 psi.

Corrosion Testing. Salt spray testing (ASTM-B-117-62,64) was used to determine durability of adhesive bond in corrosive environment. Lap shear samples were exposed to salt spray for 14 days and then immediately tested for lap shear strength.

Results and Discussion

Locus and Mechanism of Adhesion Failure during Corrosion Effects of Adhesive Chemistry. In previous studies (4-6) on the corrosion induced adhesion loss of coatings from steel surfaces, a primary mechanism for coating deadhesion was polymer degradation at the coating/metal interface by corrosion reactions that generate hydroxide ion:

$$Fe \rightarrow Fe^{+2} + 2e^-$$

$$2e^- + H_2O + 1/2\ O_2 \rightarrow 2OH^-$$

The extent of coating adhesion failure was found to be dependent upon the resistance of the polymer in the coating to hydrolysis by corrosion generated hydroxide. In this study, similar trends have been observed for adhesives. Table I shows the results of salt spray corrosion on a series of bonds between cold rolled steel adherends and adhesives of varying chemistry. The results show that there is a direct correlation between the chemistry of the adhesive polymer and the durability of the series of adhesive bonds studied. The locus of adhesion failure also appears to be related to the type of adhesive chemistry. In this study, adhesives based on polymers having a wide range of hydrolysis resistance were examined. Adhesives based on hydrolysis resistant chemistry (i.e., adhesives D and E) show a high retention of initial properties after exposure to an aggressive corrosion environment and the failure occurs cohesively within the adhesive. The cure reactions of these adhesives involve the formation of hydrolysis resistant carbon-nitrogen bonds in reactions involving the free N-H functionality of the Versamid or Genamid hardener with the oxirane functionality of the epoxy resin that is present in the adhesive formulation:

$$R-C-NH\begin{cases} R'-NH_2 \\ R'-NH_2 \end{cases} + \overset{O}{CH_2-CH}-CH_2-R''-CH_2-\overset{O}{CH-CH_2}$$

$$R-C-NH\begin{cases} R'-N'-CH_2-\underset{}{\overset{OH}{|}}{CH}-CH_2-R''-CH_2-\overset{OH}{\overset{|}{CH}}-CH_2-N-R' \\ R-N- \qquad\qquad\qquad\qquad -N-R' \end{cases} NH-\overset{O}{\overset{\|}{C}}-R$$

In contrast, adhesives based on hydrolysis prone chemistry (i.e., adhesives C, F and G where urethane and ester linkages are formed during crosslinking) are degraded by corrosion generated hydroxide and subsequently show significant reductions in adhesive bond durability. Saponfication reactions are the principal degradation processes for the polyester and polyurethane based adhesives. Evidence from interfacial analysis of surfaces generated as a result of adhesive bond failure, which we will see later in this report, indicates that degradative processes consistent with hydrolysis reactions involving corrosion generated hydroxide ion contribute significantly to bond failure. These processes result in the formation of hydrophilic ionic and polar products at the interface between the adhesive and the metal substrate:

$$R-O-\overset{O}{\overset{\|}{C}}-R' + OH^- \rightarrow ROH + {}^-O-\overset{O}{\overset{\|}{C}}-R'$$

$$R-O-\overset{O}{\overset{\|}{C}}-NH-R' + 2OH^- \rightarrow ROH + CO_3^{=} + NH_2-R'$$

Table I. Adhesive Structure/Property Studies

Adhesive	Chemistry	Strength (MPa)		Locus of Failure
		Initial	Salt Spray (14 Days)	
A	Epoxy/Dicy	11.96	5.21	Cohesive
B	Epoxy/Dicy	16.24	8.23	Cohesive
C	Urethane	8.04	3.69	Adhesive
D	Epoxy/Polyamide	9.62	7.55	90% Cohesive 10% Adhesive
E	Epoxy/Amidoamine	7.93	7.55	Cohesive
F	Vinyl Plastisol	6.27	0.69	Adhesive
G	Acrylic	8.96	1.03	Adhesive

The presence of these ionic materials at the interface, as well as the corresponding reduction of polymer crosslink density associated with this degradative process, likely contribute to the observed reduction in bond durability.

In a specific example of adhesive bonds between cold rolled steel and SMC adherends (Table II) an adhesive based on hydrolysis resistant epoxy chemistry (i.e., adhesive E) was compared with an adhesive based on hydrolysis prone urethane chemistry (i.e., adhesive C) in composite to cold rolled steel bonds. After corrosion testing, a significant difference in both retention of initial bond strength and locus of failure was observed. For bonds prepared with adhesive E, little if any reduction of the initial bond strength was observed after corrosion testing. The locus of failure for both the tested and untested bonds was largely in the

Table II. Composite to Metal Bonding: Urethane vs. Epoxy Adhesives

Test	Urethane		Urethane	
	Strength (MPa)	Failure	Strength (MPa)	Failure
Unexposed SMC/CRS	5.23	Fiber Tear	4.97	Fiber Tear
Corrosion Exposed SMC/CRS	3.21	Adhesion Loss to Steel	4.48	Fiber Tear
Untested SMC/SMC	5.36	Fiber Tear	4.87	Fiber Tear

SMC adherend (fiber tear out observed); only about 10% of the failure was at the adhesive-metal interface. The locus failure was typical of bonds involving composite materials. Bond strength values for these composite/metal bonds compared favorably with similar bonds involving composite/composite bonds where the effects of corrosion reactions are not present. For bonds prepared with adhesive C (hydrolysis prone material), nearly a 40% reduction in bond strength was observed and the failure was entirely at the metal-adhesive interface. The observed results with composite metal bonds in corrosion are consistent with a corrosion induced degradation of the adhesive at the metal-adhesive interface that reduces the overall strength of the adhesive below the cohesive strength of the composite (hence, the change in locus of bond failure).

The interfacial adhesive bond surfaces generated as a result of corrosion induced failure (for adhesives C and E) have been examined using x-ray photoelectron spectroscopy. The results of these studies are shown in Table III and Figures 1 and 2. Changes in

Table III. X-Ray Photoelectron Spectroscopy of Composite to Metal Bond Failure in Corrosion

Sample	Percent Atomic Composition				
	C	O	N	Fe	N/O
Adhesive C Unexposed	80.2	17.3	2.5	--	.145
Adhesive C After Corrosion Polymer Interface	73.0	24.9	2.1	--	.084
Adhesive C After Corrosion Metal Interface	51.7	42.5	1.7	4.1	.040
Adhesive E Unexposed	78.5	15.4	6.1	--	.396
Adhesive E After Corrosion Polymer Interface	82.0	12.6	5.4	--	.428
Adhesive E After Corrosion Metal Interface	49.9	45.4	--	4.7	0

elemental composition were observed with the urethane based adhesive after corrosion induced adhesion failure. The most notable change observed is in the N/O ratio. There is a marked reduction of nitrogen concentration (i.e., N/O ratio for untested adhesive C surface of 0.145 reduced to N/O ratio of 0.084 for interfacial adhesive surface after corrosion). The lower nitrogen concentration on the interfacial adhesive C surface after corrosion is consistent with a degradation of polymer with loss of nitrogen-containing

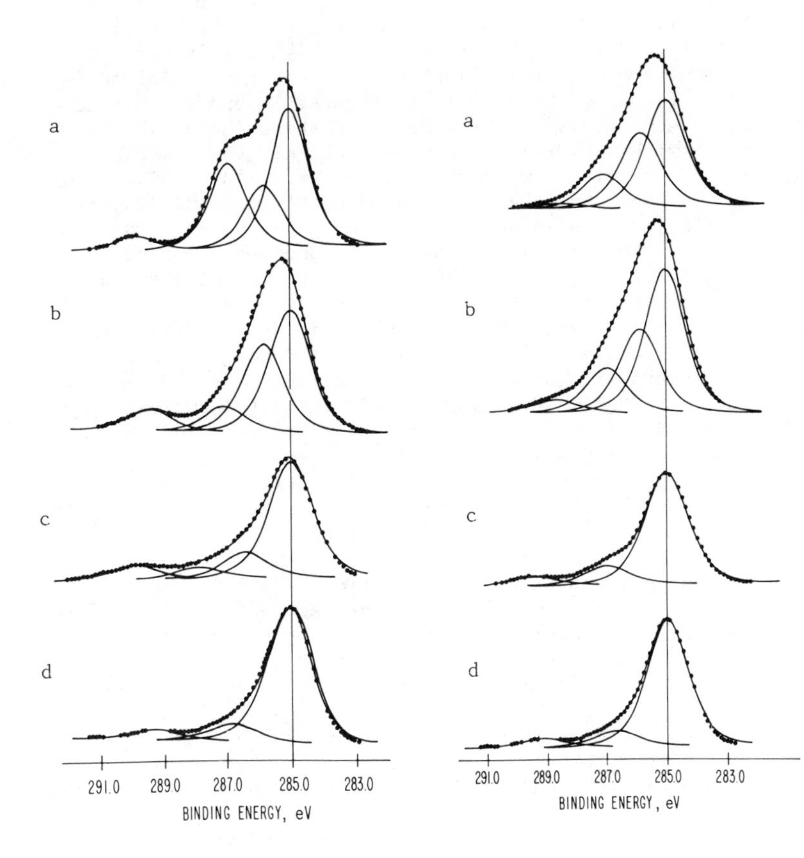

Figure 1. C 1s X-Ray Photoelectron Spectra (XPS) of interfacial surfaces of adhesive C showing (a) C 1s XPS spectrum of untested Adhesive C surface having peaks at 285.0 eV, 285.8 eV, 286.8 eV, and 289.3 eV; (b) C 1s XPS spectrum of interfacialAdhesive C polymer surface after corrosion showing peaks identical to (a); (c) C 1s XPS spectrum of interfacial Adhesive C metal surface after corrosion showing components at 285.0 eV, 286.1 eV, 287.3 eV, and 288.9 eV; (d) C 1s XPS spectrum of cold rolled steel standard showing peaks at 285.0 eV, 286.5 eV, and 288.7 eV.

Figure 2. C 1s XPS of interfacial surfaces of Adhesive E showing (a) C 1s XPS spectrum of untested Adhesive E surface having peaks at 285.0 eV, 285.7 eV, 286.7 eV, and 287.9 eV; (b) C 1s XPS spectrum of interfacial Adhesive E polymer surface after corrosion showing peaks identical to (a); (c) C 1s XPS spectrum of interfacial Adhesive E metal surface after corrosion showing components at 285.0 eV, 286.1 eV, 287.3 eV, and 288.9 eV; (d) C 1s XPS spectrum of cold rolled steel standard showing peaks at 285.0 eV, 286.5 eV, and 288.7 eV.

species during corrosion. The presence of nitrogen on the interfacial metal surface is consistent with the presence of polymer residue on that side of the interface. The high resolution C 1s XPS spectra for adhesive C (Figure 1) show that the interfacial surfaces generated as a result of bond failure (Figures 1B and 1C) differ from the untested adhesive. The observed changes in the spectra are consistent with a hydrolysis process that cleaves a water-soluble amine-containing component from the crosslinked network of adhesive C. In contrast, the delaminated portion of the adhesive E bond that failed adhesively shows only minor changes in composition (Table III). N/O ratios remain essentially unchanged for both untested adhesive E surface and the interfacial adhesive surface generated as a result of bond failure. Bond failure is likely to result from a simple displacement of the adhesive from the steel adherend by water or by a simple peel process during the bond failure that occurs without polymer degradation. High resolution C 1s spectra for adhesive E (Figure 2) show nearly identical spectra for both untested and tested surfaces, suggesting the lack of significant polymer degradation in this corrosion test. The absence of nitrogen on the interfacial metal surface as well as a C 1s nearly identical to steel standard further indicates that bond failure occurred without polymer degradation.

The Effect of Adhesive Primers. In practice, adhesive bonds involving metal adherends often use primers as pretreatments of the metal surface prior to bonding. Table IV shows the durability of composite-metal bonds prepared with adhesive C over a series of primers (of varying corrosion resistance) in 240 hour salt spray test. The results indicate that the performance of bonds is directly related to the corrosion resistance of the primer used to prepare the adherend surface. In general, the adhesion of the primer to the steel adherend, rather than the adhesive chemistry,

Table IV. Effect of Primer Chemistry on Bond Strength of SMC/Primed Steel Epoxy Adhesive Bonds

Primer Chemistry	Primer Adhesion Loss in SS*(mm)	Bond Strength (MPa) Initial	After SS	Failure
Epoxy Ecoat	0	6.27	5.86	Fiber Tear
Epoxy Ester	2-3	6.59	5.29	Fiber Tear + Primer
Urethane	4-5	6.36	2.73	Primer
Epoxy Urethane	<1	6.19	4.65	Fiber Tear + Primer

*Salt Spray

controls the durability of these adhesive bonds. XPS analysis of the interfacial primer surfaces generated as a result of adhesion failure are consistent with previously reported work on adhesion loss of coatings during corrosion (5-6).

Effects of Adhesive Chemistry on the Fatigue Resistance of Bonds. Fatigue testing was employed to study the durability of adhesive bonds under conditions more representative of service conditions (i.e., load + environment). Results (Table V) show that for composite/composite bonds exposed to fatigue and moisture environments, reductions in fatigue resistance are observed. Bond failure is also associated with the adhesive rather than the composite adherends. A range of performance, related to the adhesive chemistry, was observed. The performance of the bond during fatigue testing was found to be at least partially dependent upon the water uptake of the adhesive. The greater the water uptake, the greater the reduction in fatigue resistance. For one epoxy adhesive (adhesive D), the water uptake was 10% of the initial adhesive weight within four hours exposure to water at room temperature. In fatigue tests, where the adhesive bond was cycled between 50 and 500 psi, the fatigue resistance was reduced from 20,000 cycles to 3,300 cycles and bond failure occurred cohesively within the adhesive. In contrast, for a urethane adhesive (adhesive C), the water uptake in the identical experiment was 2-3% and the reduction in fatigue

Table V. Effects of Environment and Load on Bond Durability During Fatigue

Adhesive	Cycles to Failure
Untested	
Epoxy	20,000
Urethane	40,000
Environment then Load[a]	
Epoxy	3,293
Urethane	5,004
Environment with Load[b]	
Epoxy	1,501
Urethane	3,849

a) Sample exposed to water at room temperature for five days then fatigue tested between 50 to 500 pounds.

b) Sample exposed to water during fatigue test. Fatigue test conducted between 50 to 500 pounds. Total test time less than four hours.

resistance was from 40,000 cycles to 5,000 cycles, and bond failure occurred at the composite-adhesive interface indicating that the adhesive strength was less than the cohesive strength of the material. Equivalent reductions in fatigue life are observed when the adhesive bonds are exposed to environment (i.e., water) during fatigue. The results indicate that the effects of simultaneous exposure of adhesive bonds to both fatigue and environment has a greater effect of reducing bond durability than fatigue loading after environmental exposure.

Conclusion

The effects of adhesive polymer chemistry on the durability of adhesive bonds during corrosion has been examined. Specific efforts have been aimed at determining the effects of corrosive environments on the locus and mechanism of adhesion failure of adhesive joints. Interfacial chemistry of the adhesive has been found to strongly affect bond durability in corrosion. Adhesive chemistry resistant to degradation and swelling by water and corrosion products form the most durable adhesive bonds. The corrosion resistance of adhesive primers used to repair metal surface prior to bonding strongly influences bond durability. Primers having good corrosion resistance offer the best bond performance in corrosion environment.

The extent of adhesive bond failure under corrosive environments is greatly accelerated when cyclic mechanical stresses are imposed on the adhesive bond during exposure. The effects are greatest for adhesives having high water uptake.

Literature Cited

1. J. D. Minford, Aluminum Adhesive Bond Permanence, in Treatise on Adhesion and Adhesives, Vol. 5 (R. L. Patrick, ed.), Marcel Dekker, New York, 1981.
2. J. D. Minford, Adhesives Age, 17, March 1978.
3. O. Ishai, G. Yahiv and E. Altus, Durability of Structural Adhesively Bonded Systems, United States Army, Contract Number DAJA37-80-C-0303, November 1983.
4. R. A. Dickie, J. S. Hammond and J. W. Holubka, Ind. Eng. Chem. Prod. Res. Dev., 20, 339 (1981).
5. J. S. Hammond, J. W. Holubka and R. A. Dickie, J. Coating Technol., 51, 45 (1979).
6. J. W. Holubka, J. S. Hammond, J. E. DeVries and R. A. Dickie, J. Coating Technol., 52, 63 (1980).
7. S. A. Zahir, The Mechanism of the Cure of Epoxide Resins by Cyanamide and Dicyandiamide in Advances in Organic Coatings Science and Technology, Vol. IV, p. 83.
8. J. W. Holubka and A. R. Krause, in preparation.

RECEIVED February 14, 1986

18

New Polymeric Materials for Metal Conversion Coating Applications

A. Lindert and J. I. Maurer

Parker Chemical Company, 32100 Stephenson Highway, Madison Heights, MI 48071

In order to obtain maximum corrosion protection for painted metal articles, the metal parts are pretreated with an inorganic conversion coating prior to the painting operation. These zinc or iron phosphate coatings greatly increase both paint adhesion and corrosion protection. Traditionally, a chromic acid post-treatment has been applied to these phosphatized metal surfaces to further enhance corrosion protection. Due to environmental concerns, research efforts have been directed toward the replacement of chromate-based post-treatments. This paper focuses on a new unique chromium free post-treatment based on "Mannich derivatives" of poly-vinylphenol which have demonstrated excellent performance on both zinc and iron phosphate treatments.

A large segment of the metal parts produced by industry are painted for both decorative purposes as well as to increase the corrosion resistance and extend the useful life of the product. To obtain maximum quality from painted metal articles, it is of paramount importance to pretreat the metal parts with a conversion coating process.(1,2) Pretreatment processes contribute a significant improvement in corrosion protection and durability to metal articles by:

1. Increasing blister resistance in humid environment
2. Promoting paint adhesion
3. Decreasing the spread of corrosion of the metal substrate
4. Reducing metal-paint interactions.

A conversion coating provides an insulating, often non-conducting coating which both inhibits corrosion once painted and provides a better surface for paint adhesion.

0097-6156/86/0322-0203$06.00/0

There are essentially three main steps in a conversion coating process; cleaning, conversion coating, and post-treating. These three different, but equally important, steps in the pretreatment of metal articles will be discussed in more detail for the purpose of providing a background for the main emphasis of this paper, the post-treatment part of the conversion coating process, and more specifically chromium-free polymeric post-treatments which have been developed in recent years to replace the environmentally unacceptable chromate systems.

Background - Present Industrial Practice

Depending upon the quality level desired, the type of paint and the application method used, and the metal mix treated, a number of different types of pretreatment processes are available. For use as a paint base, the phosphate processes are either of the "iron phosphate" or the "zinc phosphate" type.

The pretreatment can be as simple as a three-step process, in which the cleaning and conversion coatings are combined in the same step as follows:

THREE STAGE PROCESS

1. Clean/Conversion Coating
2. Water Rinse
3. Post-Treatment

The above type of process is most often used in the application of an iron phosphate-iron oxide conversion coating. These amorphous coatings are generally deposited at a coating weight of from 20 to 80 mg./sq. ft. Although these iron phosphate coatings improve paint adhesion, a post-treatment is required to obtain acceptable corrosion protection, and improves the corrosion properties of the conversion coating by an order of magnitude.

On the other end of the spectrum, the pretreatment can encompass many steps:

EIGHT STAGE PROCESS

1. Clean
2. Water Rinse
3. Clean
4. Water Rinse (Conditioning Agent)
5. Conversion Coating
6. Water Rinse
7. Post-Treatment
8. Deionized Water Rinse

This more involved multi-step process can be used for both iron phosphate and zinc phosphate conversion coating processes.

In general, the zinc phosphate process requires the above multi-step procedure since, with these coatings, it is often necessary to employ a surface activating agent containing titanium phosphate. The titanium phosphate can be formulated either into the cleaner or used in the water rinse preceding the zinc phosphate coating solution. The zinc phosphate treatment baths can deposit either crystalline zinc phosphate (hopeite) or zinc-iron phosphate (phosphophylite) coatings at approximately 130 to 250 mg./sq. ft. By themselves, these coatings can improve paint adhesion and give corrosion protection to metal articles. Application of a post-treatment can further improve corrosion protection by an order of 1 to 2 times depending on the type of paint used.

Chromium Post-Treatments

The original post-treatments consisted of 0.02-0.1% operating solutions of chromic acid in water. Variations and improvements have been made in chromate post-treatments to overcome alkaline water conditions, design treatments specific for zinc phosphate coatings(3), and to develop reactive trivalent-hexavalent chromium chromate complex treatments which could be rinsed with water without decreasing the corrosion resistance.(4) A further development in chromium post-treatment technology was the use of trivalent chromium compounds.(5) These post treatments are less toxic and eliminate the reduction step employed for waste water treatment in chromate effluents. However, trivalent chromium based posttreatments are less forgiving and more care must be exercised in controlling process variables in order to maintain quality. In addition, chromium effluent is always a possibility if inadequate control of the disposal occurs.

Chrome-Free Post-Treatments

The naturally occurring tannins were the first polyphenolic oligomers and polymers employed as post-treatments on phosphatized metal surfaces. Tannin post-treatments gave reasonably good results over zinc phosphate conversion coatings where the post-treatment can increase the corrosion resistance by 1 to 2 fold.(6) Formulations of tannins/melamine-formaldehyde resins further improved the above system and were shown to increase the corrosion resistance of iron phosphate coatings(7) where the use of a post-treatment is critical for good corrosion resistance. These tannin/melamine-formaldehyde formulations gave the best results without the use of a water rinse and required an oven cure step.

The development of new paint technology has mandated further improvements in the performance of non-chrome rinses. Greater use of a final water rinse in treatment applications has taken place particularly with the advent of anodic and cathodic electro-painting which require the use of a deionized rinse to prevent paint contamination. High solids paints which tend to be more sensitive to surface residues and irregularities also benefit from a final water rinse since this extra step reduces surface impurities. These more recent developments in the conversion coating process, require that the post-treatment be reactive with the metal surface and not wash off in the water rinse step.

In the development of a reactive non-chrome post-treatment, a variety of phenolic resins were synthesized and commercial phenolic resins evaluated. It was found that phenol-formaldehyde resins, cresol-formaldehyde condensates, ortho-novolak resins, and phenol-formaldehyde emulsions gave positive results when employed as post-treatments over zinc and iron phosphate conversion coatings. The above materials all possessed drawbacks. The materials in general have poor water solubility at low concentrations used in post-treatment applications and had to be dried and baked in place in order to obtain good performance. The best results were obtained with poly-4-vinylphenol and derivatives thereof as shown in the following structure:(8,9,10)

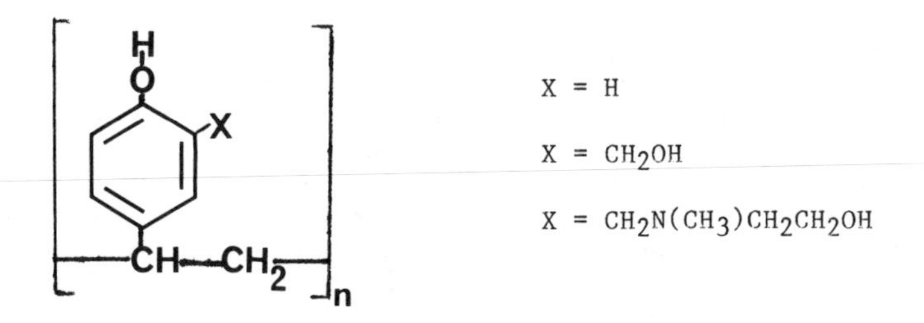

Structure 1

Poly-4-vinylphenol (X=H) was evaluated from ethanol due to poor solubility and stability in alkaline solutions. Poly-4-vinylphenol-formaldehyde condensate ($X=CH_2OH$) was applied from an alkaline solution and was soluble and stable at a pH of 9.5 or greater. The poly-2-[methyl(2-hydroxyethyl)amino]methyl-4-vinylphenol [$X=CH_2N(CH_3)CH_2CH_2OH$] derivative synthesized by the "Mannich" condensation of 2-(methylamino)ethanol, formaldehyde and the poly-4-vinylphenol was applied to the treated metal at an acidic pH as described below and was soluble in water over a wide pH range (below 1 to 7.8). Both the poly-4-vinylphenol-formaldehyde condensate and the "Mannich" derivative have demonstrated excellent performance as post-treatments over both zinc and iron phosphate conversion coatings. However, in a typical industrial treatment installation, there is considerable acid "drag-in" from the phosphatizing and subsequent rinsing stages. This acid "drag-in" will precipitate and contaminate the alkaline soluble poly-4-vinylphenol-formadlehyde condensate. The "Mannich" adduct, being acid soluble, is not affected by acid drag-in from the prior treatment stages and has found greater acceptance. Commercial products have been developed employing the poly-4-vinylphenol derivatives and are presently used in broad spectrum of industrial installations including the furniture, appliance, and automotive markets.

Polyvinylphenol Post-Treatments

The quality observed with "Mannich" derivatives of polyvinylphenol is affected by the concentration, time of treatment, temperature, pH, and whether or not a final deionized water rinse is used. The results shown in Tables I-III below represent evaluations conducted for poly-[methyl(2-hydroxyethyl)amino]methyl-4-vinylphenol, as shown in Structure I. Post-treatments based on polyvinylphenols overcome deficiencies observed with previous chrome-free rinses, since these systems are reactive and a final water rinse actually improves performance as is illustrated in Table I where the new non-chrome system is evaluated on Bonderite 1000, an iron phosphate conversion coating, as a function of concentration with and without a final water rinse. It is also

TABLE I
POLYVINYLPHENOL POST-TREATMENT IRON PHOSPHATE
BATH CONCENTRATION VS. FINAL WATER RINSE#

POST-TREATMENT CONCENTRATION (% SOLIDS)	WATER RINSE		SALT SPRAY 504 HOURS##
	NO	YES	
.01	X		5-5
.01		X	0-2^{S}
.05	X		0-1^{2S}
.05		X	0-1
.10	X		2-3
.10		X	0-1^{S}
.25	X		4-6
.25		X	0-1^{S}
CHROMIC CHROMATE*		X	0-1^{S}
NO POST-TREATMENT		X	7-9

DURACRON 200 WHITE ENAMEL, PPG INDUSTRIES, INC.
ASTM - B-117 RATING-PAINT CREEPBACK FROM SCRIBE IN 1/16TH INCH.
* FOR CHROMIC CHROMATE POST-TREATMENTS SEE REFERENCE 3

TABLE II
POLYVINYLPHENOL POST-TREATMENT IRON PHOSPHATE##
TEMPERATURE - PH (TREATMENT TIME - 30 SECONDS)

BATH TEMPERATURE	BATH PH	SALT SPRAY##
100	6.5	0-1[2S]
100	4.6	0-1
120	6.5	0-1[S]
120	4.6	0-1
140	6.5	0-1[S]
140	4.6	2-2

DURACRON 200 WHITE ENAMEL, PPG INDUSTRIES, INC.
ASTM - B-117 RATING-PAINT CREEPBACK FROM SCRIBE IN 1/16TH INCH.
* FOR CHROMIC CHROMATE POST-TREATMENTS SEE REFERENCE 3

TABLE III
POLYVINYLPHENOL POST-TREATMENT ZINC PHOSPHATE
AUTOMOTIVE BODY PAINT
SCAB OR CYCLE TEST

COLD ROLLED STEEL	RATING	AVERAGE CREEPAGE**
CHROME FREE SYSTEM	8/7	1.5 mm
CHROMIC CHROMATE*	8/6	2.0 mm
WATER ONLY	2/2	9.0 mm

GALVANIZED STEEL	RATING	AVERAGE CREEPAGE
CHROME FREE SYSTEM	9/8	0.8 mm
CHROMIC CHROMATE*	9/6	1.6 mm
WATER ONLY	5/4	5.0 mm

* FOR CHROMIC CHROMATE POST-TREATMENT - SEE REFERENCE 3.
** PAINT CREEPBACK FROM A SCRIBE IN MILLIMETER.

apparent that the concentration is not critical as excellent results are obtained from 0.05% to 0.25% solids in a working solution. For optimum cost/performance the post-treatments are operated at 0.1% solids when a final water rinse is used and at 0.05% in those installations where a final water rinse is not available.

Table II illustrates the interrelation between temperature and pH of the polyvinylphenol treatment solution. These solutions can be adequately operated over a wide range of pH, but the temperature of the solution is an important parameter and should be maintained above 100°F for optimum results. In most cases, treatment times of 15 to 30 seconds are adequate.

The "Mannich" adduct synthesized from the condensation of formaldehyde, 2-(methylamino)ethanol and poly-4-vinylphenol as shown in Structure I, has been evaluated as a function of molecular weight versus corrosion resistance as measured by salt spray and humidity tests on Bonderite 1000, an iron phosphate conversion coating. The molecular weight of the polymer was varied from approximately M_w = 2,900 to 60,000. The corrosion resistance results were essentially equivalent over the molecular weight range evaluated.

In general, paint adhesion and salt spray corrosion inhibition of iron phosphate coatings post-treated with polyvinylphenol derivatives have been equal to or superior to that obtained with chromate based systems. This has been the case particularly with many high solids paints.

On zinc phosphate conversion coatings, the "Mannich" derivatives of poly-4-vinylphenol have demonstrated performance equivalent to chromic chromate systems in salt spray, humidity, and physical testing. In addition, Table III illustrates results observed with automotive body paint systems evaluated by the "scab" or "cycle" test which causes failure more typical of actual end use conditions than do salt spray evaluations. Again, results equivalent to chromic-chromate post-treatments were obtained. In addition, the humidity resistance and adhesion tests were essentially equivalent to the chromium controls.

Surface Analysis of Polyvinylphenol Post-Treatments

A Physical Electronics Model 565 Scanning Auger Spectrometer was employed for analysis of the polyvinylphenol based rinses on Bonderite 1000 iron phosphate conversion coatings. Analysis and depth profiling of the conversion coating showed the coating to be of a granular appearance and composed of an iron phosphate-iron oxide coating of approximately 2000A thickness. A 100A layer of an organic material covered the phosphate coating and was assumed to contain the polyvinylphenol based post-treatment. This conclusion was further confirmed by XPS analysis using a Model 560 Auger/ESCA spectrometer in which the nitrogen present in the post-treatment was observed in the top 100A layer containing the organic material.

Acknowledgment

The Auger and XPS analysis and interpretation of the post-treatment on iron phosphate surfaces was conducted by Dr. J. A. Kramer of Parker Chemical Company. Auger analysis was performed at the CRISS Center, Montana State University which is supported by NSF grant CHE 7916134.

Literature Cited

1. J. I. Maurer, "Preparation of Metal Surfaces for Organic Finishes", American Society of Tool and Manufacturing Engineers, Technical Paper, FC-68-652
2. "Metal Handbook", Volume 5, 9th Edition, pp. 434-456, American Society for Metals
3. W. R. Cavanagh, U.S. Patent 2,970,935
4. J. I. Maurer, R. E. Palmer, V. D. Shah, U.S. Patent 3,279,958
5. J. I. Maurer, R. E. Palmer, V. D. Shah, U.S. Patent 3,222,226
6. L. Kulick, K. I. Saad, U.S. Patent 3,975,214
7. L. Kulick, J. K. Howell, Jr., U.S. Patent 4,039,353
8. A. Lindert, U.S. Patent 4,376,000
9. A. Lindert, U.S. Patent 4,433,015
10. A. Lindert, J. Kramer, U.S. Patent 4,457,790

RECEIVED March 14, 1986

Enhancement of Acid-Chloride Resistance in a Chromate Conversion Coating

V. S. Agarwala

Naval Air Development Center, Warminster, PA 18974

Chromate conversion coatings perform poorly in environments containing acidified chloride. In salt/SO_2 spray tests the substrate metal is heavily pitted after three to four days of exposure. In this work, a new coating was developed which improved the corrosion resistance of the conventional chromate coating remarkably. The new coating protected the substrate metal (7075-T6 Al) in 5% NaCl + SO_2 spray for up to two weeks. It even extended the stress corrosion cracking resistance of 7075-T6 Al alloy. Studies showed the leaching rate of Cr through the new coating was significantly reduced. The technique of developing the coating consisted of incorporating molybdate into the pre-existing coating containing chromium oxides on an aluminum surface.

On aluminum surfaces, chromate conversion coatings are mostly used to protect the metal from corrosion both as a pretreatment for bonding primers and paints, and for bare metal corrosion(1-2). On steels, they are applied after the surfaces have been plated with cadmium. For most military hardware (naval aircraft systems), use of chromate conversion coatings are standard practice and are specified (Qualified Product List, MIL-C-81706) for corrosion protection from industrial and marine (naval) environments. Although chromate conversion coatings offer good resistance to most ambient environments, including up to 5% NaCl, they fail to perform when the environments become acidic. Aboard aircraft carriers, the stack gases and aircraft exhausts introduce large amounts of SO_2, soot, and other carbonaceous products which are highly corrosive into the environment.

Recent studies on monitoring of a carrier environment(3) have shown that its corrosion severity parallels to a laboratory cyclic environment of 5% NaCl spray (fog) with 30ppm of SO_2. The development of a new chemical conversion coating which is also resistant to this environment would be of great importance.

Almost all aluminum structures are painted with organic polymers for corrosion protection. The purpose of incorporating an inhibitor interface (chromate conversion coating) between the substrate and the paint film is to ensure protection when paints fail to perform(4). It has been generally accepted that no matter what kind of paint system, and how well it is applied, it always will have some porosity defects and will degrade with time during service. In military aerospace vehicles, the service conditions are most severe for paints. One way to increase protecion of aluminum in acidic marine environments is to enhance the corrosion inhibiting ability of the inorganic coating. This paper describes the results of such an effort. A new coating which not only inhibits the attack of chloride ions but also that of the acidic environments such as SO_2 has been developed.

Experimental

Panels of high strength aluminum alloy (7075-T6) were used in this study. The panels were approximately 10 x 3 x 0.032 inch (25 x 7.5 x 0.08 cm) in size. The test environments for coating evaluation were: (1) a 5% NaCl spray (fog) chamber according to ASTM Standard Method of Salt Spray (Fog) Testing (B117-73), and (2) a modified 5% NaCl/SO_2 spray (fog) chamber with SO gas introduced periodically - ASTM Standard Practice for Modified Salt Spray (Fog) Testing (G85-84(A4)). In the latter case, a constant spray of 5% NaCl was maintained in the chamber and SO_2 was introduced for one hour four times a day (every 6 hours)(5). Coated test panels were examined for corrosion after one- and two-week exposure periods.

The procedure for applying coatings was almost the same as used for conventional chromate conversion coating. The new conversion coating called CMT was an additional surface application after standard chromate conversion coating. The experimental protocol used was as follows:

1. Cleaned in non-etching mild alkaline solution at 160°F for ten minutes.
2. Double rinsed in tap water.
3. Deoxidized in a 15% H_2SO_4 + 2.5% chromic acid mixture at about 150°F for 2 to 5 minutes.
4. Double rinsed in tap water and immersed in a chromate conversion coating bath (MIL-C-81706) for five minutes.
5. Washed in tap water and rinsed in distilled water.
6. Applied CMT (molybdate) treatment, rinsed in tap water and dried.

CMT treatment is a proprietary process developed at the Naval Air Development Center. It basically incorporates molybdates in the coating containing chromates.

All coatings were allowed to age for at least one week in ambient air before testing for corrosion resistance. The CMT coated panels were compared with the standard chromate conversion coated panels for their performance. The evaluation plan consisted of the following measurements: (1) visual observations; (2) weight loss; (3) electrochemical polarization; (4) stress corrosion cracking; (5) scanning electron microscopy; and (6) XPS surface analysis. Visual inspections of the specimens were made during and after the salt spray chamber exposure. They were inspected for general corrosion and pits. Conventional electrochemical measurements were made to determine their corrosion potentials and potential vs. current density (E vs Log i) relationships. In both cases, neutral and acidified salt solutions were used; the details of the procedure are described in ASTM Standard Practice for Conventions Applicable to Electrochemical Measurements in Corrosion Testing (G3-74). Stress corrosion cracking tests were performed on 1/8 inch (3mm) diameter tensile specimens fabricated from 7075-T6 Al alloy. Proving rings were used as sustained load devices (ASTM Standard Practice for Preparation and Use of Direct Tension Stress Corrosion Test Specimens, G49-76). All tensile specimens were tested with their respective chemical conversion coating. Tensile loads of 40 ksi (275 MPa) were applied on each specimen. The neck region of the tensile specimen was surrounded by a plastic wrap containing cotton soaked with 1% NaCl. In order to accelerate the test environment, acidified (pH 2) 1% NaCl solution was used. These sustained load tests were carried out for up to 300 hours.

Results

Salt Spray Test : Panels coated with the standard chromate conversion coating and CMT were compared with each other in their corrosion resistant properties in several ways. The conventional 5% NaCl/SO_2 fog chamber tests showed excessive corrosion and pitting within one week on chromate conversion coated (CCC) 7075-T6 Al alloy panels. The CMT coated panels were almost uncorroded and without any pits. The plates in Figure 1 show the conditions of the panels after 7 and 14 days' exposure in this environment. Even after 14 days' exposure the CMT panels were still far better than CCC panels.

Weight Loss : A quantitative evaluation of the coatings, both CCC and CMT, was also made. The panels exposed in salt/SO_2 fog chamber were measured and weighed before and after the test to calculate corrosion losses. All exposed panels were cleaned in 1:1 HNO_3 for after-the-test weight according to ASTM Standard Practice for Preparation, Cleaning, and Evaluating Corrosion Test Specimens (G1-81, para 7.2.2). This is to remove the corrosion products for calculating weight losses. The results of these measurements are given in Table 1. Each panel was also examined visually and the number of pits were counted in a specified area, i.e., 50cm^2 surface area; an average of 4 specimens for each coating was determined and reported in Table 1. Both the weight loss and pitting data for CMT showed significant inprovement over the CCC. In fact, the pitting resistance for the CMT coating was excellent. In a 14-day period, it showed only 10 pits/cm^2 compared to 75 pits/cm^2 for the CCC.

Table 1 - Corrosion test results of coated 7075-T6 Al alloy exposed to salt/SO_2 fog environment.

Coating	*Weight Loss, mg/cm^2/48 hrs.	# Pits/cm^2 (in 14 days)
Bare	1.10 ± 0.20	Heavily Corroded
CCC	0.65 ± 0.10	75 ± 20
CMT	0.25 ± 0.05	10 ± 3

* Based on specimen size of 8 x 6 x 0.3cm

<u>Stress Corrosion Cracking</u> : Results of the proving ring tests of the coated specimens in both the neutral and acidified salt solutions are given in Table 2. Although there is no significant difference in the failure times for both the coated specimens in neutral 1% NaCl, the differences were remarkable at low pH. The CMT coating was able to extend the stress corrosion cracking resistance of the chromate coated material from 60 to 90 hours at pH 2.

Table 2 - Stress corrosion cracking properties of coated 7075-T6 Al alloy from proving ring tests (stress = 40 ksi or 275 MPa).

Coating	*Time to Failure, Hours	
	1% NaCl	1% NaCl (pH 2)
CCC	300	60 ± 15
CMT	300	90 ± 15

* Mean average of 10 specimens.

Electrochemical Behavior : The corrosion (open circuit) potential values for the coatings were as given in Table 3. For CMT coating, the corrosion potential was least active. Generally a less active (less negative) potential indicates more corrosion resistant properties of the surface. The stability of the CMT coating was excellent since it showed no change in its corrosion potential (-0.700V) as the environment was changed from neutral to acidic or high to low chloride concentrations (cf. Table 3).

The differences in the corrosion resistant properties of the CCC and CMT coatings were better characterized from the potentiostatic anodic and cathodic polarization behaviors in acidic salt solutions. The typical plots of the potential vs current density are shown in Figure 2. The curves in Figure 2 also include a plot for untreated (bare) 7075-T6 aluminum alloy for comparison. Although both the coatings, CCC and CMT, showed significant anodic shifts in the anodic polarization curves when compared to untreated specimens, they were almost the same among themselves. The shifts in the cathodic polarization curves (cf. Figure 2) between the CMT and CCC were highly significant. The cathodic current densities for the CMT coated material were almost an order of magnitude lower than for CCC coated material. In other words, the corrosion rate (determined as the point of intersection of the anodic and cathodic Tafel slopes) for the CMT will be lower by an order of magnitude than for CCC in 1% NaCl solution of pH 2. In 3.5% NaCl solution (pH 2), the polarization plots were almost the same but indicated an even greater shift of the cathodic polarization curve toward lower current densities for the CMT coating.

Table 3 - Effect of chemical conversion coatings on corrosion potentials of 7075-T6 Al alloy.

Environment	pH	Ecorr. vs SCE, volt		
		Bare	CCC	CMT
3.5% NaCl	6	-0.725	-0.700	-0.700
3.5% NaCl	2	-0.750	-0.720	-0.700
1.0% NaCl	2	-0.760	-0.720	-0.700

Coating Stability : The stability of a chemical conversion coating is best described by the non-leaching character of its corrosion inhibiting constituents and the insoluble nature of the oxides of the substrate metal. Chromate conversion coatings suffer from the lack of these properties. Thus, a study was conducted in which all the coated panels

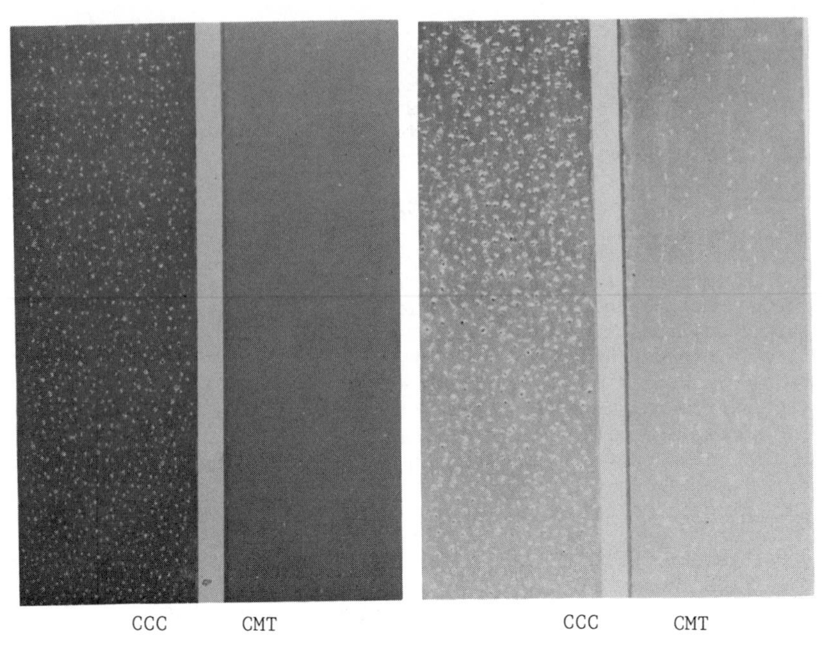

Figure 1. Effect of 5% NaCl/SO_2 fog exposure on corrosion resistance properties of chemical conversion coatings on 7075-T6 aluminum alloy at 7 days (left) and 14 days (right).

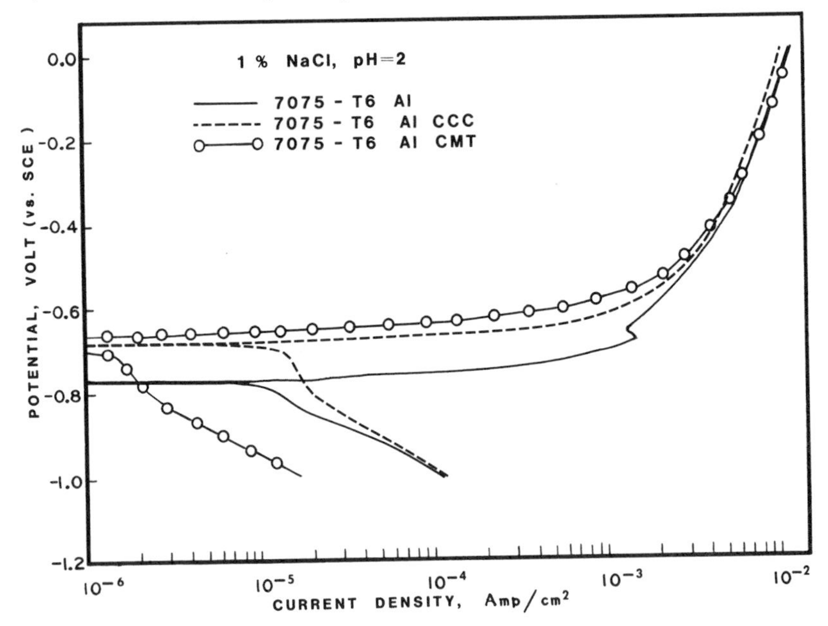

Figure 2. Electrochemical polarization behavior of chemical conversion coatings on 7075-T6 aluminum alloy in 1% NaCl solution at pH 2.

were immersed in 3.5% NaCl solutions of different pH for various exposure periods. A chemical analysis of the solutions by atomic absorption spectroscopy (AAS) showed that the amount of Cr leached out from the CMT coatings was approximately 1/3 of that from CCC. In acidified salt solutions (pH 2), leaching of Cr was much greater for CCC. Based on 14 days'exposure in 3.5% NaCl (pH 2), the CCC coating lost approximately 2.0 μg /cm^2 of Cr compared to 0.4 μg/cm^2 for CMT. The initial chromium concentration in both the CCC and CMT coatings was approximately 7-8 μg/cm^2. This was determined by stripping the coatings with 1:1 HNO_3 and analyzing the solution by AAS.

Surface Characterization : The morphology of the coating surface was examined by scanning electron microscopy and the composition of the films were determined form X-ray photoelectron microscopic (XPS) scans. Both the CCC and CMT coatings were alike in appearance but differed in their chemical contents. Well pronounced peaks for Mo and Cr were found in the CMT coating during energy dispersive X-ray analysis. Of course, there was no Mo detected in the CCC coating. Elemental distribution of Mo and Cr in the CMT was very uniform. The surface composition of coatings, determined by XPS, was as follows:

Coating	Cr	Mo	Others
CCC	Cr_2O_3	--	M_XO_Y, M(OH)
CMT	CrOOH	MoO_4^{--}	M_XO_Y

These results indicate the presence of Cr(III) in both the coatings but a less soluble hydrated oxide (CrOOH) in CMT. Presence of Cr(VI) was not detected by the XPS technique. This was in agreement with the findings of Glass (6), and Matienzo and Holub (7). In the CMT coating, the presence of Mo(VI) as molybdate was significant. It was also noted that the CCC had at least two different forms of other metal (M) oxides ,probably aluminum, while the CMT had only one. Although no quantitative estimation of the individual species present was made, the approximation was that Cr and Mo concentrations were not significantly different in the CMT coating.

Discussion

The mechanism of developing corrosion protective properties in an inorganic coating principally consists of forming insoluble oxides on the metal surface. Additionally, oxides must have certain corrosion inhibition (redox) properties which can protect the metal substrate from corrosive species like Cl^- and SO_4^{--}. In the case of chromate conversion coating, CCC, the oxides of aluminum and chromium have been responsible for their corrosion inhibitive properties which were derived from their soluble and insoluble portions of the chromate film. In acid-chloride environments, the protective properties of chromate become limited because the oxides of aluminum which hold the chromium oxides in place are no longer insoluble. In this study, these considerations were taken into account to alter the composition of the surface films. In the CMT process, the chromate conversion coating was re-inforced with other substances which contained molybdates. This surface film was five times more stable than the conventional chromate coating in both the neutral and acid-chloride environments. XPS studies and those by other investigators (7) suggest that chromium in chromate coatings is mostly present as Cr_2O_3. In the CMT coating it was found to be in CrOOH form. The hydrated oxides of chromium are known to be more stable than Cr_2O_3 ; they tend to form polymer type links (8) and most probably protect the substrate metal by forming a barrier. Additionally Mo in the CMT coating, which is present as MoO_4^{--} , is also capable of forming polymeric links on the surface (9). The presence of Mo(VI) in the film enhances the protection by counteracting the aggresive nature of Cl^- ions. Most probably Mo(VI) ties up Cl^- ions during the process. XPS analysis showed a very even distribution of Cr and Mo oxides with very little aluminum oxide on the top layers of the film. In contrast, the CCC coating contained a significant amount of both aluminum and chromium oxides. The surfaces which contain more stable and insoluble oxides can offer better resistance to acid-chloride environments. The high corrosion resistant behavior of the CMT coating supports this conclusion. Electrochemical studies showed a marked reduction in the cathodic polarization behavior which means the CMT coating offers a higher resistance barrier (potential drop) than the CCC to achieve the same cathodic reaction rates (current densities). Superiority of the new coating was exhibited in its ability to reduce stress corrosion cracking susceptibility of 7075-T6 Al alloy in acid-chloride environments.

Conclusions

The conventional chromate conversion coating is non-protective in acid-chloride environments. A new coating called CMT which contains oxides of Mo and Cr, was developed for aluminum Alloys. An evaluation of this coating was made with the following conclusions: (1) In 5% NaCl + SO_2 spray there was no corrosion of the substrate metal for up to two weeks; (2) CMT enhanced the stress corrosion cracking resistance of 7075-T6 Al alloy significantly; (3) the CMT coating was more stable than CCC and showed almost no leaching of Cr or Mo in acid-chloride environments; and (4) The composition of the CMT film was mostly Mo as MoO_4^{--} and Cr as CrOOH, and had almost no oxides of aluminum on the surface.

Acknowledgments

The author thanks the assistance of the co-op students and P. J. Sabatini of Naval Air Development Center (NADC); L. J. Matienzo of Martin-Marietta Corporation in performing XPS analysis; and support of the NADC Independent Research program for financial support.

Literature Cited

1. Wernick, S. and Pinner, R. In "Surface Treatment of Aluminum"; Draper, Robert, Ed.; Teddington, U. K., 1972; Vol. I, pp. 233-290. .
2. Katzman, H. A., Malouf, G. M. and Stupian, G. W., Applications Surf. Sci. , 1979, 2, 416-432.
3. Agarwala, V. S. In "Atmospheric Corrosion" Ailor, W. H., Ed.; Wiley Interscience, New York, N.Y., 1982; pp. 183-192.
4. Montle, J. F. and Hasser, M.D., Materials Performance , 1976, 50, 15-18.
5. Ketcham, S. J. and Jankowsky, E. J., "ASTM Symposium on Laboratory Corrosion Tests and Standards", Bal Harbor, FL, 14-17 Nov. 1983.
6. Glass, A. L., " A Radiochemical Investigation of the Leaching of Cr from Chemically Chromated Al Alloy Surfaces, Part III, The "Uptake" of the Chloride Ion Into Such Surfaces," Naval Air Development Center, Warminster, PA; Report No. NADC-MA-6702, August 1967.
7. Matienzo, L. J. and Holub, K. J., Applications Surf. Sci. , 1981, 9, 47-73.
8. Lollar, R. M. In "Chromium-Chemistry of Chromium and Its Compounds"; Udy, M. J., Ed.; Reinhold Publishing, New York, N.Y., 1956; Vol. I, p. 306.
9. Cotton, F. A. and Wilkinson, G. "Advanced Inorganic Chemistry"; Wiley Interscience, New York, N.Y., 1972; pp. 965-972.

RECEIVED February 24, 1986

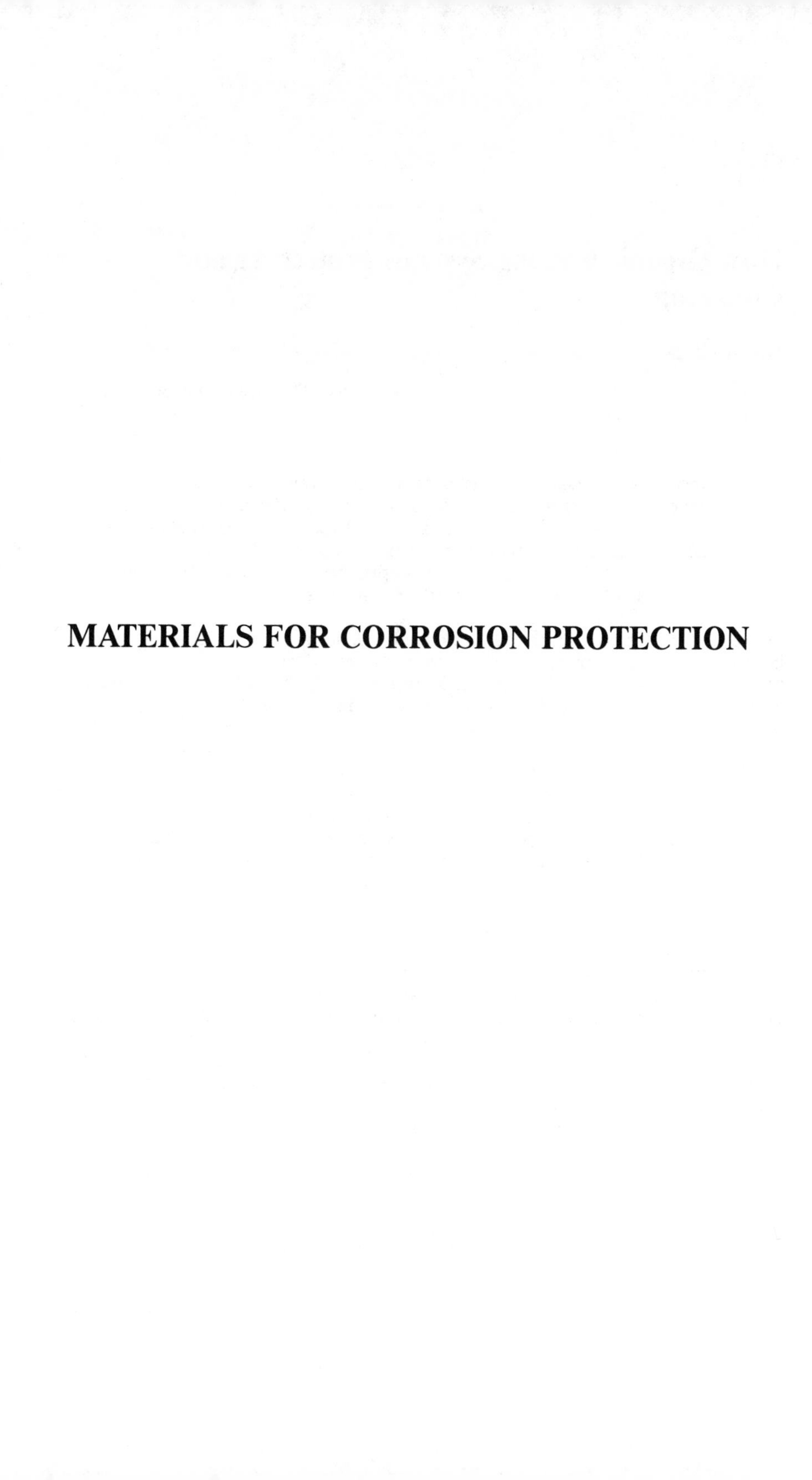

MATERIALS FOR CORROSION PROTECTION

20

How Organic Coating Systems Protect Against Corrosion

Werner Funke

Forschungsinstitut für Pigmente und Lacke, Allmandring 37, D-7000 Stuttgart 80, Federal Republic of Germany

The electrochemical, physicochemical and adhesional aspects of corrosion protection by organic coatings are shortly discussed. Attention is drawn to some inconsistancies in the interpretation of protective mechanisms and suggestions are given how protective principles may be optimally realized in practical systems.

There are essentially three important mechanisms by which organic coating systems protect against metal corrosion:
The electrochemical, the physicochemical and the adhesional mechanism. In order to obtain optimum protection, it is commonly proposed to incorporate as many as possible of these mechanisms in a coating system. It will be discussed how far this strategy is tenable in practical paint formulation and whether it is reasonable in the light of a critical judgement. For this purpose it is helpful to recall how these mechanisms work and what the requirements are for their operation. Correlations of permeability with anticorrosive action in corrosion protection by organic coatings have been recently discussed (1).

The Electrochemical Mechanism

The electrochemical mechanism is generally connected with the presence of active anticorrosive pigments, like red lead or zinc chromate, and occasionally also to corrosion inhibitors, which are added to the base coat of the system. It is a wide-spread opinion that such anticorrosive pigments are almost indispensable for a satisfactory corrosion protection because "no organic coating is impermeable to water" (2). Therefore anticorrosive pigments are considered to be an ultimate line of defense for corrosion protection. Active anticorrosive agents act only in presence of water, which dissolves a small fraction of them and makes them available at the coating/metal interface (3,4). Passivation or corrosion inhibition of metal surfaces is mostly achieved by supporting the in situ formation of thin layers of insoluble corrosion products (4,5), which cover corroding areas and stifle the action of the corrosion elements. For a continued corrosion inhibition the anticorrosive solution must keep steady contact with the metal surface. Substitution by normal water usually initiates corrosion again.

To allow diffusion of the dissolved anticorrosive agent to the coating/metal interface, the binder should be permeable to water. This requirement clearly contradicts the other requirement of corrosion protective coatings, namely to prevent the access of water as a corrosive agent to the metal surface. Frequently binders used in practical corrosion protective coating systems scarcely swell by water and are only slightly permeable to it.

Accordingly the protective agent is locked in the binder and is not available in sufficient concentration at the metal surface. Good protective properties claimed in these cases are rather due to other protective mechanisms than to the electrochemical one.

Unfortunately some corrosion stimulants, like Cl^-, SO_4^{2-} or NO_3^-, strongly oppose inhibition by anticorrosive pigments and inhibitors (6). Steel corrodes in saturated aqueous solutions of an anticorrosive pigment in presence of small amounts of these stimulants, e.g. 1% w/w NaCl is sufficient to make a saturated aqueous extract of zinc chromate corrosive (7). Therefore, irrespective of environmental requirements, the usefulness of active anticorrosive pigments and inhibitors as well has become questionable.

The Physicochemical mechanism

The physicochemical mechanism consists in blocking up diffusion of corrosive agents, like water and oxygen, and of corrosion stimulants. This barrier action of organic coatings may be enhanced significantly by pigments, fillers or extenders which, due to a flaky or platelike geometrical shape, greatly increase the length of diffusional pathways through the cross section of the coating film.

In order to avoid diffusion in the pigment/binder interface, interfacial bonds between both phases should be as water-resistant as possible. If permeability is taken as a measure, properly formulated barrier coatings may compare in corrosion protective efficiency with normal coatings, the thickness of which is two or three times as high.

The binder contributes also to the barrier effect of a coating system. Permeability of the binder depends on the rigidity and polarity of its macromolecular structure and also on the density of the molecular packing. Accordingly permeability decreases by decreasing the chain mobility, e.g. by crosslinking, by decreasing the hydrophilic character of the macromolecules and by increasing the density of molecular packing up to crystalline or crystalline-like structures.

Considering the measures to be taken in binders for ensuring an optimum barrier effect, they obviously oppose the requirements for the anticorrosive function of pigments, which need water-permeable and swellable binders. One may argue, however, that concurrently with improving the barrier properties of a coating system the anti-corrosive function of pigments or inhibitors is less challenged.

Furthermore the rigidity of the macromolecular structure of the binder may oppose the demand for mechanical strength and shock resistance of the coating film. Increasing the rigidity by crosslinking leads to internal stresses, which accumulate with increasing film thickness. A way out of this dilemma may be the use of very thin but highly crosslinked base coats (8).

The Adhesional Mechanism

The adhesional mechanism up to now has not yet received sufficient attention in corrosion protection by organic coatings. As long as adhesion of the base coat to the metal surface is unchanged no corrosion can take place below a coating.

Too much emphasis has been given to adhesion under dry conditions. However, corrosion is only possible if enough water is present in the coating/metal interface to provide the electrolyte for the corrosion elements to operate. This condition is hardly imaginable without a previous significant reduction or even the loss of adhesion. Therefore "wet adhesion" is considered to be of crucial importance to corrosion protection by organic coatings (9).

It is generally agreed that due to the polar nature of oxidic metal surfaces good dry adhesion is only possible by incorporating polar groups in the binder molecules. However, these polar groups may effect water sensivity of the coating/metal interface thus causing poor wet adhesion. That water accumulates at the interface coating/metal substantially, has been shown by comparing water absorption of free and supported films (10). One way to make water-sensitive interfaces resistant against water is to adsorb polar groups which are attached to rigid polymer backbone chains.

It is still not known for certain, whether on exposure to water adhesion is uniformly reduced over the exposed area or only locally lost at channels providing the electrolytic pathways between anodic and cathodic areas of the metal surface.

In choosing binders with good adhesion, again protective properties are encountered, which exclude each other, e.g. non-polar macromolecules with low permeability would be benificial to the barrier effect but objective to good dry as well as wet adhesion i.e. to the adhesional mechanism. On the other hand polar groups supporting dry adhesion are required despite of their weakness in presence of water. The question remains how the adhesional interaction may be stabilized to resist the attack of water.

For the sake of good adhesion a metal surface should be clean and free of water-soluble substances. On the other hand, for the protective action of anticorrosive pigments a soluble fraction, i.e. an "impurity", must be present at the interface. It is hard to reconcile this requirement with good wet adhesion of the base coat.

Corrosion protective base coats with hydrophilic binders, such as water-borne coatings drying at ambient temperatures, usually exhibit high water permeability and poor wet adhesion. In these cases active anticorrosive pigments are needed. Protective properties can be improved by delaying the access of water to the coating /metal interface, e.g. by increasing film thickness, incorporating barrier pigments or applying barrier top coats (Figure 1). However, even then these systems are latently weak and may fail on prolonged exposure to water or high humidity, especially if mechanical stressess simultaneously act on the coatings and place excessive demands on their adhesion.

The Combination Of Different Protective Mechanisms

The combination of different protective mechanisms in one coat or one coating system is frequently recommended for optimal results in corrosion protection. However, ways and measures to optimize protective mechanisms may be different and sometimes even exclude each other. For example trying to combine good wet adhesion and corrosion inhibition by an active anticorrosive pigment in the same base coat does not make much sense, despite being frequently postulated to explain protective properties of commercial paint systems (Figure 2). Binders with good wet adhesion lock in the anticorrosive pigment and therefore diminish or even prevent its corrosion inhibiting effect. Sometimes it is claimed that practical experience disproves this statement, but it cannot be excluded in these cases that protection is mostly due to good barrier properties and/or good wet adhesion. On the other hand it is advantageous to choose a primer exhibiting optimal wet adhesion and simultaneously optimal barrier properties.

The barrier mechanism not only decreases water permeating to the coating/metal interface but likewise retards the release of solvents from a coating. In order to avoid delayed film formation for this combination solventless paint systems are most suitable.

In two-layer coating systems the best choice is to endow both layers with the barrier effect and choose a binder having good wet adhesion. Other combinations are less effective or even not reasonable (Figure 3).

The use of active anticorrosive pigments is only justified to prevent corrosion at scratches, pinholes or similar coating defects and even then only in absence of virtual amounts of corrosion stimulants. It is commonly assumed by paint technologists that protective effects incorporated in each layer of a coating system add together in preventing corrosion at the coated metal surface. However, in coating systems, which base coats protect by an electrochemical mechanism (Figure 4), the successive layers including the top layer rather should prevent any physicochemical or electrochemical reaction at the base coat/metal interface. Considering the

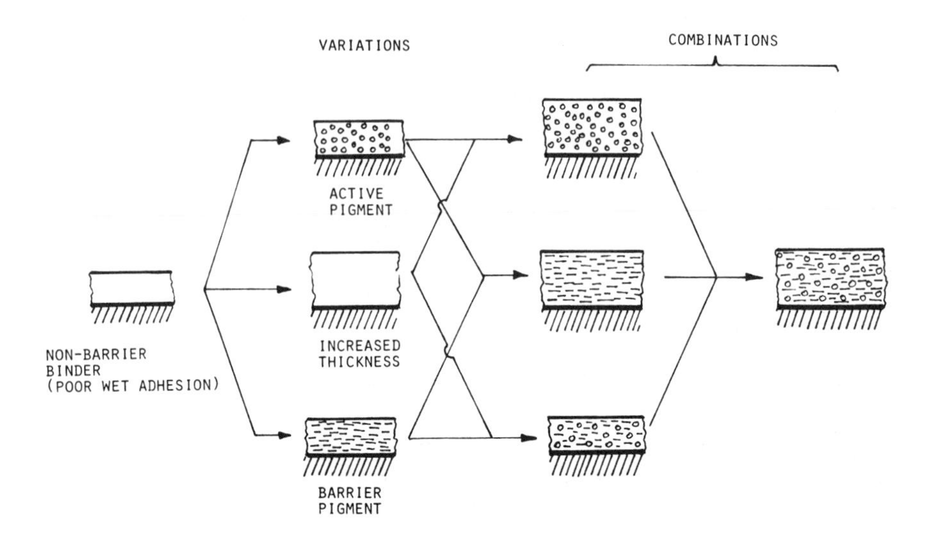

Figure 1: Variation and combination for improving corrosion protection by organic coatings composed of non-barrier binders

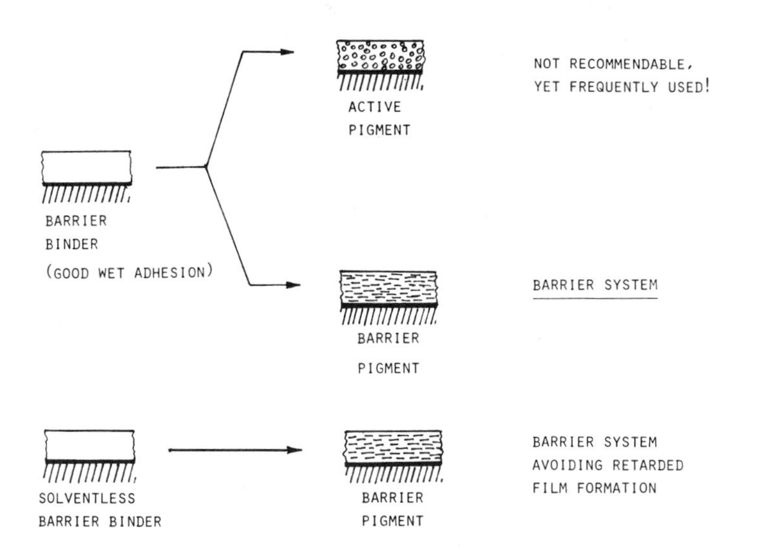

Figure 2: Variation for improving corrosion protection by organic coatings composed of barrier binders

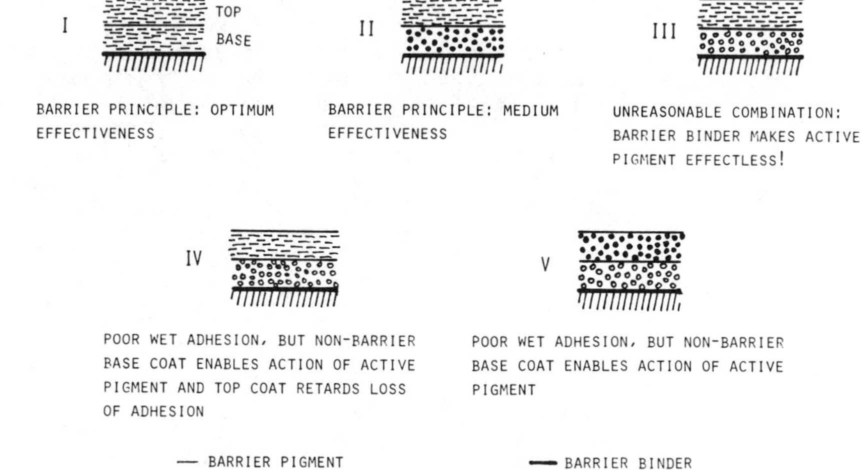

Figure 3: Barrier principle in two-layer, defectless coating systems

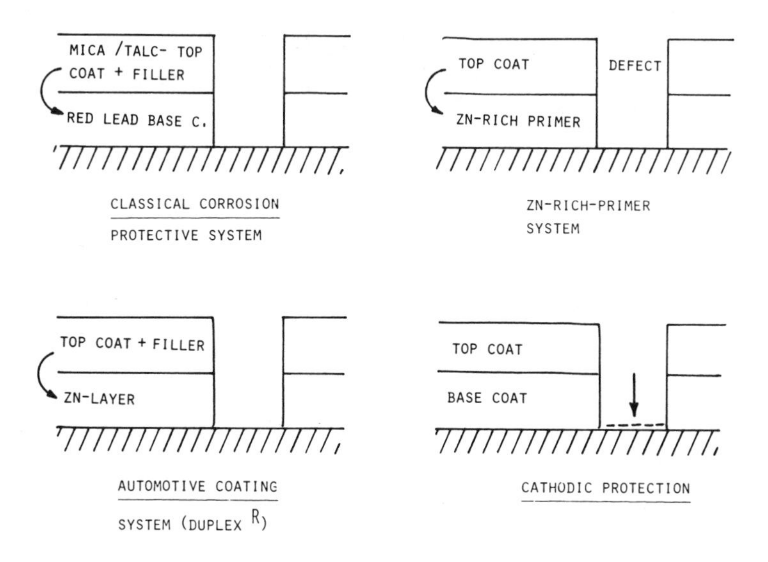

Figure 4: Two-layer system for corrosion protection at coating defects (Top coat protects base coat, base coat protects metal support at defects!)

high demands on the metal surface pretreatment to achieve good adhesion, it is not conceivable to allow this interface being kind of a reaction vessel. Layers succeeding to electrochemical protecting base coats should prevent any reaction in the base-coat and especially at its interface to the metal surface. The base coat should only come into action at coating defects extending down to the metal surface. Likewise coating systems suitable for cathodic protection should prevent any reaction at the metal surface below the intact coating system. Otherwise cathodic delamination is unavoidable.

In all these cases we actually have not a "twofold-protection" but interdependent as well as complementary mechanisms. The intact coating system protect by its barrier action and, possible, by good wet adhesion, whereas the electrochemical mechanism must be restricted to coating defects.

Literature Cited

1. F.L. Floyd, R.G. Groseclose, C.M. Frey, J. Oil Col. Chem. Assoz., 1983, 329
2. J.E.O. Mayne, Pigment Handbook Vol. III, Edited by T.C Patton, Wiley Interscience Publ. 1973, p. 459
3. J.E.O. Mayne, E.H. Ramshaw, J. Appl. Chem. 13, 1969, 553
4. H. Leidheiser, J. Coatings Technol., 53 No. 678, 1981, 29
5. J.E.O. Mayne, Pigment Handbook Vol. III, Edited by T.C. Patton, Wiley Interscience Publ. 1973, p. 457-464
6. L.A. Buckowiecke, Schweizer Archiv f. Wissenschaft u. Technik (6), 1954, 1
7. W. Funke, unpublished results
8. W. Funke, J. Oil Col. Chem. Assoz., 1985, 229
9. W. Funke, J. Coatings Technol., 55 No. 705, 1983, 31
10. W. Funke, Fette, Seifen, Anstrichmittel, 64, 1962, 714

RECEIVED March 5, 1986

21

Improving the Performance of Zinc-Pigmented Coatings

T. Szauer and A. Miszczyk

Institute of Inorganic Chemistry and Technology, Technical University of Gdánsk, 80–952 Gdánsk, Majakowskiego 11, Poland

The investigations comprised zinc pigmented coatings with the aim of testing the mechanism of their action and looking for ways of improving protective properties. With the use of impedance technique the mechanism for two periods of activity has been proven with these coatings. Zn pigmented coatings impose the full cathodic protection during the first period of action, while in the second period the sealing and inhibiting properties of Zn pigment corrosion products are the main factors influencing the protection. The work has been focused on looking for ways of improving the coatings by influencing both periods. Positive results have been obtained with zinc phosphate used as an coating additive to modify the second period of the protection.

Organic and inorganic coatings containing metallic zinc dust are extensively and successfully used for anticorrosion steel protection in various agressive media such as sea water, industrial and sea atmosphere (1). They comprise frequently a prime layer in multilayer coatings or a paint for temporary protection. Their unique feature is steel protection even in the course of minor damages. Standard layers contain from 85 to 95 % of zinc with respect to dry mass of the coating. Such a large content of zinc in the form of dust makes the coating electrically conducting and porous (2,3).

According to generally accepted concepts, the protective action of a highly zinc pigmented coating has a two step mechanism which can be distinguished (1-6). The first one, which is relatively short, is a period of cathodic protection. In zinc-steel microcells, dissolution of zinc at the anode is accompanied by protection of steel at the cathode. For its existence several factors are necessary: the contact between zinc particles as well as between zinc particles and steel substra-

0097–6156/86/0322–0229$06.00/0

tum and wetting of zinc and steel surface. These conditions are fulfilled due to the large zinc content and porosity of the coating. The decrease of effectiveness of cathodic protection is associated with the loss of contact between steel and zinc dust (7) or between particles of zinc due to formation of the corrosion products of low conductivity. The second stage, several times longer than the first one (6), is associated with the blocking action of zinc corrosion products in pores existing in the coating (4,8,9) or with inhibiting action of zinc salts depositing on the steel surface and forming a tight layer, impeding significantly the corrosion of steel(10). Thus, one of the important factors for the long-term protection is the formation of proper quality and amount of zinc corrosion products during the step of cathodic protection. Therefore, the ratio of zinc and steel surface areas being in contact with the electrolyte during the initial period is significant (11).

In order to change definite properties of the coating and/or to reduce its coast, various actions were undertaken to substitute part of zinc dust by other pigments (12,13). However, in general this led to the decrease in protective properties of the modified coating.

Another question is the proper technique for the evaluation of protective properties of zinc pigmented coatings. Until now, the most popular one is the test in a salt chamber (12,14). This technique does not allow a complete understanding of the complex phenomena occuring in the coating. It seems that the impedance technique selected in this work provides more adequate and useful data.

Experimental

An epoxy paint for temporary protection of high zinc content 88.3 % relative to dry mass of the coating was investigated on mild steel wire electrodes of 5 mm diameter. The coatings of 27 ± 2 µm in thickness were studied. The measurements were carried out in 3 % non - - deaerated NaCl solution at room temperature in the frequency range from 1 Hz to 60 kHz using a sine signal of 10 mV amplitude. The measurements were performed in a three-electrode system with the corrosion potential measured vs. the saturated calomel electrode.

Results and Discussion

It was decided to carry out modifications of the highly zinc pigmented paint using the impedance technique to test the protective properties. Steel electrodes with the following coatings were tested:

A - standard highly zinc pigmented paint,

B - standard highly zinc pigmented paint modified by a conductive substitution of 1 % of zinc by carbon black,

C - standard highly zinc pigmented paint modified by substitution of 3 % of zinc by zinc phosphate.

In the case of electrode B the conductive carbon black was used with the purpose to influence the cathodic protection period. The modification C was made with zinc phosphate to influence the second step in protective action of zinc pigmented coatings.

The obtained results are shown in Figures 1 through 3. The impedance diagrams obtained after 24, 140 and 900 hrs from the moment of immersion are presented in relation to the changes of corrosion potential, E, vs. time of immersion, t, in 3 % NaCl solution. Other symbols on figures are: R_s - series resistance and $\frac{1}{\omega C_s}$ - - series reactance, expressed in $\Omega \cdot m^2$. Frequency in the diagrams is given in Hz. Since there is no significant difference in the duration of full cathodic protection period in Figures 1 and 2, the addition of carbon black can be regarded as ineffective. Despite that fact, the smallest increase in diffusion resistance during this period has been noticed for the paint with the addition of carbon black.

In the final step of protection, the largest diffusion resistance has been found for the paint modified with zinc phosphate. It reflects the highest difficulties in the mass transport throughout this coating. Also observed is a positive effect of zinc phosphate on the formation of sealing and inhibiting zinc compounds

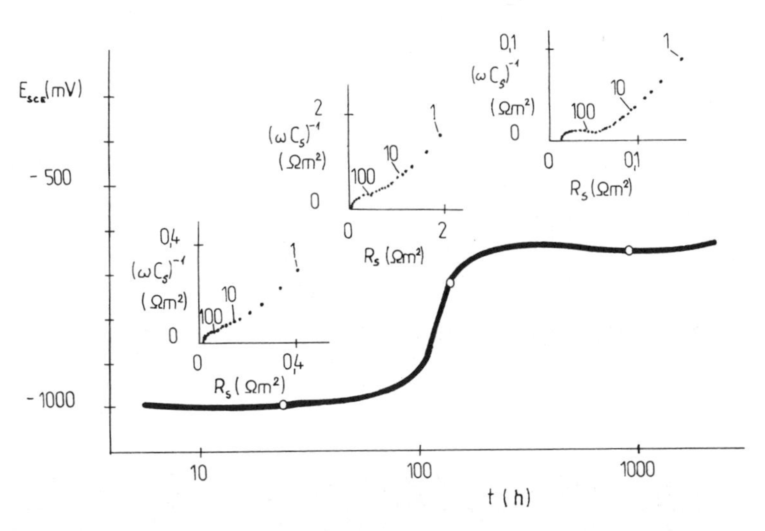

Figure 1. Electrode potential vs. immersion time in 3 % NaCl solution and impedance diagrams for steel electrodes with unmodified zinc pigmented coatings.

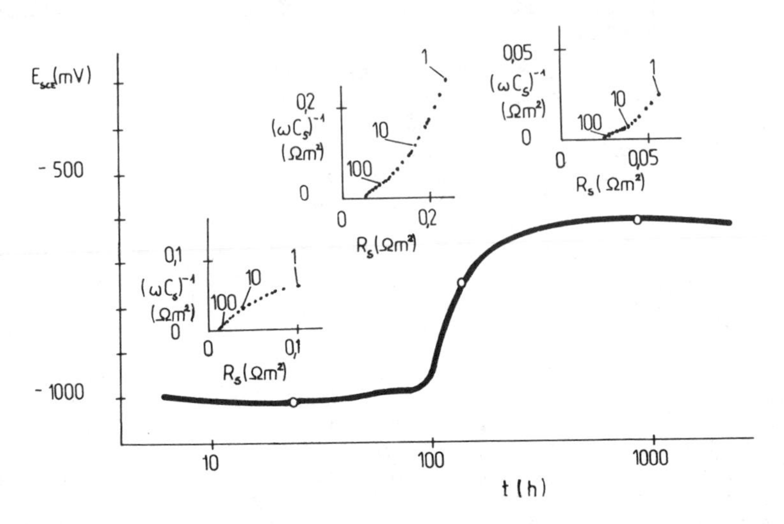

Figure 2. Electrode potential vs. immersion time in 3 % NaCl solution and impedance diagrams for steel electrodes with zinc pigmented coating modified by substitution of 1 % of zinc with conducting carbon black.

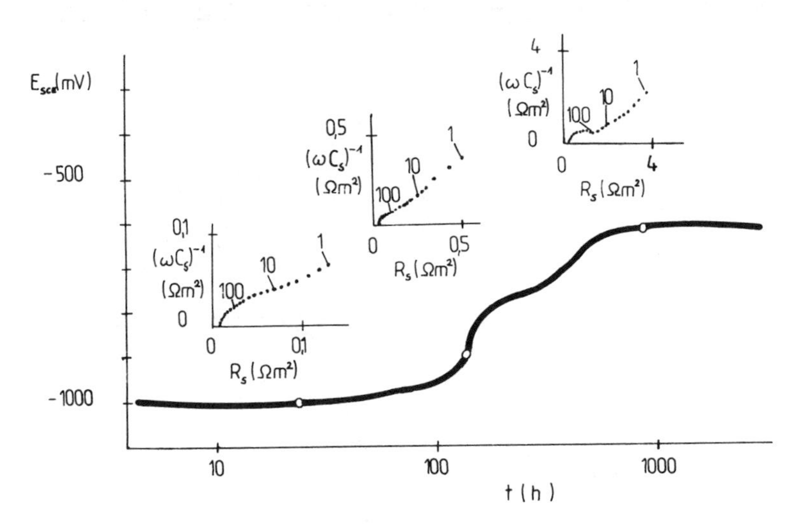

Figure 3. Electrode potential vs. immersion time in 3 % NaCl solution and impedance diagrams for steel electrodes with zinc pigmented coating modified by substitution of 3 % of zinc with zinc phosphate.

within the phase of the coating and on the steel surface. The attempt to modify the highly zinc pigmented paint with zinc phosphate can thus be regarded as promising such that results of test in a salt chamber evidenced better performance of this paint.

Conclusions

1. The impedance technique permits observation of the effect of modifying additives on protective properties of highly zinc pigmented coatings.
2. It has been found that protective properties of zinc pigmented coatings can be enhanced by using zinc phosphate as an additive influencing the performance of the coatings in the second stage of their protective action.
3. Negative results have been obtained with carbon black as an additive intended to improve protection properties of zinc pigmented coatings by influencing the first stage of their protective action.

Literature Cited

1. Munger, Ch.G. Good Painting Practice; Steel Structures Painting Council: Pittsburg, 1982; Vol. I, p. 125.
2. Henning, H.J. Farbe und Lack 1980, 86, 798.
3. Schmid, E.V. Farbe und Lack 1982, 88, 435.
4. Pass, A.; Meason M.J.F. JOCCA 1961, 44, 417.
5. Van Oeteren, K.A. Fette, Seifen, Anstrichmittel 1974, 76, 72.
6. Szauer, T.; Brandt, A. JOCCA 1984, 67, 13.
7. Mayne, J.E.O. J. Iron Steel Inst. 1954, 176, 140.
8. Schuster, H.J. Werskstoffe und Korrosion 1959, 10, 49.
9. Newton, D.S.; Sampson, F.G. JOCCA 1965, 48, 382.
10. Lincke, G.; Immenroth, R. Farbe und Lack 1979, 85, 733.
11. Theiler, F. Corrosion Science 1974, 14, 405.
12. Newton, D.S. JOCCA 1962, 45, 657.
13. Simpson, V.P.; Simko, F.A. JOCCA 1973, 56, 491.
14. Dickman, H.T. Modern Paint and Coatings 1983, 73,32.

RECEIVED February 24, 1986

22

Organic Corrosion Inhibitors to Improve the Durability of Adhesion Between Aluminum and Polymeric Coatings

L. J. Matienzo[1], D. K. Shaffer, W. C. Moshier, and G. D. Davis

Martin Marietta Laboratories, 1450 South Rolling Road, Baltimore, MD 21227

The application of selected organophosphonates and organosilanes onto pre-treated aluminum surfaces improves their environmental corrosion resistance and bond durability with an external polymeric coating. Ionizable phosphonates, such as nitrilotris methylene phosphonic acid (NTMP), adsorbed at monolayer concentrations, are effective inhibitors against hydration and are compatible with a nitrile-modified epoxy adhesive material. Aqueous solutions of selected organosilane compounds containing reactive side chains (e.g., epoxy, mercapto) render protection against both hydration and localized corrosion, and provide good adhesive bond durability with both nitrile-modified and polyamide (primer) epoxy resin systems. Wedge test results suggest that the curing process (e.g., percent crosslinking) of the epoxy-polyamide primer system is not affected by the addition of organosilanes, but may be affected by NTMP. The results of substrate surface characterization, adsorption behavior of applied films, and evaluation of candidate inhibitors by chemical, mechanical, and electrochemical test methods are presented. Mechanisms to explain the observed behavior of the various phosphonate and silane polymer systems are discussed.

The environmental durability of adhesively-bonded aluminum structures is of prime importance in the aircraft industry. Whether the adhesive function is structural or protective, proper pretreatment of the aluminum prior to epoxy bonding remains essential for developing high bond strengths. (1-8) The incentive to eliminate environmentally undesirable materials, such as chromates, has led to the consideration of organic inhibitor compounds as anticorrosion additives.

[1]Current address: IBM Corporation, Systems Technology Division, Endicott, NY 13760

0097-6156/86/0322-0234$06.00/0

The overall performance of a polymer-metal bond system is affected by:

1. Surface roughness of the substrate, needed to provide good mechanical interlocking with the polymer (6-8);

2. Long-term stability of the Al_2O_3 in a humid environment (9-12);

3. Selection of corrosion inhibitors which slow down the transformation of Al_2O_3 into AlOOH (12-14);

4. Effect of a selected corrosion inhibitor on the curing of the polymeric coating.

Typically, the first three effects, which deal with metal-polymer and metal-inhibitor interactions, are studied with respect to corrosion control processes. In this paper, we have also examined the relationship between the inhibitor compounds and the polymeric top coat from both a chemical and a physical point of view.

The results of our investigations on metal-polymer systems treated with selected inhibitors are presented. Using etched Al substrates, organosiloxane and organophosphonate compounds are evaluated with respect to their resistance to environmental degradation and overall bond durability. In addition to their ionizable (RO^-) moieties for bonding to the metal oxide surface, many of these compounds contain functional groups (X) designed to couple chemically with applied polymeric epoxy systems. This scheme is illustrated in Equation (1).

CORROSION-RESISTANCE FUNCTION OF INHIBITOR

$$\begin{array}{l} -OH \\ -OH \\ -OH \end{array} + (RO)_n\text{-}I\text{-}X \longrightarrow \begin{array}{l} -OH \\ -O\text{-}I\text{-}X \\ -OH \end{array} + ROH\uparrow \qquad (1)$$

Oxide surface — Inhibitor

Phosphonates: R=H, Me, Et; n=2

Silanes: R=Me, Et; n=3

Studies for the organophosphonates include the effects of solution pH on their adsorption onto the anodized aluminum substrates. Mechanisms are discussed for the respective interactions of the ionizable phosphonate and neutral silane compounds with two different polymeric epoxy systems.

Experimental

Substrate Characterization. Test coupons and panels of 7075-T6 aluminum, an alloy used extensively for aircraft structures, were degreased in a commercial alkaline cleaning solution and rinsed in distilled, deionized water. The samples were then subjected to either a standard Forest Products Laboratories (FPL) treatment (1) or to a sulfuric acid anodization (SAA) process (10% H_2SO_4, v/v; 15V; 20 min), two methods used for surface preparation of aircraft structural components. The metal surfaces were examined by scanning transmission electron microscopy (STEM) in the SEM mode and by X-ray photoelectron spectroscopy (XPS).

Epoxy Systems

Two conventional epoxy-based products were used for these studies:

1. A nitrile-modified epoxy structural adhesive (American Cyanamid FM 123-2)

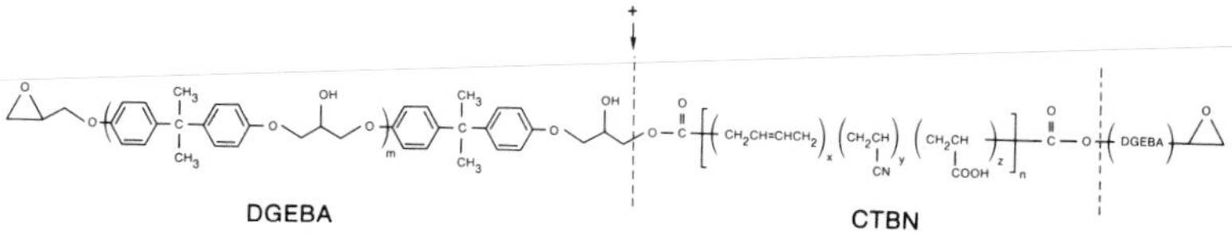

DGEBA CTBN

2. An unpigmented epoxy-polyamide top coat (Shell EPON 1001-T75 epoxy and Versamid 115 amidopolyamine).

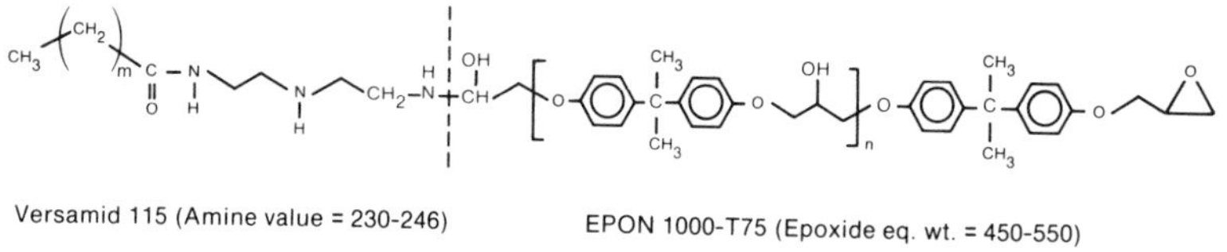

Versamid 115 (Amine value = 230-246) EPON 1000-T75 (Epoxide eq. wt. = 450-550)

The FM 123-2, supplied as a supported film on a knitted micro-filament nylon carrier, was applied directly to one pretreated (SAA or FPL) Al panel (15 cm x 15 cm x 0.3 cm). A second panel was then pressed onto the exposed adhesive side and the "sandwich" structure was subsequently cured at 120°C under 40 psi for 1 hr. A 1:1:1 mixture of the epoxy-polyamide formulation (EPON resin : Versamid curing agent : thinner, MIL-T-81772) was sprayed onto prepared 7075-T6 specimens to a thickness (dry) of 0.015 - 0.023 cm.

Inhibitor Coverage

The FPL- or SAA-prepared coupons and panels were immersed for 30 min in an aqueous (or aqueous/alcoholic) inhibitor solution at room temperature, followed by rinsing in distilled, deionized water and forced air drying. Coverage levels of phosphonate and silane inhibitors were determined by XPS from the surface concentration ratios of their characteristic elements, P/Al or Si/Al, respectively. Selected organosilanes were applied to the metal surface by spraying, after dissolution (0.1 - 0.5%, v/v) in the EPON-Versamid primer formulation.

Corrosion Testing

Hydration. The treated aluminum specimens were placed in a Blue M Humidity chamber maintained at 65°C and 95% relative humidity for specified time periods, removed and then dried.

Wedge Test. The adhesive bond durabilities of the inhibitor-treated 7075-T6 surfaces were evaluated by wedge tests (ASTM D-3762) on bonded specimens using the FM 123-2 epoxy adhesive to simulate the epoxy primer. The specimens were placed in a humidity chamber at 65°C and 95% relative humidity and removed at specified time intervals to record the crack tip locations; after each examination, they were returned to the humidity chamber.

Electrochemistry. The electrochemical behavior of the treated specimens was analyzed by anodically polarizing the specimens with a PAR Model 273 potentiostat/galvanostat to assess the ability of the inhibitors to promote passivation in chloride-containing electrolytes. The Al surfaces were polished to a 4000 grit finish with SiC paper, degreased with a solvent, and washed in an alkaline detergent solution. Each sample to be polarized was then placed for 30 min in deaerated electrolytic solutions containing chloride (0.002 N KCl) as the aggressive ion species and 0.1 N Na_2SO_4 to minimize the impedance of the electrolyte. Inhibitors were added to the solution in concentrations known to provide the optimum surface coverage. (15) Each sample was scanned anodically from the equilibrium corrosion potential at a rate of 0.5 mV/s in order to determine the propensity of the alloy to pit in a specific electrolytic solution.

Results

Substrate Characterization. Venables et al. (7) have described the FPL oxide morphology using STEM in the SEM mode. Figure 1 is an isometric representation of the FPL surface. In contrast, the SAA process produces a much thicker oxide layer, isometrically represented in Fig. 2. Quantitative XPS analysis indicates that a constant

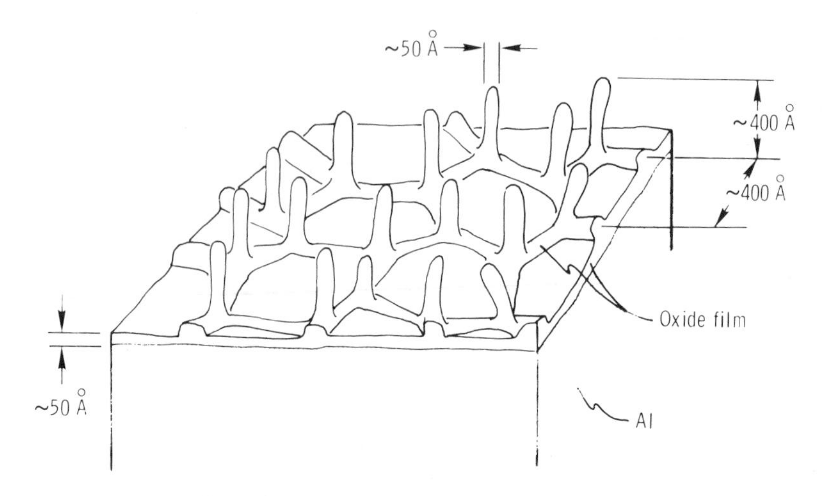

Figure 1. Structure of a Forest Products Laboratories (FPL) prepared aluminum surface (7).

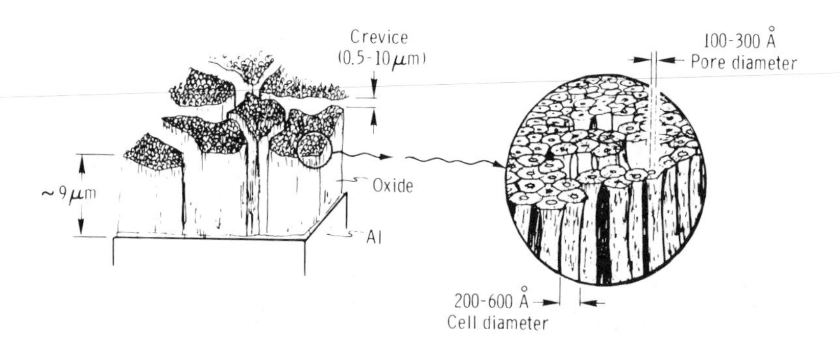

Figure 2. Structure of a sulfuric acid-anodized (SAA) aluminum surface.

composition (30% Al, 60% O, 3.5% S) is obtained after five minutes of anodization. The surface layer is $Al_2(SO_4)_3$ on top of the Al_2O_3 matrix.

Inhibitor Coverage. The adsorption behavior of phosphonate compounds, including those shown in Fig. 3, on FPL-prepared aluminum surfaces, has been reported. (11,12,14) The respective concentrations of P or Si and Al were measured by XPS. Figure 4 shows the correlation of P coverage on SAA surfaces following treatment in NTMP solution over a concentration range of 10 to 10^5 ppm. Using a constant immersion time of 30 minutes, the adsorption maximum observed was attributed to a multilayer buildup of the inhibitor compound in the pH 2-4 region. (15)

The relationship between the level of NTMP coverage and the solution pH has been reported previously. (15) Essentially, if the NTMP solution concentration is maintained at 100 ppm, an adsorption maximum is observed between pH 2-4.

The surface coverages for three organosilanes using different solution adsorption conditions, including type of solvent system, solution pH, and total immersion time, are shown in Table I. The results indicate that the extent to which these materials are adsorbed onto the metal surface depends primarily upon the aqueous composition and pH of the inhibitor solution. Both factors influence the hydrolysis of silicon alkoxy groups to silanols, which are the moieties that actually bond to the metal surface.

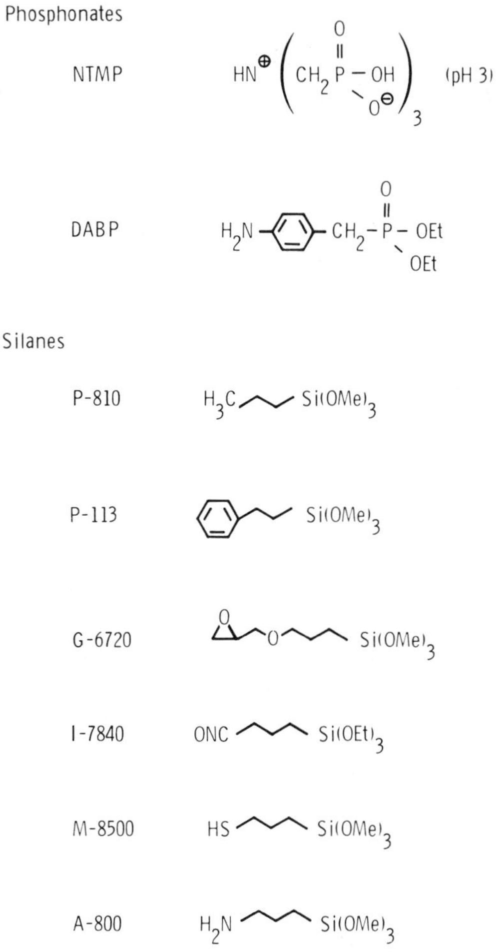

Phosphonates

NTMP $HN^{\oplus}(CH_2P(=O)(OH)O^{\ominus})_3$ (pH 3)

DABP $H_2N-C_6H_4-CH_2-P(=O)(OEt)OEt$

Silanes

P-810 $H_3C\sim Si(OMe)_3$

P-113 $C_6H_5\sim Si(OMe)_3$

G-6720 (epoxy)$\sim O \sim Si(OMe)_3$

I-7840 $ONC\sim Si(OEt)_3$

M-8500 $HS\sim Si(OMe)_3$

A-800 $H_2N\sim Si(OMe)_3$

Figure 3. Chemical structures of corrosion inhibitor compounds.

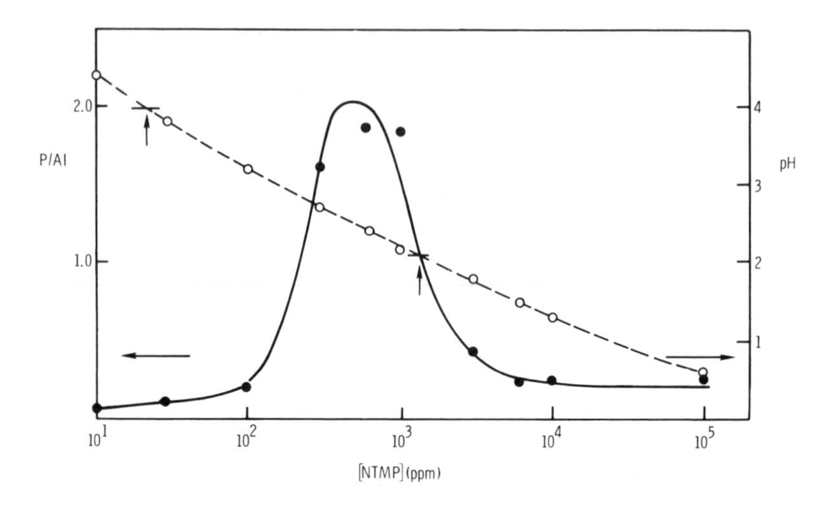

Figure 4. NTMP coverage as a function of concentration on SAA-prepared 7075-T6 aluminum surface.

TABLE I Silane Adsorption onto FPL 7075-T6 Surfaces

Silane	Concentration (ppm)	Solvent System	pH	Immersion Time	Si/Al
P-810	1000	W (water)	4	40 min	0.040
P-810	1000	W (water)	7	40 min	0.070
P-810	1000	M/W (methanol-water, 1:1)	4	40 min	0.076
P-810	1000	W	4	4 hr	0.100
G-6720	1000	W	4	40 min	0.148
G-6720	1000	W	7	40 min	0.083
G-6720	1000	M/W	4	40 min	0.127
G-6720	1000	W	4	4 hr	0.158
A-800	1000	W	4	40 min	0.364
A-800	1000	W	7	40 min	0.460
A-800	1000	M/W	4	40 min	0.330

Hydration Resistance. Visual examination of a series of inhibitor-treated FPL-prepared 7075-T6 Al coupons exposed to high humidity conditions for specified time intervals indicated good short-term hydration resistance for several phosphonate and silane compounds (no visible discoloration). The most effective silane compound tested contained the mercapto (-SH) functional group.

Adhesive Bond Durability

SAA Surface. The corrosion resistance and adhesive coupling capabilities of the inhibitors were evaluated by the wedge test. Results obtained using SAA-prepared surfaces (Fig. 5) indicated that two inhibitor treatments (100 ppm NTMP and 5000 ppm epoxysilane) provided systems which outperformed those treated with higher concentrations of NTMP and by FPL alone, but were apparently no more effective than the SAA control with respect to overall adhesive bond durability. However, XPS analysis (Fig. 6), in conjunction with SEM examination of the failed debonded sides, identified the true modes of failure. The SAA control (hydrated oxide on both sides under SEM; high Al and O levels on both sides) failed within the oxide. Examination of the specimen treated with multilayer-forming 5000 ppm NTMP solution (distinct "metal" and "adhesive" sides under SEM; high Al and O, low C levels on "metal" side; high C, low Al and O levels on "adhesive" side) indicated that the failure occurred between the metal and the adhesive (i.e., adhesive failure).

Although distinct "metal" and "adhesive" sides were apparent upon visual examination of the debonded surfaces treated with 100 ppm NTMP, SEM analysis showed the presence of an adhesive layer on the "metal" side. XPS analysis indicated low Al and O and identical high C levels on both debonded sides, confirming a failure within the adhesive layer (cohesive failure), i.e., the best possible performance in a given adherend-adhesive system. This result is similar to that obtained using a 2024 Al alloy prepared by the phosphoric acid-anodization (PAA) process (16) and indicates the importance of monolayer NTMP coverage for good bond durability (Fig. 4).

FPL Surface. A second wedge test was performed to evaluate six silanes and NTMP using the thinner, more sensitive FPL oxide (Fig. 7). The results indicated that four silanes performed better than the FPL control but not as well as NTMP; one silane (an aminopropyl derivative) performed very poorly with respect to the control; and one silane (M-8500, mercaptopropyl derivative) outperformed all of the other inhibitor systems, including NTMP and organosilanes containing methyl, phenyl, isocyanate, and epoxide side chains. Subsequent XPS analysis of the adsorbed mercaptosilane inhibitor showed high concentrations of Si and S near the surface, with corresponding low Al, the S being primarily in the reduced "inhibitor" (i.e., R-SH) form relative to sulfate ($-SO_4^{2-}$) which resulted from the FPL pretreatment.

Primer Epoxy vs. Nitrile-Modified Epoxy. The compatibility of the epoxy-polyamide primer with the nitrile-modified epoxy adhesive facsimile and the aluminum oxide surface was also evaluated by the wedge test, since earlier tests using the primer as the adhesive had failed immediately. As shown in Fig. 8, the addition of the primer directly to the prepared

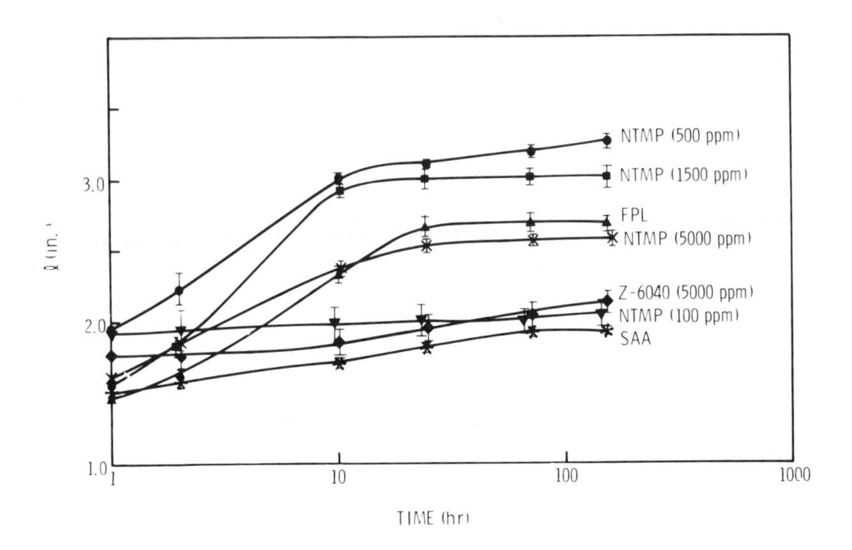

Figure 5. Wedge test results for inhibitor-treated SAA 7075-T6 aluminum specimens.

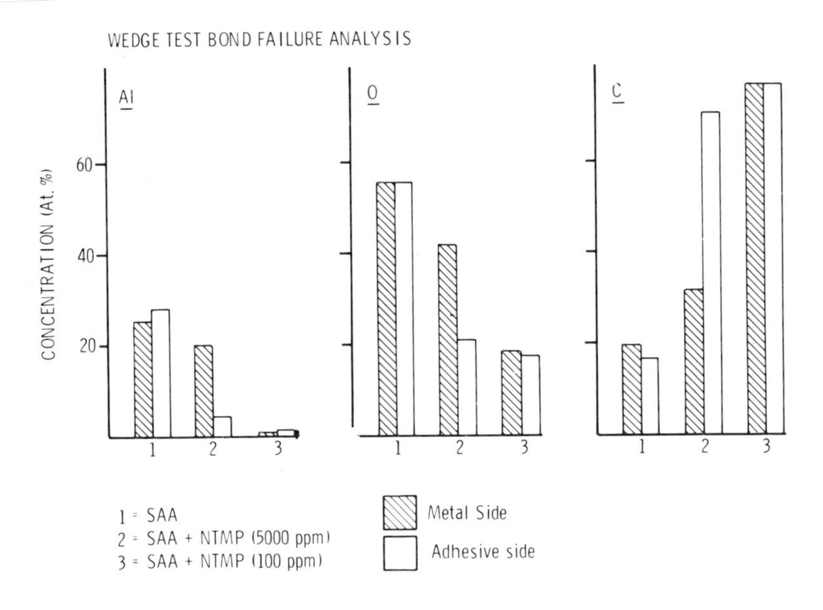

Figure 6. XPS surface analysis results for debonded SAA 7075-T6 aluminum specimens.

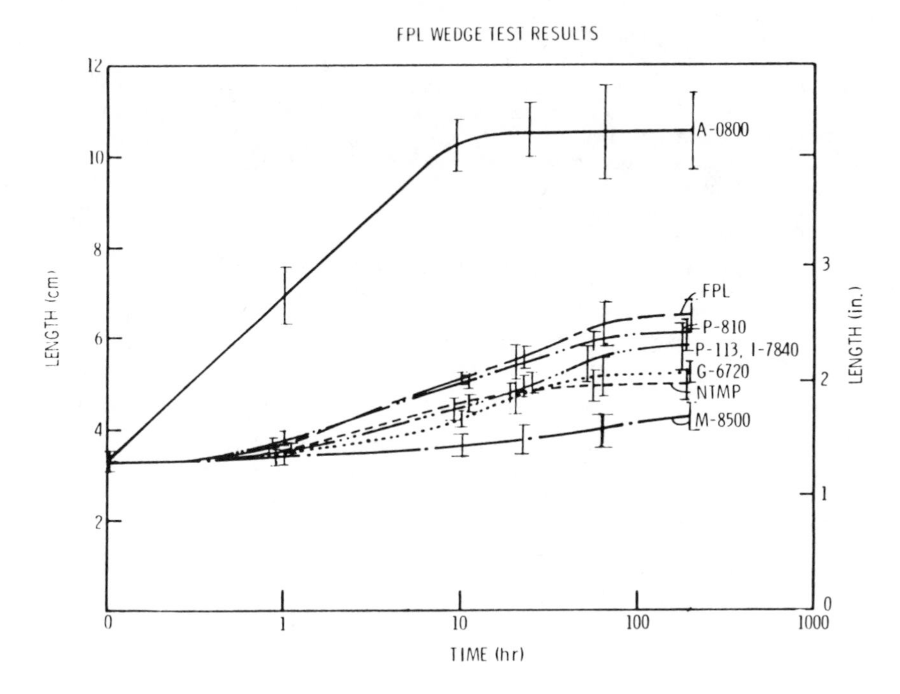

Figure 7. Wedge test results for inhibitor-treated FPL 7075-T6 aluminum specimens.

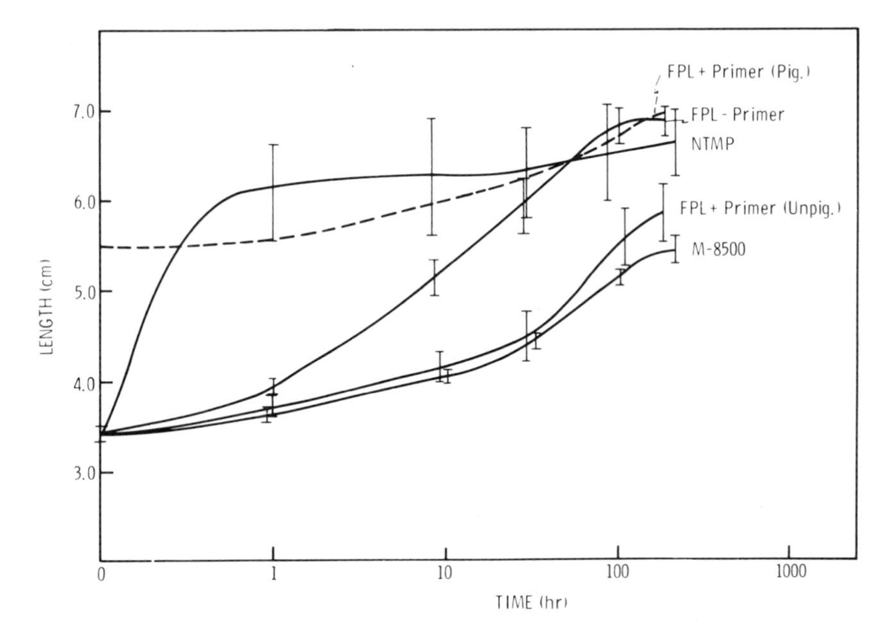

Figure 8. Wedge test results for primed FPL 7075-T6 specimens.

metal surface improved the overall durability of the aluminum-adhesive system. The application of the mercaptopropylsilane derivative by preadsorption onto the substrate surface effectively maintained or slightly enhanced the bond strength between the primer and the FPL oxide. In contrast, adsorbed NTMP, which preserves the integrity of the (nitrile-modified, high-temperature cured) adhesive-metal bond, failed to prevent rapid deterioration of the (epoxy-polyamide, room-temperature cured) primer-metal bond.

XPS analysis of the debonded specimens showed that the unprimed and primed FPL control and mercaptosilane-treated specimens failed primarily within the oxide, which represented the weakest layer in the system. On the other hand, the NTMP-treated sample debonded between the oxide and the polyamide primer.

Electrochemical Testing. Potentiodynamic polarization measurements provided a sensitive means of evaluating the inhibitors with respect to environmental (Cl^-) corrosion protection. The results obtained from anodically polarizing polished 7075-T6 Al samples are presented in Fig. 9. For the control electrolyte (0.1N Na_2SO_4, 0.002N KCl, no inhibitor), pitting was observed almost immediately on the surface, and the aluminum showed no evidence of passivation. The addition of NTMP to the solution did not appear to protect the metal

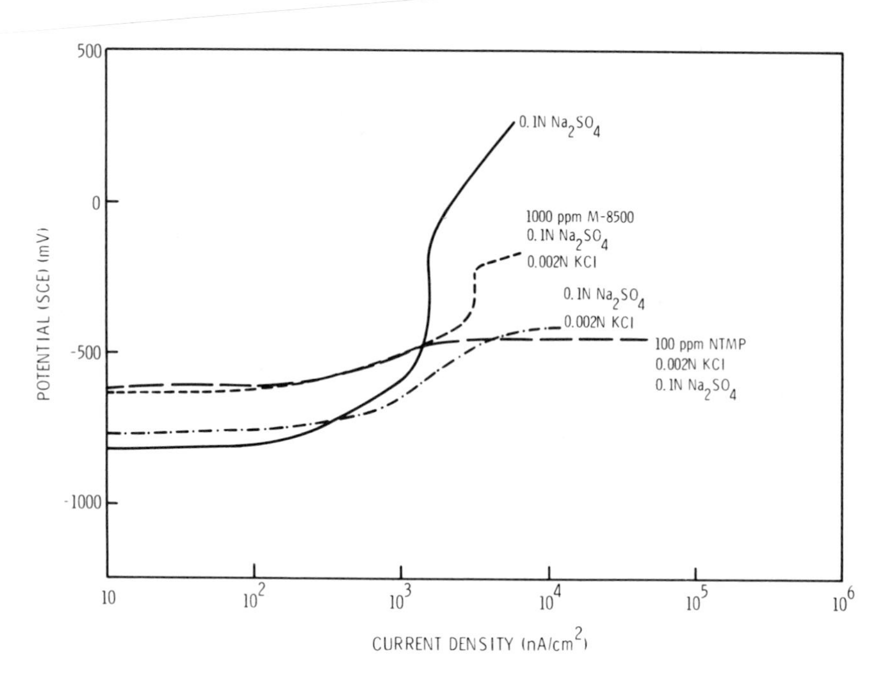

Figure 9. Electrochemical test results for two different inhibitors in 0.1N Na_2SO_4 electrolyte with added KCl.

surface, when compared to the solutions containing no inhibitors. In contrast, the sample polarized in the electrolyte solution containing 1000 ppm mercaptopropyl silane (M-8500) did passivate, and had a distinct pitting potential well above (more noble than) the potential where pitting initiates in NTMP-containing electrolytes. The Al was passivated up to -200 mV (SCE), and the alloy surface remained free from pits. Above this potential, small pits began to form.

Discussion

The corrosion resistance and polymer-bonding compatibilities of the ionizable organophosphonates and the neutral organosilanes are directly related to their inherent chemical properties. Specifically, NTMP inhibits the hydration of Al_2O_3 and maintains or improves bond durability with a nitrile-modified epoxy adhesive which is cured at an elevated temperature. The mercaptopropyl silane, in addition to these properties, is compatible with a room temperature-cured epoxy-polyamide primer and also exhibits resistance to localized environmental corrosion. These results, in conjunction with the adsorbed inhibitor films and the metal substrate surfaces, are subsequently discussed.

Surface Morphology. The initial integrity of an adhesively bonded system depends on the surface oxide porosity and microscopic roughness features resulting from etching or anodization pretreatments. (17) The SAA surface characterized in this study consists of a thick (9 μm), porous columnar layer which provides excellent corrosion resistance in both humid and aggressive (i.e., Cl^-) media. I The thinner FPL oxide (7) provides a suitable substrate surface for evaluating the candidate inhibitors.

Inhibitor Adsorption. There are some interesting differences between the adsorption of ionizable (aminophosphate) and neutral (organosilane) compounds onto FPL- and SAA-prepared aluminum surfaces. The acidic NTMP species, consisting of a quaternary $\equiv N^+H$ and unprotonated $-O^-$ groups in solution, (18) exhibits a pH-dependent multilayer adsorption maximum in an aqueous solution. At low solution concentrations ($\leq$ 100 ppm), the coverage is essentially monolayer, corresponding to a P/Al ratio of 0.15. (12) Wedge test results confirmed that this monolayer of NTMP leads to better long-term bond durability than multilayer films, which are probably formed by weak intermolecular hydrogen bonding and fail by a mixed-mode process.

On the other hand, alkoxysilane coupling agents produce structures which may consist of a fused network of polymeric five- or six-membered Si--O rings. (19,20) Although multiple layers of adsorbed silane (siloxane) films can be obtained, only the first few layers are bound to the metal by Si-O-Al bonds, through the silanol (hydroxyl) groups, and cannot be removed by aqueous rinsing. (21) The remaining film is available for general corrosion protection and possible coupling to the adhesive primer.

Bonding Mechanisms. The corrosion behavior of inhibitor-treated Al oxide systems provides clues to the mechanism of action for the phosphonates and silanes. In a purely hydrating environment, in which the Al_2O_3 progresses from the AlOOH (boehmite) to $Al(OH)_3$ (bayerite) stages, the hydrophilic NTMP provides short-term protection relative to untreated controls, probably by displacing H_2O in the oxide and forming a more stable complex. (14,22,23) Low levels of silanes applied to an aluminum substrate can likewise protect the surface from hydration (24), as we observed with the mercaptosilane compound. This may be associated with the coordination ability of the -SH groups, resulting, for example, in the formation of an oligometic $(S\text{-}S)_n$ film.

The compatibility of the quaternary amine-containing NTMP with the nitrile-modified epoxy adhesive, which leads to a bond that fails only within the adhesive (i.e., cohesive failure) can be explained. The FM 123-2 adhesive contains the storage-stable curing agent, dicyanodiamide, which does not release low-molecular weight amines until it becomes soluble in the resin above 90°C. (Equation 2) (25)

$$\underset{\text{solid}}{H_2N\text{-}\underset{\substack{|\\NH_2}}{C}\text{=N-CN}} \quad \underset{\text{solution}}{H_2N\text{-}\underset{\substack{\|\\NH}}{C}\text{-}\overset{\substack{H\\|}}{N}\text{-CN}} \xrightarrow[>90^\circ C]{} \text{LMW Amines} \qquad (2)$$

Tautomeric forms of dicyanodiamide

The curing sequence and kinetics of this adhesive system prevent the NTMP from inhibiting the reactive amines until the curing reaction is well under way. In contrast, the epoxy-polyamide primer contains free amino groups at room temperature and may be inhibited by the electrophilic NTMP species prior to curing. (Equation 3)

$$\text{Epoxide + Polyamine} \xrightarrow[25^\circ C]{\oplus NTMP} \text{Cured primer} \qquad (3)$$

A similar example of curing inhibition in an acidic medium has been observed when moderate amounts of salicylic acid were added to epoxy-amine matrix systems. (26) Such behavior may beresponsible for the poor bond strengths observed with the NTMP-containing oxide-primer specimens. In environments containing an aggressive species (e.g., Cl^-), the anion may interact with and become incorporated into the NTMP-oxide matrix, whereby it can attack the metal surface.

The hydration resistance of the organosilane compounds was reflected by the wedge test performances of our silane-treated systems. Electrochemistry results further indicated that the adsorbed silane film maintained the resistance in environments containing aggressive (Cl^-) species and oxidizing conditions. In contrast to NTMP, however, certain organosilanes are compatible and reactive with the epoxy polyamide primer as well as with the nitrile-modified adhesive, which, in most cases, strengthens the oxide-epoxy bond. This result is presumably due to a chemical coupling of functional epoxide or mercapto side chains on the organosilane with the epoxy coatings during the curing process.

The amine-catalyzed mercaptan-epoxide reaction (Equation 4) proceeds exothermally at room temperature (27, 28). The order of average relative nucleophile-displacement rates (Table II) further suggests that mercaptans react significantly faster than amines and that the addition of the mercaptide (RS^-) ion to the epoxide group is the rate determining step (30).

$$R\text{-}SH + \overset{O}{CH_2\text{-}CH}\text{-}CH_2OR' \xrightarrow{R''_3N} R\text{-}S\text{-}CH_2\text{-}\overset{OH}{CH}\text{-}CH_2OR' \qquad (4)$$

The mercaptan-epoxide coupling reaction indicates the true bi-functional nature of such inhibitors. The poor performance of the aminosilane compound [expected to strengthen the metal-adhesive bond (20)] is not completely understood at this time. Using

TABLE II Average Relative Nucleophile-Displacement Rates[a]

Nucleophile	Relative Rate
$C_4H_9S^-$	680,000
$C_6H_5S^-$	470,000
$S_2O_3^{2-}$	3,000
$C_2H_5O^-$	1,000
$C_6H_5O^-$	400
$(CH_3)_3N$	30
NO_3^-	1

[a] Data from Ref. (29).

similar concentrations, Walker (31) has shown that some silanes are effective adhesion promoters for urethane and epoxy paints on aluminum and mild steel surfaces and significantly improve the initial, wet, and recovered bond strengths.

Conclusions

The ionic phosphonates like NTMP are effective hydration inhibitors because they can form an insoluble complex with the oxide surface. They are useful as epoxy adhesive couplers in cases where the adhesive and its curing cycle are compatible with the adsorbed phosphonate molecule. (14) Wedge test results indicate that in two epoxy-aluminum systems studied, certain organosilanes tend to both increase the epoxy-metal bond durability and maintain hydration resistance. The results of anodic polarization experiments further suggest that these silane films are effective against localized pitting.

Adsorbed NTMP exhibits a pH-dependent surface coverage on anodized aluminum, which includes a region characterized by a multilayer of hydrogen-bonded phosphonate molecules. These thick layers are weak and fail to provide good bond durability in a humid environment. NTMP monolayers are protective against hydration and are compatible with a nitrile-modified epoxy adhesive, but not with an epoxy-polyamide primer topcoat.

In contrast, hydrolyzed silane compounds, presumably adsorbed as oligomeric films, confer corrosion resistance in both hydrating and Cl^- environments. These inhibitors can also couple with applied epoxy primer or adhesive formulations to further protect the metal against corrosion by strengthening the metal-epoxide bond. The organosilanes do not appear to affect the curing process, e.g., % crosslinking, of the polymeric epoxy systems.

Acknowledgments

The authors wish to thank Drs. John D. Venables and John S. Ahearn for their valuable discussions in conjunction with this work. The support of the Naval Air Development Center, which sponsored these studies under contract N00019-82-C-0439, is also gratefully acknowledged.

Literature Cited

1. H.W. Eichner and W.E. Schowalter, Forest Products Laboratory, Madison, WI, Report No. 1813 (1950).
2. G.S. Kabayashi and D.J. Donnelly, Boeing Co., Seattle, WA, Report No. DG-41517 (Feb. 1974).
3. Fokker-VFW, Amsterdam, Process Specification TH 6.785 (August, 1978).
4. W.M. McCracken and R.E. Sanders, SAMPE J., 5, 37 (1969).
5. J.C. McMillian, J.T. Quinlivan, and R.A. Davis, SAMPE, 7, 13 (1976).
6. J.M. Chen, T.S. Sun, J.D. Venables, and R. Hopping, Proc. 22nd National SAMPE Symposium, 25 (April 1977).

7. J.D. Venables, D.K. McNamara, J.M. Chen, and T.S. Sun, Appl. Surf. Sci., 3, 88 (1979).
8. W. Brockman in Adhesion Aspects of Polymeric Coatings, K.L. Mittal, ed., 265 (Plenum, New York, 1983).
9. J.D. Venables, D.K. McNamara, J.M. Chen, B.M. Ditchek, T.I. Morgenthaler, T.S. Sun, and R.L. Hopping, Proc. 12th Nat. SAMPE Symposium, Seattle, 909 (October 1980).
10. G.D. Davis and J.D. Venables in Durability of Structural Adhesives, A.J. Kinloch, ed., 43 (Applied Science, Essex, 1983).
11. D.A. Hardwick, J.S. Ahearn, and J.D. Venables, J. Mater. Sci., 19, 223 (1984).
12. J.S. Ahearn, G.D. Davis, T.S. Sun, and J.D. Venables, op. cit. Ref. 8, 281.
13. J.D. Venables, M.E. Tadros, and B.M. Ditchek, U.S. Pat. 4,308,079 (1981).
14. G.D. Davis, J.S. Ahearn, L.J. Matienzo, and J.D. Venables (accepted by J. Mater. Sci.).
15. Matienzo, L. J.; Shaffer, D. K.; Moshier, W. C.; Davis, G. D., J. Mater. Sci. 21, 1601, 1986.
16. D.A. Hardwick, J.S. Ahearn, and J.D. Venables, J. Mater. Sci. (in press).
17. J.D. Venables, J. Mater. Sci., 19, 2431 (1984).
18. R.P. Carter, M.M. Crutchfield, and R.R. Irani, J. Inorg. Chem., 6, 943 (1967).
19. K.W. Allen, A.K. Hansrani, and W.C. Wake, J. Adhes., 12, 199 (1981).
20. E.P. Plueddemann, Proc. 24th Ann. Tech. Conf., Reinf. Plastics/Composites Div., SPI, Sec. 19-A (1969).
21. K.W. Allen and M.G. Stevens, J. Adhes., 14, 137 (1982).
22. F.I. Belskii, I.B. Goryunova, P.V. Petrovskii, T.Y. Medved, and M.I. Kabachnik, Academia Nauk USSR, 1, 103 (1982).
23. J.C. Bolger, op. cit. Ref. 8, 8.
24. F.J. Boerio, R.G. Dillingham, and R.C. Bozian, 39th Annual Conference, Reinforced Plastics/Composites Institute, The Society of the Plastics Industry, Inc., Session 4-A, 1 (January 1984).
25. A.F. Lewis and R. Saxon in Epoxy Resins: Chemistry and Technology. C.A. May and Y. Tanaka, eds., 413 (Marcel Dekker, New York, 1973).
26. A.M. Ibrahim and J.C. Seferis, Polym. Compos., 6, 47 (1985).
27. K.R. Kramker and A.J. Breslau, Ind. Eng. Chem., 47, 98 (1956).
28. E.H. Sorg and C.A. McBurney, Mod. Plastics, 34, 187 (1956).
29. C.A. Stretweiser, Chem. Rev., 56, 571 (1956).
30. Y. Tanaka and T.F. Mika, op. cit. Ref. 25, 169.
31. P. Walker, J. Oil Colour Chem. Assoc., 65, 436 (1982).

RECEIVED January 27, 1986

23

Inhibition of Copper Corrosion by Azole Compounds in Acidic Aqueous Solutions

R. Johnson[1,2], M. Daroux[3], E. Yeager[3], and Hatsuo Ishida[1,4]

[1]Macromolecular Science Department, Case Western Reserve University, Cleveland, OH 44106

[2]Department of Chemistry, Case Western Reserve University, Cleveland, OH 44106

Azole compounds, poly-N-vinylimidazole (PVI-1) and 2-undecylimidazole (UDI), are studied as alternative inhibitors to benzotriazole (BTA) for copper corrosion in aqueous systems using electrochemical techniques. It is shown that UDI, either as a cast film or dissolved in solution at concentrations as low as 7 x 10^{-5}M, inhibits oxygen reduction on copper in acidic solutions of 0.1M $HClO_4$ (pH=1) while BTA and PVI-1 do not. In phosphate buffer (pH=5.6), UDI and BTA inhibit oxygen reduction on copper, but PVI-1 does not. All three compounds suppress the anodic dissolution of copper at pH=1. At pH=5.6, both UDI and PVI-1 cast films produce passivation, UDI reducing the anodic currents by an order of magnitude more than PVI-1. Linear polarization measurements on Cu in UDI solution (pH=5.6) indicate that UDI may have practical application as a corrosion inhibitor for copper.

In the past, heterocycles containing nitrogen have been examined for use as inhibitors for copper (Cu) corrosion [1-25]. To date benzotriazole (BTA), figure 1A, has been the most successful compound and is widely used in industry. Recently there has been interest, especially in the printed circuit board industry [26], in alternative inhibitors to BTA that not only inhibit corrosion but also promote adhesion. Imidazoles such as poly-N-vinylimidazole (PVI-1), figure 1B and 2-undecylimidazole (UDI), figure 1C are potential candidates because imidazoles are commonly used as curing agents for epoxy resins and some are reported to inhibit Cu corrosion [26,27].

BTA inhibits Cu corrosion in many aqueous environments [1-22]. The partial anodic or cathodic process or processes that are inhibited by BTA is controversial; however, most researchers

[3]Current address: Automotive Research Center, Inmont, 26701 Telegraph Road, Southfield, MI 48086-5009

[4]To whom correspondence should be addressed

0097-6156/86/0322-0250$06.00/0

agree that the mechanism by which BTA acts involves reinforcement of the oxide film. This mechanism is supported by the poor inhibition of BTA observed at pH<3 [4,6,13,19], where copper oxides are not thermodynamically stable [28]. Acidic conditions, however, are often met in practice; for example in atmospheric corrosion where evaporation concentration can occur and in treating baths for processing. Inhibitors which are effective at pH<3 would therefore be valuable.

PVI-1 and UDI have been investigated as anti-oxidation agents for Cu in dry air at elevated temperatures [29-30]. Fourier transform infrared reflection-absorption spectroscopy (FT-IR RAS) was used to monitor the onset and intensity of the Cu_2O band as a measure of the surface oxidation of copper substrates with 150nm thick films of either BTA, UDI, or PVI-1 cast onto the surface. The polymer was found to be superior to either of the small molecules. PVI-1 protected the copper from oxidation up to 400^oC while UDI and BTA lost their protective action after short exposure at 150^oC [30].

It has not been reported whether UDI or PVI-1 inhibit Cu corrosion in aqueous environments. In the present work the electrochemical behavior of cast films of BTA, PVI-1, and UDI on Cu has been studied in acidic solutions. The compounds were studied primarily as cast films to correlate with the above research on PVI-1 and UDI. UDI, which was found to be the most promising inhibitor was also studied as a solution species. This paper deals with the overall effects and compares the behavior of the different compounds.

Experimental

Mass transport and the electrode kinetics of copper electrodes were studied using the rotating disk electrode (RDE) technique in conjunction with potential sweep and steady-state potentiostatic measurements. The RDE was made by press-fitting a spectro-grade Cu rod, area $0.2cm^2$, into a cylindrical poly(tetrafluoroethylene) (PTFE) holder. Before testing, the RDE was polished with 1um diamond paste, rinsed with ethanol, and cleaned in distilled water in an ultrasonic bath. Cast films were made by placing calculated amounts of UDI, PVI-1, or BTA methanol solutions (approximately 1% by weight.) dropwise by a microsyringe onto the electrode surface. Each drop wetted the entire electrode and was allowed to dry slightly before the next drop was applied. When the last drop was deposited the film was allowed to dry for at least 5 minutes before testing. The resultant films were nonuniform. Unless otherwise stated, the thickness of the films was either 15um or 150nm. The thickness was calculated from the surface area of the electrode and the concentration of the inhibitor solution by assuming a film density of $1g/cm^3$.

The experiments were conducted in a one-compartment cell with the reference electrode separated from the main compartment by a Luggin capillary and a closed electrolyte-wetted stopcock. The counter electrode was a gold wire loop and the reference electrode was a saturated calomel electrode. The potential of the RDE was controlled by a Princeton Applied Research Corporation (PARC) model 176 potentiostat. The potentiostat was coupled to a PARC model 175 universal programmer to continuously vary the potential at a constant rate. The electrode was rotated at controlled frequencies by a Pine Instruments ASR2 analytical rotator.

The potential sweep method used was cyclic voltammetry. Unless otherwise stated, the cycles were begun at -100mV (SCE) and cycled between -750 and +150mV (SCE) in 0.1M $HClO_4$ solution and between -1100 and +250mV (SCE) in phosphate buffer, pH=5.6. The sweep rate was 50mV/s in the perchloric acid solution and 20mV/s in the phosphate buffer solution. All measurements were made at room temperature and the electrode was rotated at 15Hz. In the perchloric acid solution, the steady-state potentiostatic measurements were made by sweeping the potential at 50mV/s between -100 and -600mV (SCE) three times then sweeping at 5mV/s and holding the potential constant at certain intervals until the current remained constant. These results were used to calculate the Tafel slopes and exchange current densities. In the phosphate buffer, these measurements were made at 20mV/s in order to obtain reproducible results. Linear polarization measurements were made by equilibrating the Cu electrode without stirring until a steady-state open-circuit potential was reached (after 4-8 hours). The polarization curve was obtained by stepping from the open-circuit potential to random potentials within 10mV. The current was taken after 3 minutes. This current was found to be close to the steady-state value. Between each potential step, the electrode was allowed to rest at open-circuit for 3 minutes before the next potential step was applied.

The water used to prepare solutions was purified by means of a commercial reverse osmosis system followed by distillation with a counter current of nitrogen to remove volatiles. The method is a modified version of a system reported by Gilmont and Silvus [31]. The nitrogen used to deoxygenate the electrochemical cell was purified by passing it through a solution of Cr(II) and amalgamated Zn (to continously regenerate Cr(II)), then washing with distilled water. The oxygen gas used to saturate the electrochemical cell with O_2 was purified by passing it through a purification train similar to that reported by [B Amadelli [32]. The train contained traps of Hopcalyte (Mine Safety Appliance, 14-20 mesh), an active mixed oxide catalyst that oxidizes CO to CO_2, and NaOH on asbestos which removes the CO_2.

The BTA (Aldrich) and UDI (Shikoku Chemicals) were recrystallized before use and the PVI-1 was synthesized using the method reported by Eng [30]. The phosphate buffer was prepared by mixing 13.8g analytical grade monobasic sodium phosphate (Fisher Scientific) and 6.7g analytical grade dibasic sodium phosphate (Fisher Scientific) into purified water, then titrating with 0.2M analytical grade H_3PO_4 (Fisher Scientific) to pH=5.6. The reagent grade perchloric acid (Fisher Scientific) was diluted to a 0.1M aqueous solution.

Results and Discussion

Cyclic voltammetry was employed to determine the nature of the anodic and cathodic reactions of bare and coated copper in aerated 0.1M $HClO_4$ (pH=1) and phosphate buffer (pH=5.6) solutions. Steady-state measurements were made in air to obtain a more quantitative picture of the effect of the inhibitor films on the oxygen reduction reaction.

Cyclic voltammetry The steady-state cyclic voltammogram of Cu in 0.1M $HClO_4$, shown in figure 2A, shows a steep rise in the anodic current beginning at -60mV (SCE). According to the Pourbaix diagram [28] of Cu in aqueous solutions, the thermodynamically stable oxidation product under these conditions is Cu^{2+}. No copper oxides are stable at this pH at any potential. Etching of the electrode surface was observed upon removal from the electrolyte and no visible coatings were formed, therefore, this anodic process is attributed to copper dissolution:

$$Cu = Cu^{2+} + 2e^- . \qquad (1)$$

Two cathodic processes can also be observed in the voltammogram. One is characterized by a current plateau between -600 and -750mV (SCE) and the other by a sharp increase of the current in the cathodic direction at potentials more negative than -850mV (SCE). The former process is expected to be oxygen reduction:

$$O_2 + 4H^+ + 4e^- = 2H_2O \qquad (2)$$

$$\text{or } O_2 + 2H^+ + 2e^- = H_2O_2 \qquad (3)$$

which becomes mass transport limited at potentials more negative than -500mV (SCE). This was confirmed by performing experiments in deaerated solution. The cyclic-voltammogram in deaerated solution in figure 4 shows no O_2 reduction current plateau. This wave, however, reappeared upon reintroduction of air. This suggests that the second cathodic process can be assigned to hydrogen evolution:

$$2H^+ + 2e^- = H_2\ (g) \qquad (4)$$

The predominant cathodic process for Cu corrosion will therefore be oxygen reduction in aerated solutions and hydrogen evolution in deaerated solutions.

The steady-state voltammogram of UDI-coated Cu in 0.1M $HClO_4$ (figure 2D) is quite featureless, lying close to the baseline until -300mV (SCE). The initial sweeps on UDI-coated Cu show much larger currents which decay to the steady-state curve only after cycling (figure 3). The steady-state condition was achieved after approximately 25 cycles (50mV/s). Similar results were observed upon testing bare Cu electrodes in solutions of dissolved UDI. The currents for the oxygen reduction reaction decreased to low steady-state values, as seen above, after approximately 7 cycles (5mV/s). The suppression of the oxygen reduction reaction on Cu in UDI solutions at steady-state was evident for concentrations down to 7×10^{-5}M, but not at 7×10^{-6}M. This indicates that there is a threshold concentration for this effect between these

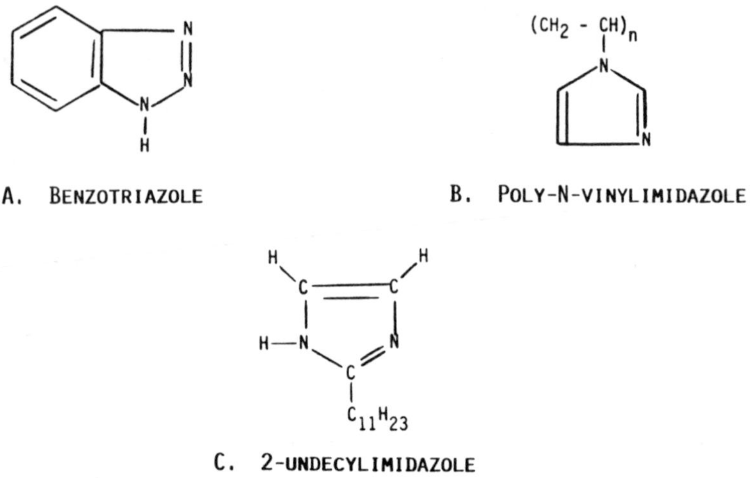

A. BENZOTRIAZOLE
B. POLY-N-VINYLIMIDAZOLE
C. 2-UNDECYLIMIDAZOLE

Figure 1. Chemical structures of potential inhibitors.

i(mA/cm2)
0.4
0
-0.4
-0.8
-1.2
-0.8
-0.6
-0.4
-0.2
0
0.2
E(V) vs. SCE
A ----- Cu
B -·-·- BTA
C ········ PVI-I
D ——— UDI

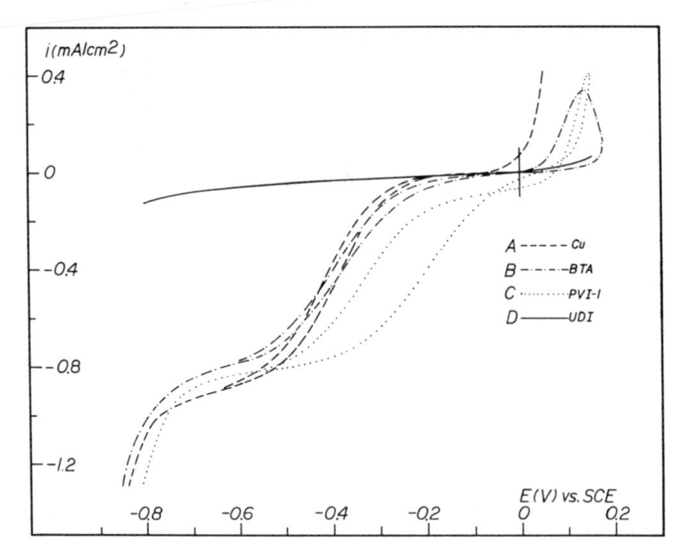

Figure 2. Steady-state cyclic voltammograms of Cu electrodes, both bare and coated with thick films (approximately 15 μm) of BTA, PVI-1, or UDI in 0.1 M $HClO_4$ in air, v = 50 mV/s, w = 15 Hz, 25 °C.

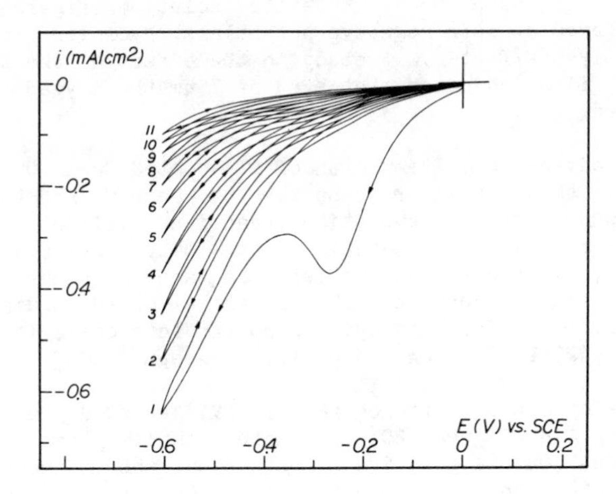

Figure 3. Current decay of a UDI-coated Cu electrode in 0.1 M $HClO_4$ in air, v = 50 mV/s, w = 15 Hz, 25 °C.

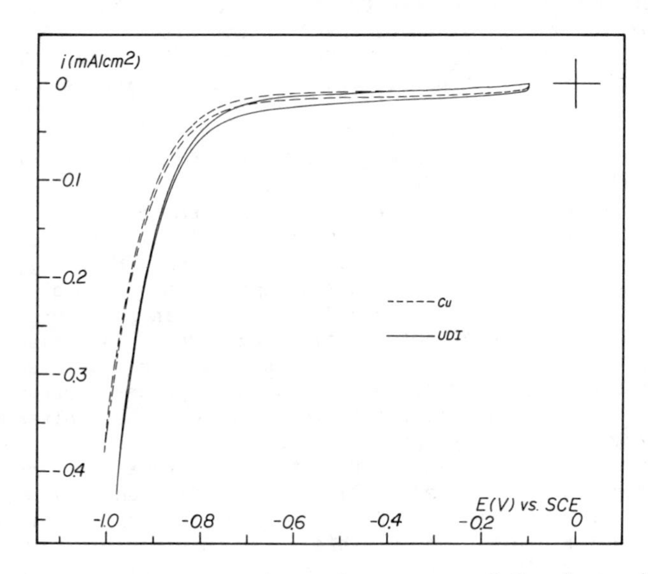

Figure 4. Steady-state cyclic voltammogram of Cu electrodes, bare and coated with UDI film (approximately 15 μm) in deaerated 0.1 M $HClO_4$, v = 5 mV/s, w = 15 Hz, 25 °C.

concentrations. The voltammogram of UDI-coated Cu is the same in both aerated and dearated solutions and is identical to the voltammogram of bare Cu in dearated solution (figure 4). The process observed at more negative potentials, hydrogen evolution, is apparently unaffected. A study on the effect of the thickness of UDI cast films indicated that even at 75nm the oxygen reduction was inhibited.

The steady-state voltammogram of BTA-coated Cu in 0.1M $HClO_4$ (figure 2B) shows similar behavior to bare Cu for the oxygen reduction and hydrogen evolution reactions (slightly higher currents for oxygen reduction at lower potentials), but significantly different behavior for the anodic reaction. The current is nearly zero at the potential where Cu dissolution becomes significant for bare Cu, and no anodic process is observed below about +125mV (SCE) on the positive sweep.

The steady-state voltammogram of PVI-1-coated Cu in 0.1M $HClO_4$, given in figure 2C, is more complex than that for BTA-coated Cu. On the positive sweep the current becomes anodic at +45mV (SCE) and as with BTA-coated Cu, the Cu oxidation is inhibited, but to a lesser extent. The initial cathodic currents are enhanced in comparison to bare Cu and BTA-coated Cu, but the limiting oxygen reduction current is close to that for bare Cu.

Upon removing the coated Cu electrodes from the electrolyte after treatment, it was observed that PVI-1 and BTA films dissolved in 0.1M $HClO_4$ while UDI films did not. The UDI, however, showed similar anodic and cathodic behavior as a solution species.

The steady-state voltammogram of bare Cu in the phosphate buffer, pH=5.6, (figure 5A) is very similar to that in 0.1M $HClO_4$ at the negative potentials. An anodic process begins at -100mV (SCE) and a current plateau is observed indicating that a process such as film formation becomes current limiting. However, on the reverse sweep the currents remain high indicating that any film that forms is either immediately dissolved or in poor contact with the Cu electrode and not passivating [28]. At this pH, the formation of Cu oxides and/or hydroxides is expected on a thermodynamic basis [28], but no reduction process corresponding to the reduction of anodic films is seen on the voltammogram. Upon sweeping to very positive potentials (figure 6), a white powder could been seen falling from the electrode and reduction peaks appeared at -170mV (SCE), -215mV (SCE), and -410mV (SCE); presumably arising from the formation and reduction of oxides and/or hydroxides. The visible loss of material from the electrode along with the fact that no visible film was observed after testing supports the argument that these materials do not form adherent films under these conditions.

The steady-state voltammogram of UDI-coated Cu in the phosphate buffer (figure 5C) is similar to the result in pH=1 solution; the curve is indistinguishable from the baseline until approximately -300mV (SCE). In the steady-state a UDI film

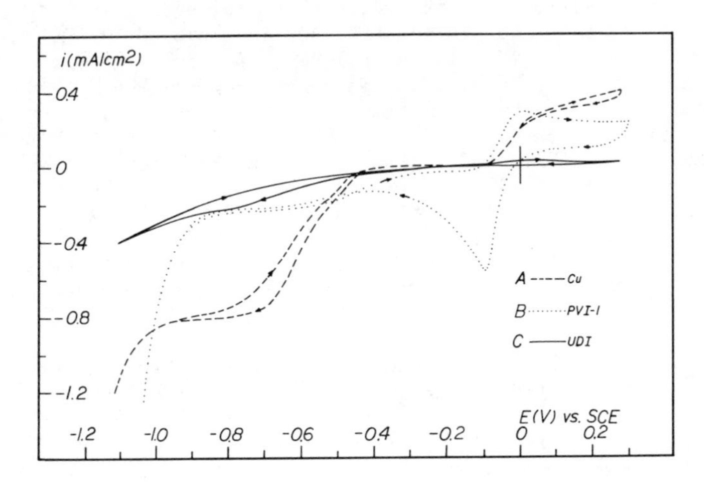

Figure 5. Steady-state cyclic voltammogram of BTA, PVI-1, or UDI in phosphate buffer, pH = 5.6, in air, v = 20 mV/s, w = 15 Hz, 25 °C.

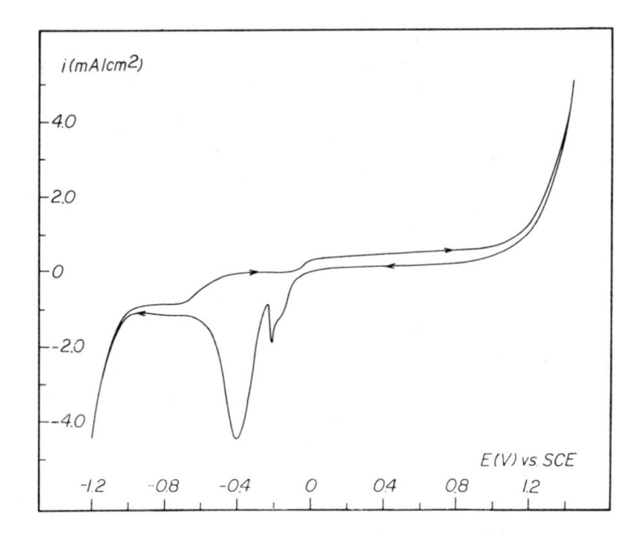

Figure 6. Cyclic voltammograms of bare Cu electrode in phosphate buffer, pH = 5.6, in air, v = 20 mV/s, w = 15 Hz, 25 °C.

strongly inhibits the oxygen reduction reaction. UDI, at this pH strongly inhibits the anodic process, at least up to +300mV (SCE) causing passivation of the electrode surface. The hydrogen evolution appears to be unaffected in both 0.1M $HClO_4$ and phosphate buffer, pH=5.6, solutions. A 150nm film., cycled between -300 and -1100mV (SCE), did not suppress the O_2 reduction currents as well as the 15um film, but the steady-state currents were still much lower than those for bare Cu. A saturated solution of UDI (UDI powder observed in the solution) showed the same effect as a 15um film on the O_2 reduction and anodic oxidation of copper. These films were found to dissolve in the phosphate buffer, unlike UDI films in 0.1M $HClO_4$, indicating that the effect of film thickness merely reflects the effect of differing concentrations of UDI in the solution produced by the dissolution of the films.

A 150nm BTA film cast onto Cu and cycled between -300 and -110mV (SCE) initially inhibited the O_2 reduction on Cu, but within 30 minutes the current had increased to those currents observed for untreated Cu. A 1mM solution of BTA showed similar behavior. A 0.01M solution, however, exhibited a lasting effect. These results are consistent with those found by McCrory-Joy et. al. [8], in an acetate buffer of pH=6.

They found that a Cu electrode, pretreated by immersing it in a 0.1M BTA solution for 15 seconds, inhibited the O_2 reduction reaction initially and that on subsequent cycles the currents increased to that of bare Cu in a short time. A similar effect was observed when a Cu electrode was cycled in a 1mM solution of BTA. They discovered that a solution of 0.1M BTA produced a lasting effect, indicating that a reservoir of BTA is necessary for continuous protection of the copper against corrosion. We found that bare Cu gives the same voltammogram in the O_2 reduction region in both acetate buffer and phosphate buffer; therefore, McCrory-Joy et. al.'s results can be directly compared to the results reported here.

The steady-state voltammogram of PVI-1-coated Cu in the phosphate buffer is given in figure 5B. The initial oxygen reduction currents on the PVI-1-coated Cu are higher than for bare Cu, but the limiting current is less by approximately 50%. This result is consistent with blocking of transport by the polymer film. A 150nm film shows the same limiting current for O_2 reduction as bare, supporting this hypothesis. The positive sweep for both 15um and 150nm films shows a current plateau commencing at approximately -100mV (SCE) and passivation is observed, although the anodic currents are at least ten times those seen with UDI. On the reverse sweep a corresponding reduction peak appears at -100mV (SCE) which does not coincide with the Cu oxide/hydroxide reduction peaks (figure 6). The charge involved in the oxidation process for the 15um film was calculated to be 6.02mC while that of the reduction process was 4.24mC. This suggests that 20% of the anodic products was not reduced and that material is lost on each cycle by dissolution. It is expected, therefore, that corrosion occurs over time.

The above results show that BTA, PVI-1 and UDI behave differently in the two pH solutions tested in this study. The differences are probably related to the form of the inhibitor, for example BTA takes on a cationic, anionic, or neutral form depending on the pH [33]. This may affect the solubility of the cast films, adsorption behavior, reactions of the inhibitor at the electrode, and the solubility of any subsequent reaction products. The BTA molecule (pK_b=0.44, pK_a=8.2) is soluble in all pH solutions in all forms [33]. The neutral form is predominant in both 0.1M $HClO_4$ and phosphate buffer (pH=5.6), but there are substantial amounts of the cation present at pH=1 and small amounts of the anion present at pH=5.6. The PVI-1 molecule (pK_a=4.7) is predominantly a soluble cation in pH=1 [34] and a neutral species at pH=5.6. The pK_a nor the pK_b are known for the UDI molecule. The solubility behavior of UDI indicates that it may be more acidic than BTA.

All of the inhibitor compounds affected the the Cu oxidation processes in both 0.1M $HClO_4$ and phosphate buffer. The type of effect depended on the pH and the magnitude depended on the inhibitor. In 0.1M $HClO_4$ the potential of observed Cu dissolution is shifted in a positive direction. Quantitative measurements were not made but the inhibitors can be ranked in order of decreasing potential shift as follows:

UDI > BTA > PVI-1.

Once dissolution commenced the anodic currents continued to increase on the reverse sweep, resulting in the hysteresis loops observed for BTA and PVI-1 in figure 2B,C. A possible explanation of the loss of the inhibiting condition is that an adsorbed layer of the inhibitor is in equilibrium with a dissolved species in solution. When dissolution occurs the resulting Cu ions consume the free inhibitor by forming Cu complexes. This depletes the concentration of the solution species near the electrode and the inhibiting layer is desorbed. On cycling to negative potentials the adsorbed layer is re-established. This is consistent with the observation that inhibition occurs on each subsequent cycle.

In phosphate buffer (pH=5.6), passivation is observed rather than activation. The effect is again believed to be caused by an adsorbed layer of inhibitor and not by the bulk cast film because UDI, only present as a solution species at this pH, shows the most pronounced effect. The position of the reduction peak seen in the case of PVI-1 (figure 5C) suggests that the reaction is not simply the reduction of Cu oxides or hydroxides but involves the reaction of the inhibitor. The resulting film is more than a monolayer because the charge consumed is of the order of millicoulombs per square centimeter. The anodic passivating species is probably a copper-inhibitor compound formed by an anodic reaction of Cu with the solution inhibitor species. At this pH the final reaction product may have oxide or hydroxide incorporated as ligands. The results in figure 5 and those in the literature [8] suggest the following ranking of the studied inhibitors in order of decreasing passivation:

UDI > BTA > PVI-1

The different degrees of passivation may be related to the solubility of Cu(inhibitor) complexes. Cu is known to readily form complexes with BTA, PVI-1, and UDI [3, 29, 30]. The charge balance for PVI-1 indicates that the passivating layer, at least for PVI-1, is partially soluble. It is known that Cu-PVI-1 and Cu-UDI complexes are soluble in 0.3M HCl [29,30]. This may explain why no such passivation is observed at pH=1, although the absence of copper oxide formation processes is also relevant.

Steady-state Measurements. In order to obtain a more quantitative understanding of the process resulting in the inhibition of the O_2 reduction reaction, Tafel measurements were made. The mass transport corrected Tafel equation [35]:

$$E = a[\log i_o - \log (i_L \times i/ i_L - i)] \quad (5)$$

where E = the measured potential, a = the Tafel slope, i_o = the exchange current density, i_L = the mass transport limiting current, and i = the measured current density) was utilized, where applicable, to calculate the Tafel slopes and exchange current densities of the oxygen reduction reaction on bare and coated Cu electrodes. The uncorrected Tafel equation [35] was used when no mass transport limitation was observed; UDI in both solutions and BTA in phosphate buffer. The i_L values for each sample were obtained by extrapolating from plots of the Levich relation:

$$1/i = (1/i_k) + 1/BCw^{1/2} \quad (6)$$

(where i = measured current density, i_k = kinetic component of the total current density, B = constant, C = bulk concentration, and w = rotation rate of the electrode). A typical Levich plot obtained in this study is represented by the bare Cu sample given in figure 7. The Tafel plots, using the corresponding limiting currents listed in Table I, are shown for measurements in 0.1M $HClO_4$ in figure 8 and for measurements in phosphate buffer in figure 9. The observed mass transport limited current densities in air and in oxygen saturated solutions (pH=1) for bare Cu are half those reported for gold (Au) in 0.1M $HClO_4$ [36], indicating that oxygen reduction on Cu follows a mechanism that involves twice the number of electrons than Au. Oxygen reduction on Au is a 2 electron process in perchloric acid [36]; therefore, the reaction on Cu must be a 4 electron process (equation (3)).

At pH=1, the Tafel slopes (Table I) for bare Cu and Cu coated with PVI-1 or BTA are the same within experimental error. PVI-1 and BTA do not appear to affect the mechanism of O_2 reduction on Cu. These results are supported by Heakal and Haruyama [13] who also found the Tafel slope of Cu with BTA present in a pH=3 solution of HNO_3 and 3% NaCl to be -0.18V/dec. These authors, however, do not cite a value for bare Cu. UDI-coated Cu shows a significantly different Tafel slope, -320mV/dec, indicating that a different mechanism applies.

In the phosphate buffer, pH=5.6, Tafel measurements were made for bare Cu, Cu coated with a 150nm cast film of PVI-1, for Cu

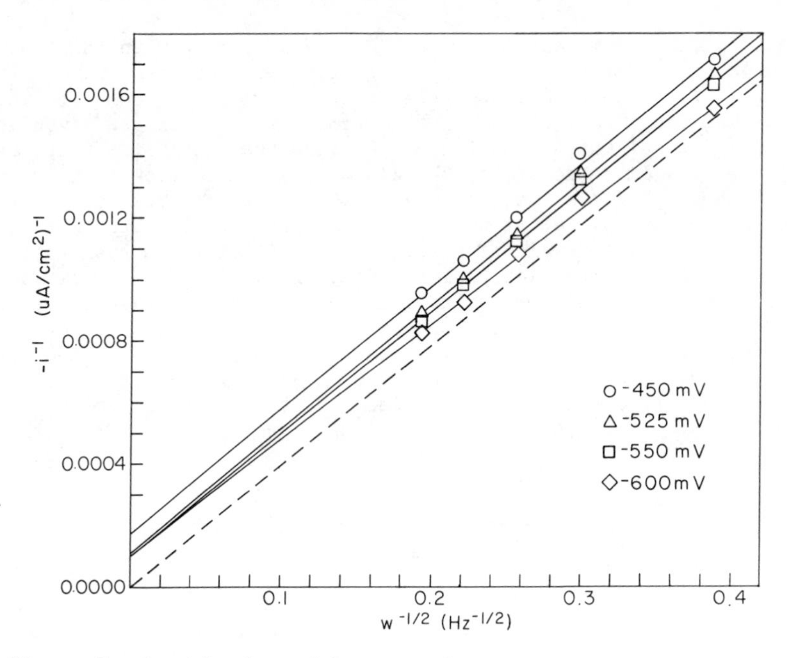

Figure 7. Levich plot of bare Cu electrode to obtain through i_L values in 0.1 M $HClO_4$ in air, v = 5 mV/s, 25 °C.

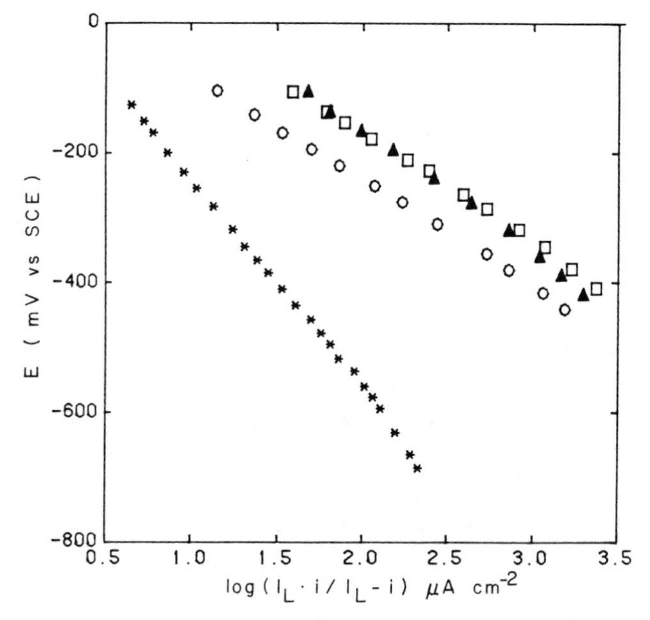

Figure 8. Tafel plots of oxygen reduction on Cu electrodes in 0.1 M $HClO_4$ in air, w - 15 Hz, 25 °C. □ bare Cu, o BTA-coated Cu, ▲ PVI-1-coated Cu, * UDI-coated Cu. $^+$log i was used since no mass transport limited behavior was observed.

coated with a 15um cast film of UDI, and for Cu in a 0.01M solution of BTA. Mass transport limited currents were only observed for bare Cu and the PVI-1 coated Cu. The values of the true limiting currents calculated from Levich plots are given in Table I. A saturated solution of UDI gave the same Tafel slope as the 15um film; therefore, UDI can be used as a thick film or in solution at pH=5.6 as well as pH=1. The inhibiting species in the two cases are different because in phosphate buffer the UDI film dissolves and the volume of solution will affect the concentration of UDI, thus, the inhibition efficiency. At pH=5.6, all four samples exhibited different Tafel slopes, but the PVI-1-coated Cu is close to that of bare Cu while UDI is distinctly different. The Tafel slope of Cu in 0.01M BTA solution lies intermediate between PVI-1 and UDI treated Cu (figure 9). The UDI and probably BTA cause a change in the mechanism of the cathodic reaction, presumably O_2 reduction. PVI-1 films do not greatly affect this process.

Table I. Tafel parameters obtained from figures 7,8 and the mass transport limited currents.

Inhibitor	Tafel Slope (V)	i_o (uA/cm^2)	i_L (uA/cm^2)
0.1M $HClO_4$, pH=1			
none	-0.19	12.62	-967
PVI-1	-0.20	11.05	-922
BTA [B	-0.18	12.88	-948
UDI	-0.32	4.01	----
phosphate buffer, pH=5.6			
none	-0.25	7.24	-1034
PVI-1	-0.30	5.23	-1102
BTA	-0.34	3.18	-----
UDI	-0.47	3.27	-----

There is clearly a change in the mechanism when UDI is present in both 0.1M $HClO_4$ and phosphate buffer, pH=5.6. The observation in 0.1M $HClO_4$ that the suppression of the O_2 reduction currents occurs only after cycling whether or not a cast film was initially present, indicates that inhibition is produced by a reaction. The reaction probably involves the inhibitor species and an intermediate formed in the O_2 reduction process, superoxide or peroxide. The anion present may also play a role. At pH=5.6, a similar effect is observed, but the change in the Tafel slope indicates that a different mechanism is operating.

Extrapolation of the linear portion of the curves to the standard oxygen reduction potential gives the corresponding exchange current density, a measure of the equilibrium rate of the reaction [35]. The results are given in Table I. Within experimental error, BTA- and PVI-1-coated Cu show the same exchange current density for oxygen reduction as bare Cu in pH=1 solution. This result supports the conclusion that at this pH, BTA and PVI-1 do not affect the O_2 reduction process. These compounds do affect the anodic process indicating that they may only act as anodic inhibitors. In the pH=5.6 solution, the i_o values cannot be directly compared because different mechanisms apply in each case.

The measurements discussed above describe the effects of the BTA, PVI-1, and UDI on the electrochemical processes believed to be involved in Cu corrosion at relatively high overpotentials. Open-circuit potential and linear polarization measurements were made in UDI solutions in order to correlate the high overpotential measurements to practical corrosion potentials.

Open-circuit potentials in the 0.1M $HClO_4$ solution were made in 0.1M $HClO_4$. The steady-state potential of bare Cu was -38mV (SCE) and this value shifted to +71mV (SCE) when UDI was added. This implies that the effect of the anodic process on the mixed potential is dominant. The effects of cycling anodically and cathodically on the open-circuit potential was also measured. After cycling between -38mV (SCE) and +225mV (SCE) for 1 hour, the potential was found to be +88mV (SCE). Subsequent cathodic cycling between -100 and -750mV (SCE) revealed that the anodic treatment reduced the inhibition of the O_2 reduction reaction usually obtained. A possible explanation of this phenomenon is that the concentration of free UDI forms Cu-UDI complexes during the anodic process. The cathodic treatment decreased the open-circuit potential to +50mV (SCE), but the original value, +71 mV (SCE) was reached after the electrode was allowed to rest at open-circuit for 30 minutes. Cycling cathodically did not affect the anodic inhibition.

Linear polarization measurements were found by Simmons [37], to be a good method for evaluation of inhibitor compounds. He found that the polarization resistance were directly correlated to weight-loss measurements, i.e. inhibitors that exhibit good weight-loss suppression possess high polarization resistance values. Linear polarization results in phosphate buffer, pH=5.6, are shown in figure 10. The polarization resistance of bare Cu was found to be 1.1Kohm cm^2 and that of a saturated solution of UDI after cycling between -300 and -1100mV (SCE) was found to be 4.8Kohm cm^2. The open-circuit potential fluctuated by a few millivolts between measurements and this is probably responsible for the scatter in the data points, but there is a clear difference in the two curves. BTA, the most commonly used Cu corrosion inhibitor, has been found by Thierry and Leygraf to have high values of polarization resistance. They found that 5 mM BTA in 0.1M and 0.1M NaCl gave values of 42 and 245 Kohms cm^2 respectively. They did not, however, report their value for bare Cu in the same solutions. A direct comparison between the values found for UDI and those reported for BTA is impossible because of differences in inhibitor concentration and differences in testing medium. In general, based on the increase of polarization

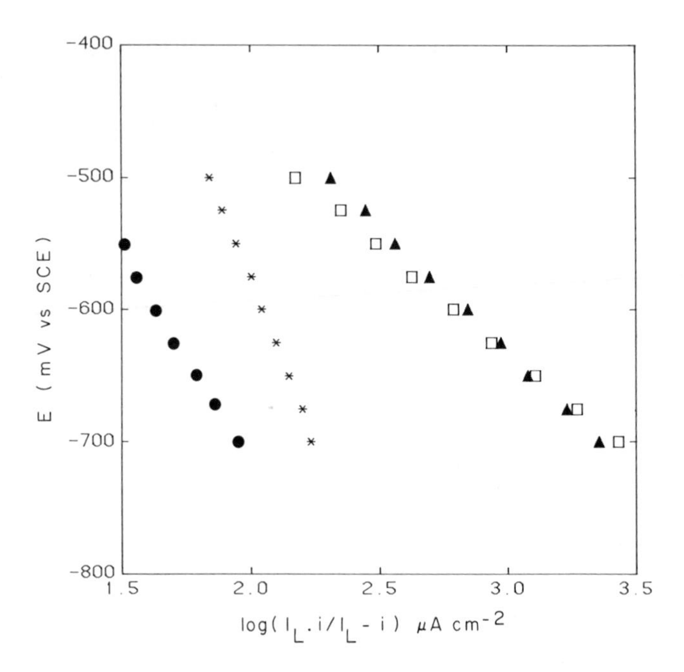

Figure 9. Tafel plots of oxygen reduction Cu electrodes in phosphate buffer, pH = 5.6, w = 15 Hz, 25 °C. □ bare Cu, ● BTA-coated Cu, ▲ PVI-1-coated Cu, and * UDI-coated Cu. [+]log i was used since no mass transport limited behavior was observed.

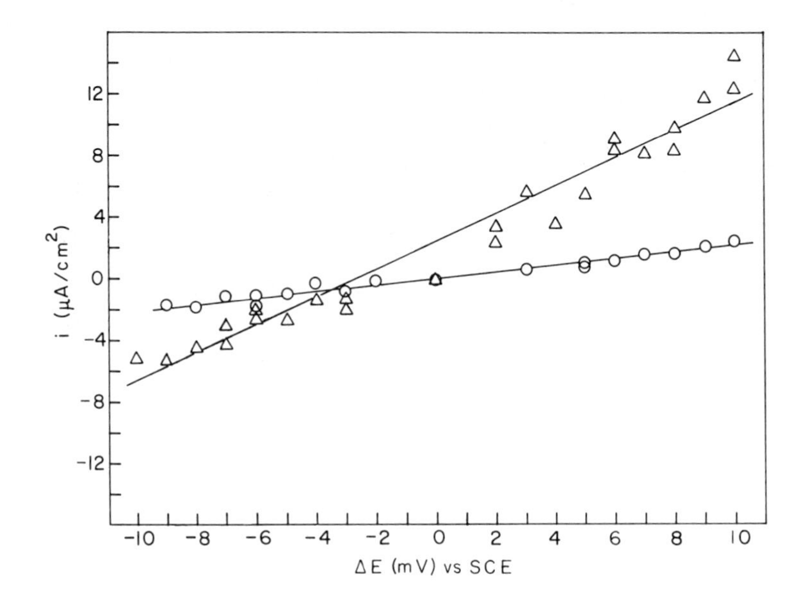

Figure 10. Linear polarization curves in phosphate buffer for △ bare Cu, o Cu in a saturated solution of UDI.

resistance by BTA, UDI is a corrosion inhibitor at least in phosphate buffer.

It should be emphasized that the results presented in this paper were obtained only in 0.1M $HClO_4$ and phosphate buffer (pH=5.6). Work currently in progress indicates that anions in the electrolyte play a role in the electrochemical processes leading to inhibition. The effectiveness of the inhibitors has been found to vary depending on the testing solution.

Summary

The electrochemical studies of the partial reactions likely to be important in copper corrosion, Tafel and polarization resistance measurements indicate that UDI should be an effective corrosion inhibitor in oxygenated aqueous solutions. The predominant cathodic process is oxygen reduction. UDI as both a thin cast film and in solution of concentrations as low as 7 x 10^{-5}M almost completely suppresses this process, apparently by reaction of the UDI with intermediate species generated in the oxygen reduction reaction. with BTA or PVI-1 present have the reaction mechanism but with UDI present a different mechanism or completely different process is operating. In the phosphate buffer, pH=5.6, all of the samples possessed different Tafel slopes (Table I) indicating different mechanisms for each case. BTA must be present in solution to provide continuous inhibition while UDI is effective as a thick, 15um, cast film. A 15um PVI-1 cast film suppressed the O_2 reduction currents, but a 150nm film did not.

The polarization resistance of Cu in a UDI/phosphate buffer solution was found to be four times that of bare Cu. The proven inhibitor, BTA, also increases the polarization resistance. This result indicates that UDI may have practical applications.

The thermal oxidation mechanism results [29,30] differ from the electrochemical results because different mechanisms apply. PVI-1 was found to be an excellent anti-oxidant in dry conditions, while it has no effect on the O_2 reduction and has less effect on the anodic processes than the other inhibitors studied. The thermal stability of the Cu(inhibitor) complex is probably important for the thermal oxidation whereas the preformed complex has little importance in the electrochemical processes.

Acknowledgments

The authors gratefully acknowledge the financial support of the International Copper Research Association, Inc.

Literature Cited

1. J.B. Cotton, "Proc. 2nd Intl. Cong. on Metallic Corrosion", NACE, New York, 590 (1963).

2. I. Dugdale and J.B. Cotton, Corr. Sci., 3, 69 (1963).

3. G.W. Poling, Corr. Sci., 10, 359 (1970).

4. T.K. Ross and M.R. Berry, Corr. Sci., 11, 273 (1971).

5. D. Altura and K. Nobe, Corrosion-NACE, 28:9, 345 (1972).

6. R. Walker, Corrosion-NACE, 29:7, 290 (1973).

7. T. Notoya and G.W. Poling, Corrosion-NACE, 35:5, 193 (1979).

8. C. McCrory-Joy and J.M. Rosamilla, J Electroanal. Chem., 136, 105 (1982).

9. F. Mansfeld, T. Smith, and E.P. Parry, Corrosion-NACE, 27:7, 289 (1971).

10. F. Mansfeld and T. Smith, Corrosion-NACE, 29:3, 105 (1973).

11. K.H. Wall and I. Davies, J. Appl. Chem., 15, 389 (1965).

12. R. Walker, Anti-Corrosion, 9 (1970).

13. F.E. Heakal and S. Haruyama, Corr. Sci., 20, 887 (1980).

14. G. Lewis, Corrosion-NACE, 34:12, 424 (1978).

15. R. Walker, J. Chem. Ed., 57:11, 789 (1980).

16. J.S. Wu and K. Nobe, Corrosion-NACE, 37:4, 223 (1981).

17. D. Chadwick and T. Hashemi, Corr. Sci., 18, 39 (1978).

18. R.J. Chin, D. Altura, and K. Nobe, Corrosion-NACE, 29:5, 185 (1973).

19. P.G. Fox, G. Lewis, and P.J. Boden, Corr. Sci., 19, 457 (1979).

20. M. Ohsawa and W. Suetaka, Corr. Sci., 19, 709 (1979).

21. P.G. Fox and P.A. Bradley, Corr. Sci., 20, 643 (1980).

22. R. Walker, Corrosion-NACE, 31:3, 97 (1975).

23. S.N. Prajapati, I.M. Bhatt, K.P. Soni, and J.C. Vora, J. Indian Chem. Soc., 5, 723 (1976).

24. D. Niki, F.M. Delnick, and N. Hackerman, J. Electrochem. Soc: Electrochemical Science and Technology, 122:7, 855 (1975).

25. D. Thierry and C. Leygraf, J. Electrochem Soc.: Electrochemical Science and Technology, 132:5, 1009 (1985).

26. Shikoku Chemicals Co., Technical Report, "Glicoat T and L", 1979.

27. N. Sawa and M. Hoda, U.S.P., 3,933,531.

28. M. Pourbaix, "Atlas of Electrochemical Equilibria in Aqueous Solutions", Pergamon Press, New York (1966), p. 384.

29. S. Yoshida, PhD Dissertation, Case Western Reserve University, Cleveland, Ohio (1983).

30. F. Eng, PhD dissertation, Case Western Reserve University, Cleveland, Ohio (1985).

31. R. Gilmont and S.J. Silvus, Am. Lab., 6, 46 (1974).

32. R. Amadelli, PhD Dissertation, Case Western Reserve University, Clevelan, Ohio (1982).

33. J.E. Fagel Jr. and G.W. Ewing, J. Am. Chem. Soc., 73, 4360 (1951).

34. M. Sato, K. Kondo, and K. Takemoto, Makromol. Chem., 179, 601 (1978).

35. A.J. Bard and L.R. Faulkner, "Electrochemical Methods, Fundamentals and Applications", John Wiley and Sons, New York (1980), p. 110.

36. R.W. Zurilla, PhD Dissertation, Case Western Reserve University, Cleveland, Ohio (1969).

37. E.J. Simmons, Corrosion-NACE, 11, 225t (1955).

38. M. Stearn, Corrosion-NACE, 8, 440t (1958).

RECEIVED March 14, 1986

24

Corrosion Protection on Copper by Polyvinylimidazole

Fred P. Eng and Hatsuo Ishida

Macromolecular Science Department, Case Western Reserve University, Cleveland, OH 44106

Fourier transform infrared reflection-absorption spectroscopy (FT-IRRAS) is applied to the study of corrosion protection of copper by an organic coating. Poly-N-vinylimidazole (PVI(1)) and poly-4(5)-vinylimidazole (PVI(4)) are demonstrated to be effective new polymeric anti-corrosion agents for copper at elevated temperatures. Oxidation of copper is suppressed even at 400° C. PVI(1) and PVI(4) are more effective anti-oxidants than the most commonly used corrosion inhibitors, benzotriazole and undecylimidazole, at elevated temperatures. These new polymeric agents are water soluble and easy to treat the metal surface.

Azole compounds such as benzotriazole, benzimidazole, indazole and imidazoles are efficient anti-corrosion agents for copper and copper-base alloys [1-10]. Many experimental techniques [11-15] have been used to study the corrosion inhibition mechanisms, however, the mechanisms are still not well understood. It is believed that the complex formation between copper and nitrogen atoms would inhibit oxygen adsorption on copper surface [16-20].

Inhibitors mentioned above are small molecules in nature. Recently, there have been great emphases in using polymers as corrosion inhibitors [21-37]. However, these inhibitors, which included different heterocyclic polymers, polythiopropinate, polymaleic acid and polyalkylolamide/alkenyl copolymers, were developed for protecting steel in sea water [21,29], tap water [22] and in acidic environments [30,31,33]. Relatively very few or no polymeric inhibitors for copper, aluminum, and iron [28,34-37] have been reported, especially in high temperature studies. In this study, we will present poly-N-vinylimidazole (PVI(1)) and poly-4(5)-vinylimidazole (PVI(4)) (Figure 1.) as new polymeric anti-corrosion agents for copper in elevated temperature

0097-6156/86/0322-0268$06.00/0

environments. Polyvinylimidazoles (PVIs) are prefered because of the following reasons. PVIs have the imidazole ring as their pendant group which would lead to complex formation with copper. Furthermore, the polymers can easily form thin films of relatively higher ductility than small molecules on copper surfaces. It is known that the higher ductility would enhance the adhesion of the film to the substrate. Lastly, amorphous polymer/copper complex is expected to have less defects which may be important in influencing the the rate of oxidation of copper.

In this elevated temperature study, commonly used copper corrosion inhibitors for copper, benzotriazole and undecylimidazole will be used for comparisons. Benzotriazole is one of the most effective and widely used corrosion inhibitors specifically for copper in both atmospheric and immersed environments for over 35 years. At the same time, undecylimidazole also exhibits superior anti-corrosion property. The reactivity of imidazole with copper is very high, forming a thick imidazole/copper complex film which makes undecylimidazole an attractive corrosion inhibitor. Fourier transform infrared reflection-absorption spectroscopy (FT-IRRAS) is utilized in the experiment. FT-IRRAS is an external reflection technique which is useful for studying thin films on metal surfaces by reflecting infrared radiation from the metal surfaces at high,[A nearly grazing angles of incidence. The theory of the technique was developed by Francis and Ellison [38] and Greenler [39]. Cuprous oxide formation is used to follow the corrosion kinetics.

Experimental

Both benzotriazole and urocanic acid were purchased from Aldrich Chemical Co. and azobis(isobutyronitrile) was from Eastman Kodak Co. Undecylimidazole and N-vinylimidazole were supplied by Shikoko Chemical Co. and BASF Co., respectively. Copper plates (2.5 x 5.0 x 0.2 cm, ASTM B 125, type ETP) were mechanically polished with No.5 chromeoxide, ultrasonically washed with acetone, rinsed with dilute hydrochloric acid and distilled water, and dried with a stream of nitrogen gas. Corrosion inhibitors were dissolved in either ethanol or methanol, solution cast onto copper substrates and air dried. Film thickness was calculated based on the concentration of the solution, density of the sample and the area of the copper surface. In this study, 150 nm thick films were used. The reflection-absorption (R-A) attachment (Harrick Scientific) along with a gold wire grid polarizer (Perkin-Elmer) were mounted in a Digilab FTS-14 Fourier transform infrared spectrometer equipped with a triglycine sulfate detector and purged with dry air. Spectra collected were the average of 200 scans at 4 cm^{-1} resolution using an optical velocity of 0.3 cm/s. The angle of incidence used was 75°.

Purification of Azobis(isobutyronitrile) - Crude AIBN was first dissolved in warm methanol ($35^{o}C$), then recrystallized in ice bath and finally dried in a vacuum oven at room temperature for two days.

Purification of N-Vinylimidazole - Crude brownish N-vinylimidazole was distilled in vacuo ($51^{o}C/2.5$ mm Hg) to yield a pure and colorless liquid.

Synthesis of Poly-N-Vinylimidazole [40] - A solution of N-vinylimidazole (30 g, 0.32 mol) and azobis(isobutyronitrile) (0.26 g, 0.0016 mol) in benzene (200 ml) was heated at $68^{o}C$ with stirring under nitrogen for two days. The white precipitated polymer was collected by filtration, washed four times with benzene (20 ml) and dried in a vacuum oven (30 mm Hg) at $40^{o}C$ for three days. The yield was 30 g (100% conversion). The weight-average molecular weights of the polymer ranged from $5.5x10^{4}$ to $1.3x10^{6}$ [40]. The polymer, as suggested by NMR studies, was atactic [41]. Density measured was 1.246 g/ml.

Synthesis of 4(5)-Vinylimidazole [42] - Urocanic acid (7.6 g, 0.055 mol) was decarboxylated at the melting point in vacuo in a distilling apparatus with a silicone oil bath of temperature $280^{o}C$. A heating tape with temperature set at $140^{o}C$ was wrapped around the condenser to prevent the distillate from solidifying before reaching the receiver. A viscous and colorless liquid was collected; the yield was 2.0 g (26%). The product solidified upon cooling to room temperature.

Synthesis of Poly-4(5)-Vinylimidazole [42] - A solution of 4(5)-vinylimidazole (1.50 g, 0.016 mol) and azobis(isobutyronitrile) (5 mg, 0.030 mmol) in benzene (200 ml) was heated at reflux with stirring under nitrogen for three days. Afterward, the white polymer was collected by filtration, washed with benzene (20 ml) four times and dried in vacuum oven (30 mm Hg) at $40^{o}C$ for three days. The yield was 0.9 g (60% conversion). Density measured was 1.246 g/ml.

Synthesis of Poly-N-Vinylimidazole/Copper(II) Complex - To a PVI(1) (1.7 mg)/methanol (5 ml) solution was added cupric chloride (0.7 mg, 0.0041 mmol). After the mixture solution was allowed to stand at room temperature overnight, blue crstals appeared at the bottom of the reaction flask. Blue crystals of PVI(1)/copper(II) complex were then collected and washed repeatedly with methanol.

Synthesis of Poly-4(5)-Vinylimidazole/Copper(II) Complex - To a PVI(4) (2.0 mg)/methanol (5 ml) solution was added cupric chloride (1.0 mg, 0.0059 mmol). After the mixture solution was allowed to stand at room temperature overnight, green crystals appeared at the bottom of the beaker. Green crystals of PVI(4)/copper(II) complex were then collected and washed repeatedly with methanol.

Results and Discussion

Molecular Structure of PVI/Copper(II) Complexes

Figure 2 shows the reflection spectra of PVI(1) and PVI(4) coated on copper mirrors. The peak at 3140 cm^{-1} is assigned to the NH stretching mode. Peaks at 3115, 2950, 2940 and 2850 cm^{-1}

$(CH_2 - CH)_n$ — $(CH_2 - CH)_n$

POLY-N-VINYLIMIDAZOLE — POLY-4(5)-VINYLIMIDAZOLE

PVI(1) — PVI(4)

Fig. 1. Polyvinylimidazoles.

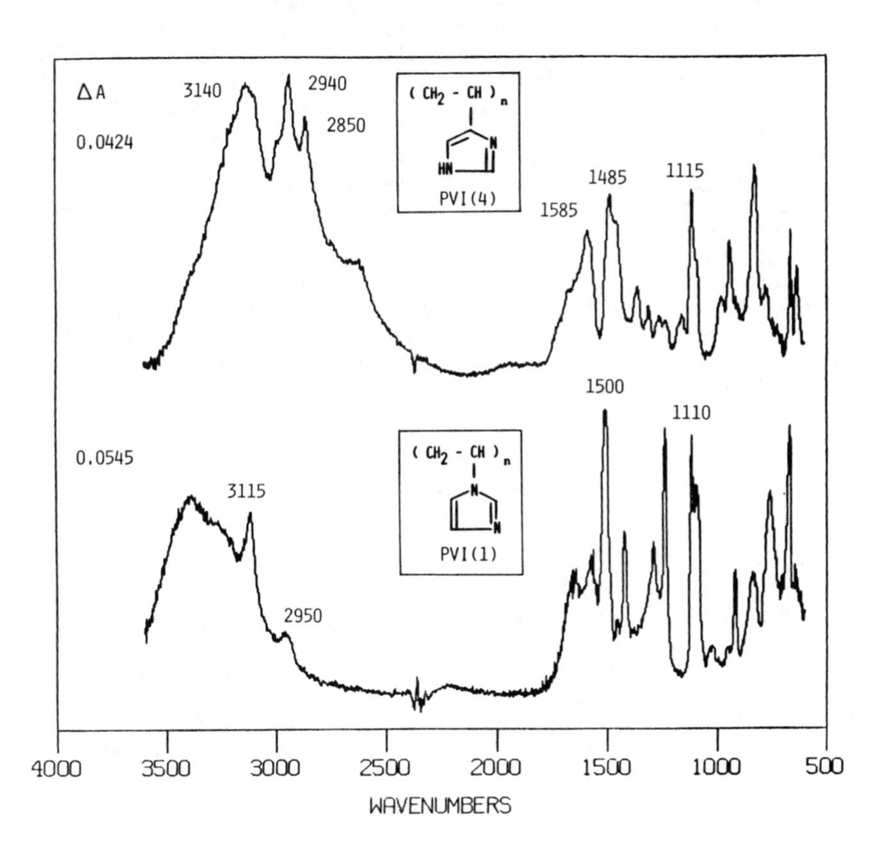

Fig. 2. R-A Spectra of PVI(1) and PVI(4).

are the CH stretching modes of the imidazole ring and the aliphatic chain. Peaks at 1585, 1500 and 1485 cm^{-1} are due to the ring stretching. Finally, peaks at 1115 and 1110 cm^{-1} are assigned to the CH in-plane bending of the imidazole ring.

From the study of the synthesized model compounds of PVI(1)/copper(II) and PVI(4)/copper(II), it was found that both polymers were capable of forming complex with copper. Solubility tests indicated that the complexes were not soluble in water, methanol, or any other common organic solvents. Then the complex formation between the PVI thin film and copper mirror surface was studied both at room and elevated temperatures. First, 150 nm PVI films were deposited on the copper mirrors and the reflection spectra were taken. After various temperature treatments, the copper mirrors were washed with methanol, dried and the reflection spectra were taken again. Area of peaks at 1115 and 1110 cm^{-1} was used to follow the kinetics of the complex formation. The assumption of this experiment was that since PVI/copper(II) complexes were not soluble in methanol, therefore the amount of material remained on the copper surface would be due to the complex formed. At room temperature, PVI(1) formed complex instantly with copper once the polymeric film was solution cast onto the copper surface and dried. However, PVI(4) did not immediately complex with the copper. Figure 3 shows that after heating at 60° C for 15 min, only 7 % of the PVI(4) film complexed with the copper. It was not until the temperature reached 120° C before 92 % of the PVI(4) complexed with copper. This phenomenon is likely caused by the conformational effects of the polymer. In each of the imidazole rings, there is a NH group causing the imidazole rings to interact with one another through hydrogen bonding intra- and intermolecularly. Such interactions cause a shrinkage, resulting in the exposure of aliphatic chains and the burying of the imidazole groups. As a consequence, steric hindrance prevents nitrogen atom from complexing with copper. Thus, the rate of complex formation with copper ions is strongly influenced by the strength of hydrogen bonding.

<u>Comparisons of Benzotriazole, Undecylimidazole and PVIs (150 to 300°C)</u>

Figure 4 shows the reflection spectra of undecylimidazole on copper heated at 150° C for various lengths of time. Complex formation was observed without any heat treatment (spectrum at 0 min) [43]. Degradation of undecylimidazole was observed as the intensity of the peaks in the 2900 cm^{-1} region decreased gradually with prolonged heating. At the same time, new bands appeared around 1600 cm^{-1}. Further degradation also resulted in nitrile formation as indicated by the peak at 2183 cm^{-1}. Increase of copper oxide formation with heating time was observed at 650 cm^{-1}. At 150° C, benzotriazole showed complete degradation after only 15 min of heating (Figure 5.). Degraded product showed an intense peak at 740 cm^{-1}. Copper oxide formation also occurred after 15 min and was even more prominent after 2 hours. As for PVI(1) and PVI(4) (Figures 6 and 7.), no degradation was detected after 1

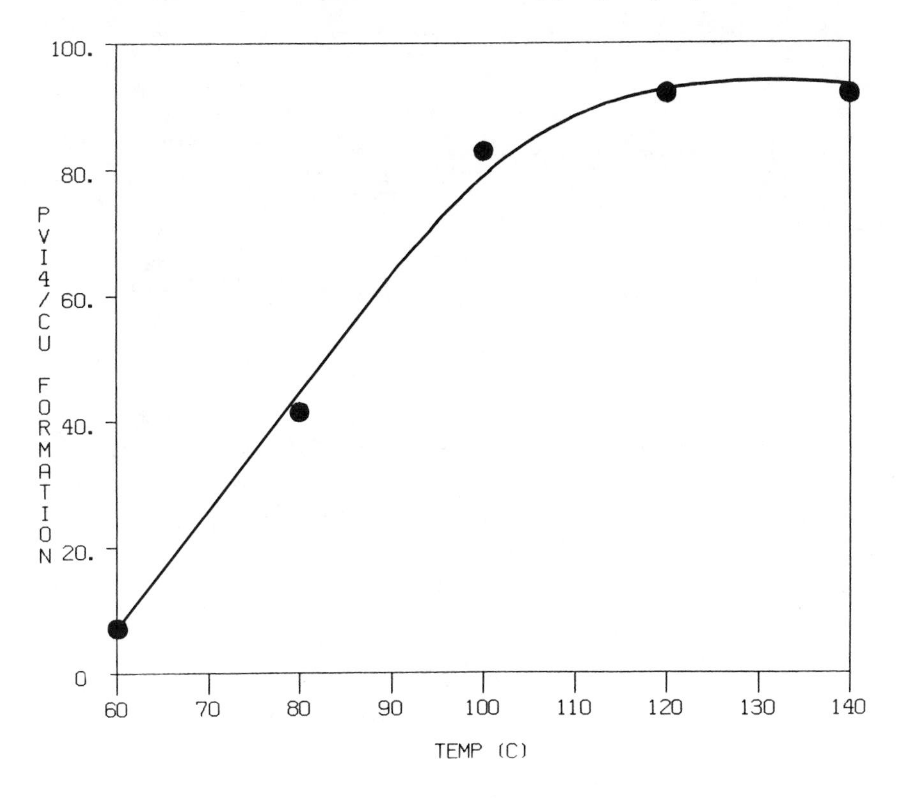

Fig. 3. PVI(4)/CU(II) Formation versus Various Temperatures Heated for 15 Minutes.

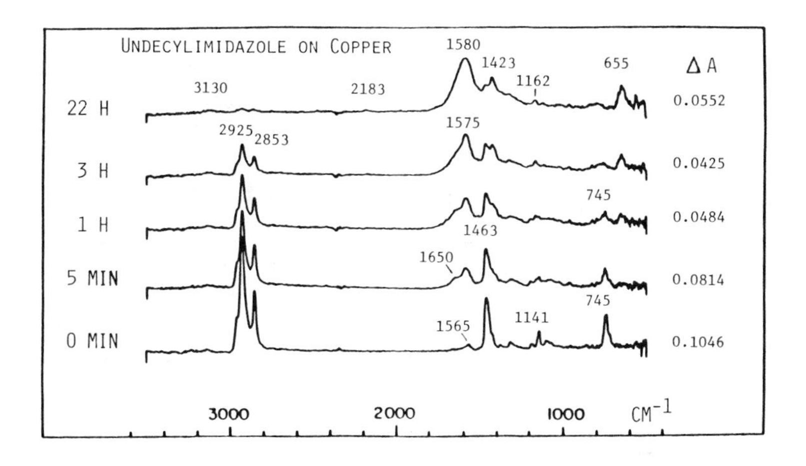

Fig. 4. R-A Spectra of Undecylimidazole with Heat Treatment at 150° C.

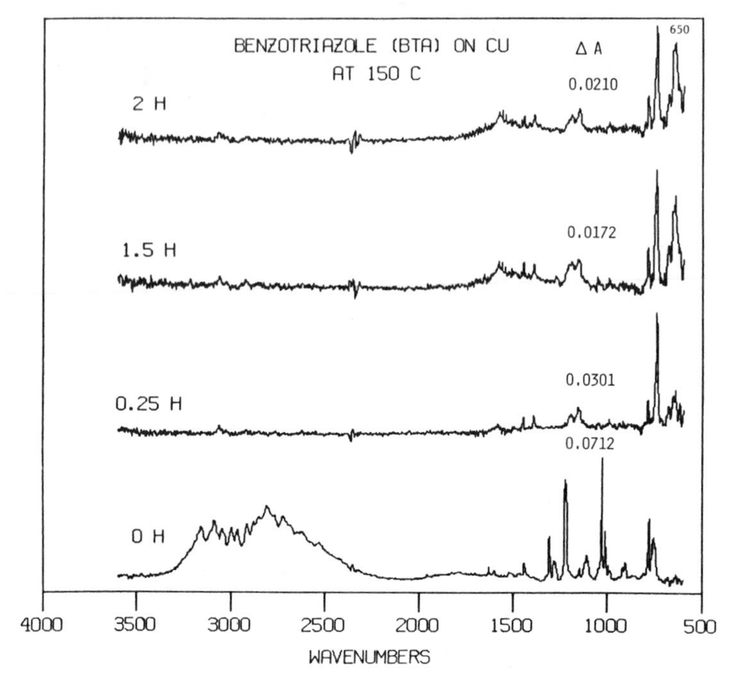

Fig. 5. R-A Spectra of BTA with Heat Treatment at 150° C.

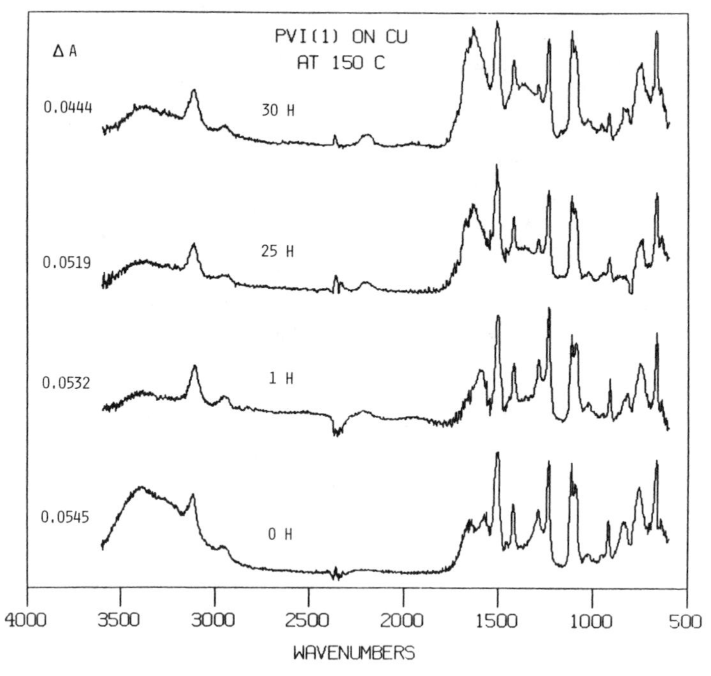

Fig. 6. R-A Spectra of PVI(1) with Heat Treatment at 150° C.

hour of heating. Even after 27 or 30 hours, little degradation was seen as the emergence of the nitrile peak at 2200 cm^{-1} and carbonyl peaks at 1600 cm^{-1} region was relatively small. No copper oxide was observed at any time for either polymer.

When the temperature was raised to 210° C, undecylimidazole was completely degradated after 15 min (Figure 8.). No imidazole ring structure was observed. The nitrile peak at 2190 cm^{-1} was pronounced, and the copper oxide formation was intense. On the other hand, PVI(1) and PVI(4) degradation was relatively mild at 210°C after 15 min (Figures 9 and 10.). It was not until the temperature was raised to 250° C and higher that major degradation of the polymers was observed. Even in such degrading conditions, no copper oxide was detected at 210 or 250° C. At 300° C, where the polymers suffered relatively severe degradation, oxidation of copper was still suppressed.

High Temperature Study of PVIs (330 to 450°C)

Figure 11 shows the reflection spectra of bare copper mirrors heated at high temperatures for 15 min. At 330°C, cuprous oxide was detected by the band absorbed at 655 cm^{-1}. However, at 350 and 400°C, two bands were observed near 611 and 655 cm^{-1}. According to the theory of reflection-absorption infrared spectroscopy developed by Greenler et al. [44], bands 611 and 655 cm^{-1} are assigned to the transverse optical and longitudinal optical modes of the high frequency phonon observed nearC 609 cm^{-1} in dielectric spectra of cuprous oxide [45]. The band around 611 cm^{-1} was observed by Boerio and Armogan [46] with oxides having a thickness of about 200 nm or thicker.

Figure 12 shows the results of PVI(1) after being heated at various high temperatures for 15 min. The bands of cuprous oxides formed on bare copper mirrors at corresponding temperatures are superimposed on the PVI(1) spectra for direct comparisons. The scale of the two types of spectra is shown by the difference between the maximum and minimum absorbance, A. Note that no cuprous oxide formation was observed at 330 or 400°C. In fact, at 400°C the polymer coated surface of the copper mirrors remained mostly shiny whereas the bare reference copper surface turned dull with a layer of reddish-black scale on it. At 410°C or higher, the polymer was no longer protecting the copper surface. The two cuprous bands near 611 and 655 cm^{-1} began to emerge. Even so, at 450°C, the amount of oxides formed was much less than that of bare copper. Similar results were obtained for PVI(4) (Figure 13.) where no significant cuprous oxide was observed at or below 400°C. Figure 14 summarizes the high temperature study with a plot of relative cuprous oxide formation versus temperatures. As the temperature went up, the amount of cuprous oxide formed on bare copper also increased. However, for the polymer samples, it is interesting to note that the transition from no oxide to oxide formation at 410°C is relatively sharp.

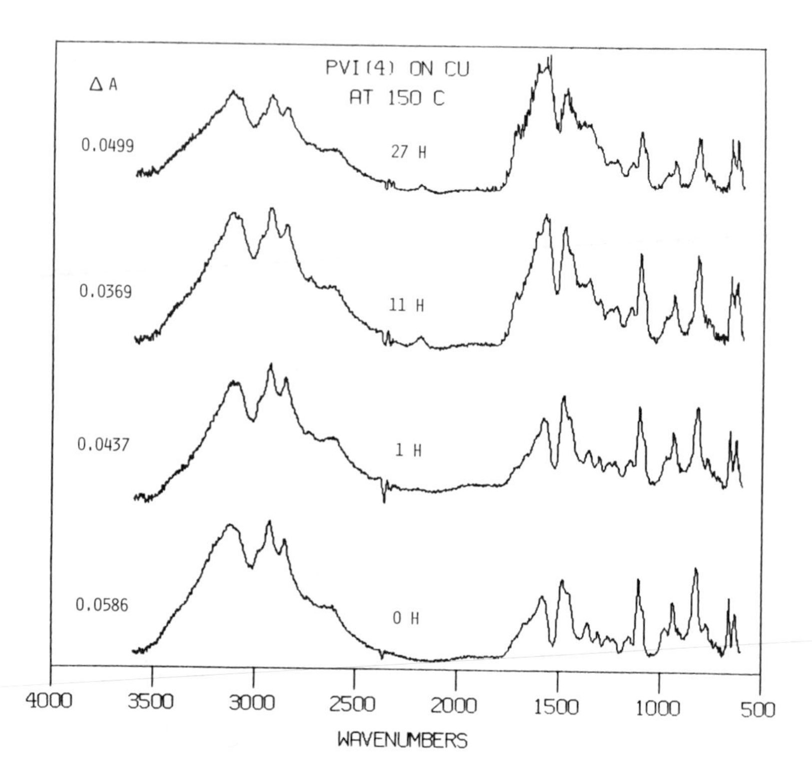

Fig. 7. R-A Spectra of PVI(4) with Heat Treatment at 150° C.

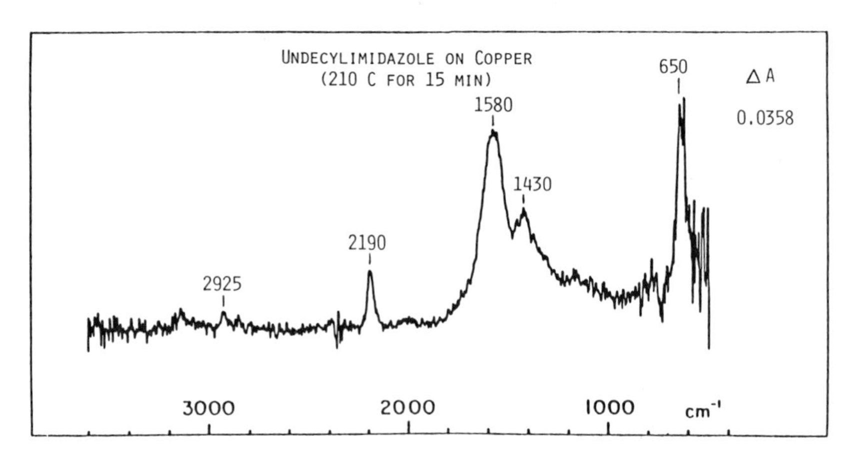

Fig. 8. R-A Spectra of Undecylimidazole with Heat Treatment at 210° C.

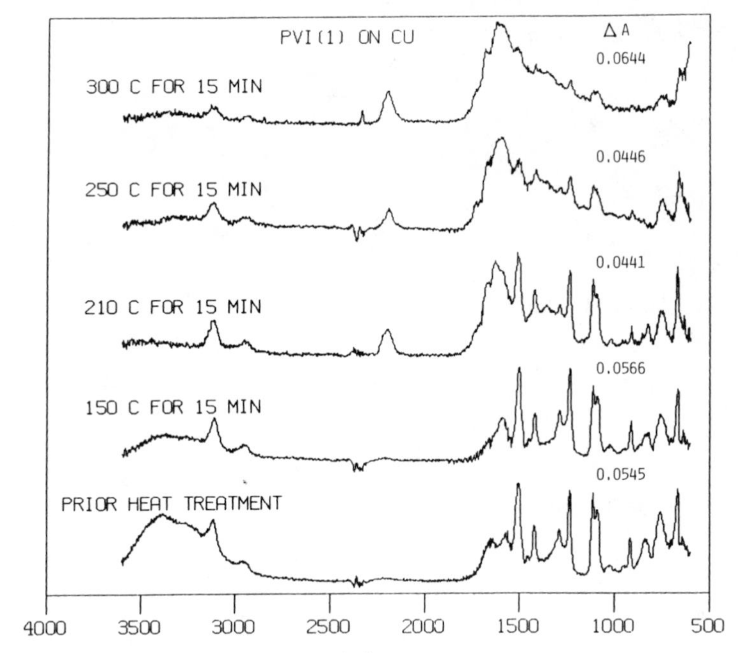

Fig. 9. R-A Spectra of PVI(1) with Heat Treatment at Various Temperatures.

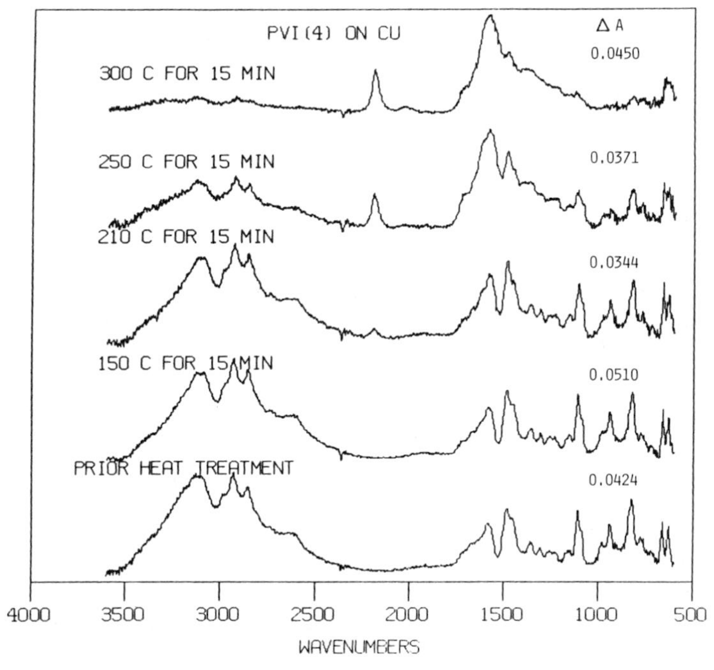

Fig. 10. R-A Spectra of PVI(4) with Heat Treatment at Various Temperatures.

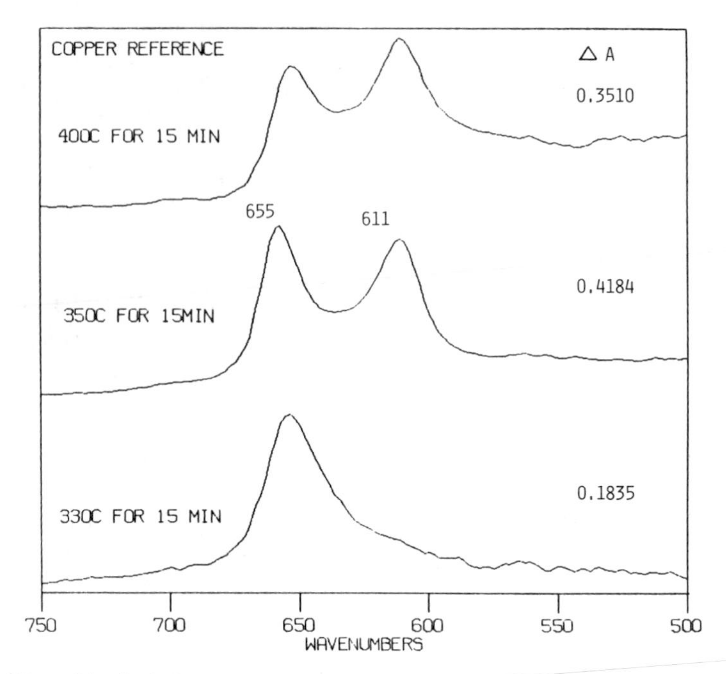

Fig. 11. R-A Spectra of Bare Copper with High Temperature Treatments.

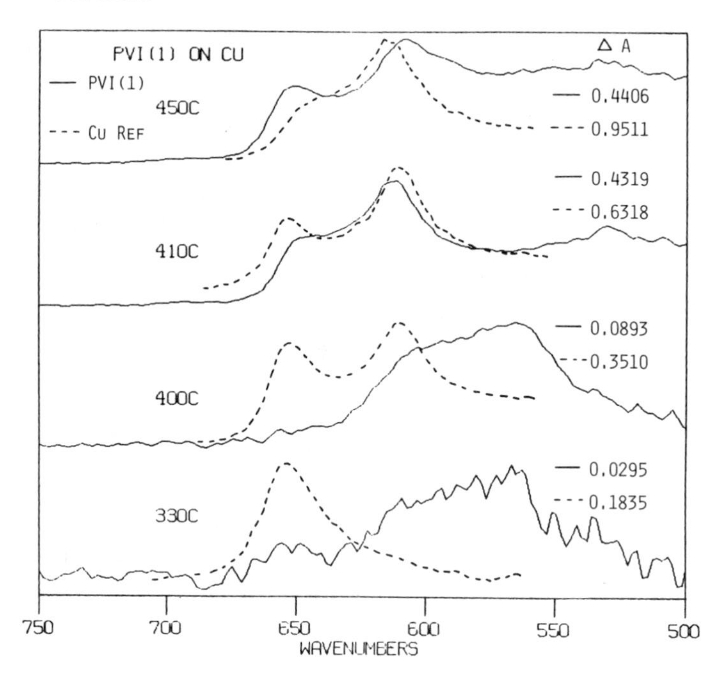

Fig. 12. R-A Spectra of PVI(1) with High Temperature Treatments.

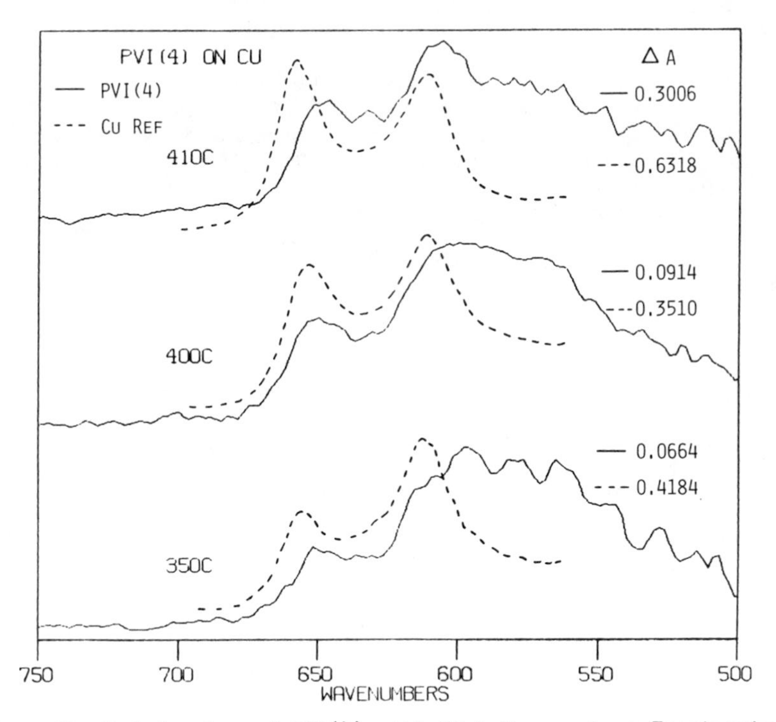

Fig. 13. R-A Spectra of PVI(4) with High Temperature Treatments.

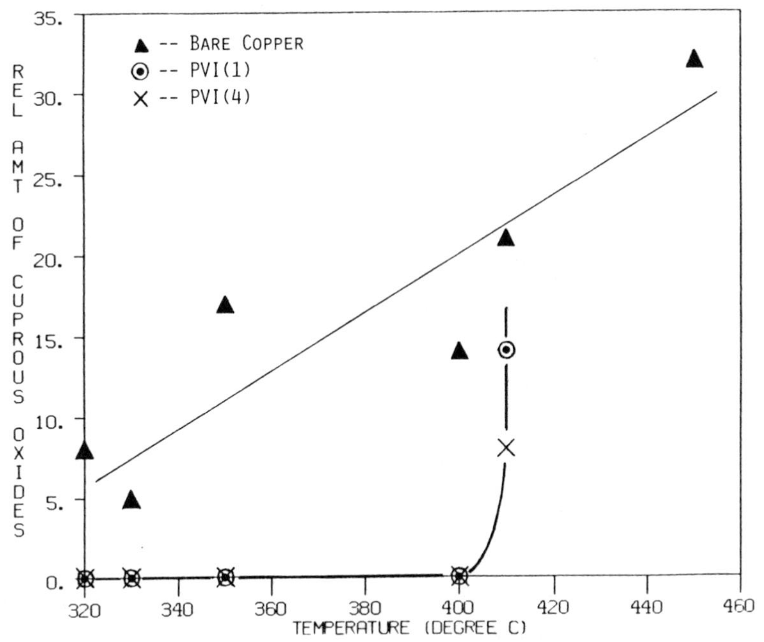

Fig. 14. Relative Amount of Oxides Formed on Bare, PVI(1) and PVI(4) Copper *versus* Various High Temperatures.

Summary

Polyvinylimidazoles are effective anti-oxidants for copper at elevated temperatures. Below 250° C, there is no major degradation of the coated polyvinylimidazole films on copper. Furthermore, degraded polyvinylimidazole films can suppress oxidation even at 400°C. Finally, polyvinylimidazoles are more effective anti-oxidants than benzotriazole and imidazoles at elevated temperatures. It is also demonstrated that FT-IRRAS is a useful technique to study degradation of polymeric coatings and corrosion of metal simultaneously.

Acknowledgment

The authors gratefully acknowledge the financial support of the International Copper Research Association, Inc. and the International Business Machines Corporation.

Literature Cited

1. J.B. Cotton, Proc. 2nd Int. Congr. Metallic Corrosion, NACE, New York (1963).

2. J.B. Cotton and I.R. Scholes, Brit. Corros. J., 2, 1 (1967).

3. S. Yoshida and H. Ishida, J. Materials Sci., 19, 2323 (1984).

4. I. Dugdale and J.B. Cotton, Corros. Sci., 3, 69 (1963).

5. R. Walker, Anti-Corrosion, 17, 9 (1970).

6. S.M. Mayanna and T.H.V. Setty, Corros. Sci., 15, 625 (1975).

7. S. Yoshida and H. Ishida, J. Chem. Phys., 78, 6960 (1983).

8. N.K. Patel, J. Franco and I.S. Patel, J. Ind. Chem. Soc., 54, 815 (1977).

9. G.W. Poling, INCRA Project No. 185, February (1979).

10. G.N. Ekilik, V.P. Grigorev and V.V. Ekilik, Zashch. Met., 14, 357 (1978).

11. D. Chadwick and T. Hashemi, J. Electron. Spectrosc. Relat. Phenom., 10, 79 (1977).

12. D. Chadwick and T. Hashemi, Corros. Sci., 18, 39 (1978).

13. A.R. Siedle, R.A. Velapoldi and N. Erickson, Appl. Surf. Sci., 3, 229 (1979).

14. S. Thiboult, Corros. Sci., 17, 701 (1977).

15. N.D. Hobbins and R.F. Roberts, Surf. Technol., 9, 235 (1979).

16. G.W. Poling, Corros. Sci., 10, 359 (1970).

17. J. Benard, Acta Metal., 8, 272 (1960).

18. F. Mansfeld, T. Smith and E.P. Parry, Corros., 27, 289 (1971).

19. R.F. Roberts, J. Electron Spectrosc. Relat. Phenom., 4, 273 (1974).

20. D. Chadwick and T. Hashimi, Corros. Sci., 18, 457 (1979).

21. J.W. Truesdell and M.R. Van de Mark, J. Electrochem. Soc., 129, 2673 (1982).

22. Katayama Chemical Works Co., Ltd., Tokkyo Koho JP 57,185,988 (1982).

23. S.D. Zeinalov, M.M. Talybov, S.A. Mamedov and N.M. Gasanov, Korroz. Zashch. Neftegazov. Prom-sti., 3, 7 (1983).

24. P. Guerit, P. Du Manoir and J. Oliver, Ger. Offen. DE 3,220,931 (1982).

25. Katayama Chemical Works Co., Ltd., Jpn. Kokai Tokkyo Koho JP 58 58,285 (1983).

26. Katayama Chemical Works Co., Ltd., Jpn. Kokai Tokkyo Koho JP 58 67,873 (1983).

27. A. Lupu, R. Avram, P. Popescu, M.V. Spiliadis and V.P. Pietris, Rom. RO 79,222 (1982).

28. M. Fradique, Eur. Pat. Appl. EP 79,236 (1983).

29. J.C. Lumaret, S. Gosset, M. Huchette, Ger. Offen. DE 3,232,396 (1983).

30. S.S. Abd El Rehim, F.M. Tohamy and M.M. Seleet, Surf. Technol., 21, 169 (1984).

31. A.P. Brynza, L.I. Gerasyutina, V.P. Fedash and E. Ya. Baibarova, Zashch. Met., 19, 961 (1983).

32. Kurita Water Industries, Ltd., Jpn. Kokai Tokkyo Koho JP 58 164,790 (1983).

33. Sanyo Chemical Industries, Ltd., Jpn. Kokai Tokkyo Koho JP 59 23,885 (1984).

34. R. Annand, D. Redmore and B. Rushton, US 3,450,646 (1969).

35. R. Annand, D. Redmore and B. Rushton, US 3,509,046 (1970).

36. J. Ehreke and W. Stichel, GWF, Gas-Wasserfach: Wasser/Abwasser, 124, 473 (1983).

37. M.G. Bondar, N. Ya. Kirillova, L.I. Makhonina, V.I. Martynenko, V.I. Nazarova, A.G. Ovcharov and V.P. Reshetov, USSR SU 1,067,086 (1984).

38. S.A. Francis and A.H. Ellison, J. Opt. Soc. Am., 49, 130 (1959).

39. R.G. Greenler, J. Chem. Phys., 44, 310 (1966).

40. J.S. Tan and A.R. Sochor, Macromolecules, 14, 1700 (1981).

41. P.M. Henrichs, L.R. Whitlock, A.R. Sochor and J.S. Tan, Macromolecules, 13, 1375 (1980).

42. C.G. Overberger and N. Vorchheimer, J. Am. Chem. Soc., 85, 951 (1963).

43. S. Yoshida and H. Ishida, J. Chem. Phys., 78, 6960 (1983).

44. R.G. Greenler, R.R. Rahn and J.P. Schwartz, J. Catal., 23, 42 (1971).

45. P. Dawson, M.M. Hargreave and G.R. Wilkinson, J. Phys. Chem. Solids, 34, 2201 (1973).

46. F.J. Boerio and L. Armogan, Appl. Spectrosc., 32, 509 (1978).

RECEIVED March 14, 1986

25

N-(Hydroxyalkyl)acrylamide Copolymers for Corrosion Control

F. Chen

Betz Laboratories, Inc., Somerton Road, Trevose, PA 19047

The study was an attempt to correlate polymer structure to its activity in water treatment applications. Copolymers of acrylic acid with N-(hydroxymethyl)-, N-(2-hydroxyethyl)-, and N-(2-hydroxypropyl) acrylamide were prepared. The resulting acrylic acid/N-(hydroxyalkyl) acrylamide copolymers were evaluated for their deposit control and dispersant activities as compared to the homopolymer of acrylic acid. Differences in the activities could be attributed to the incorporation of the N-hydroxylalkylacrylamide moiety into the polymer chain.

Corrosion is the destruction of a metal by chemical or electrochemical reaction with its environment. To increase equipment reliability and plant efficiency, corrosion inhibitors are used in boiler and cooling water programs to control fouling and deposition on critical heat-transfer surfaces. In cooling systems, corrosion inhibition is commonly achieved through the use of passivators, which encourage the formation of a protective metal oxide film on the metal surface (1).

Although chromate is the best aqueous corrosion inhibitor available, its use has been severely curtailed due to toxicity and environmental concerns (2). One of the more successful non-chromate treatments involves the use of phosphate/phosphonate combinations. This treatment employs high levels of orthophosphate to promote passivation of the metal surfaces. Therefore, it is important to control calcium phosphate crystallization so that high levels of orthophosphate may be maintained in the system without fouling or impeding heat-transfer functions.

Low-molecular-weight carboxyl-containing polymers have been used to control the deposition of calcium phosphate (2-8). These polymers also function as

0097-6156/86/0322-0283$06.00/0

dispersants to reduce the formation of scale from calcium carbonate, calcium sulfate, iron oxide, clay, etc. Other acrylamide-based copolymers, such as dicarboxymethylacrylamide (9) and alkylol amide, have been reported but are not yet commercially significant (10-11).

Copolymers of acrylic acid with N-(hydroxymethyl)-, N-(2-hydroxyethyl)-, and N-(2-hydroxypropyl)acrylamide were prepared in mole ratios of 3:1 [acrylic acid/ N-(hydroxyalkyl)-acrylamide] and similar molecular weights in order to study the effect that the variations in polymer structure have on activity. The resulting copolymers were then evaluated for their deposit control activity by comparing them to the homopolymer of acrylic acid (also in a similar molecular weight range). Thus, the difference in activity may be attributed to the particular functional group of the N-(hydroxyalkyl)-acrylamide moiety.

Experimental Section

Materials: Commercial grade acrylic acid (Rohm and Haas) and N-(hydroxymethyl)acrylamide (N-methylol acrylamide, American Cyanamid) were used without purification. Acetonitrile was dried over molecular sieves (4A) and distilled prior to use. Ethanolamine and 1-amino-2-propanol were vacuum distilled. Reagent grade acryloyl chloride, 2-propanol, and sodium persulfate were used as received.

Monomer Synthesis: N-(2-hydroxyethyl)- and N-(2-hydroxypropyl)acrylamide were prepared from the reaction of acryloyl chloride with ethanolamine and 1-amino-2-propanol, respectively, in acetonitrile:

$$CH_2{=}CHCOCl + 2NH_2CH_2CH_2OH \longrightarrow CH_2{=}CHCONHCH_2CH_2OH + HCl.NH_2CH_2CH_2OH\downarrow$$

$$CH_2{=}CHCOCl + 2NH_2CH_2CH(OH)CH_3 \longrightarrow CH_2{=}CHCONHCH_2CH(OH)CH_3 + HCl.NH_2CH_2CH(OH)CH_3\downarrow$$

1-Amino-2-propanol was used instead of 3-amino-1-propanol so that the hydroxy group would be on the 2-position, as in N-(2-hydroxyethyl)acrylamide. Reactions were carried out at -15°C instead of -5 to -10°C as reported in the literature (9) to obtain a higher yield. After the reaction, the mixture was stirred overnight and then filtered. The filtered hydroxyalkylamine hydrochloride salt was washed several times with chilled acetonitrile. The filtrates were combined, and then the acetonitrile was removed by vacuum distillation. Trace amounts of p-methoxyphenol

(300 ppm) were added to prevent self-polymerization during the distillation. Structure and purity of the distilled products were verified by ^{13}C NMR and IR spectroscopy.

Copolymerization: N-(Hydroxymethyl)-, N-(2-hydroxyethyl)-, and N-(2-hydroxypropyl)acrylamide were copolymerized with acrylic acid, in an aqueous medium. Sodium persulfate was used as the initiator. Acrylic acid was partially neutralized (ca. 80 mol %) with sodium hydroxide solution. After the neutralization, the respective N-(hydroxyalkyl)acrylamide was added to the solution and stirred. The monomer and initiator solutions were then added to a reaction flask containing water and 2-propanol at 85 to 90°C. Addition time was approximately 2 hours. After the addition, the reaction mixture was heated for an additional hour and, subsequently, a specific amount of 2-propanol/water was stripped off. The resulting polymer solution was cooled and filtered. The filtrate was clear and stable at room temperature. The detailed properties of the respective polymer solutions are shown in Table I. As analyzed by ^{13}C NMR, the composition (mol ratio) of the resulting copolymers was also close to that of 3:1. This could indicate that under the reaction conditions used the

Table I: Physical Properties

Sample No.	Comp[a]	Molar[b] Ratio	Brookfield[c] Visc.cps,25°C	pH	Mol Wt[d] Mn
	PA	-	200-500 (50.0% solids)	-	2,100
1	AA/HMAMD	3:1	20.3 (22.3% solids)	5.9	3,600
2	AA/HEAMD	3:1	18.9 (24.7% solids)	5.6	2,900
3	AA/HEAMD	3:1	18.0 (23.6% solids)	5.9	2,340
4	AA/HPAMD	3:1	16.0 (24.1% solids)	5.9	2,400

[a] Abbreviations used: PA, polyacrylic acid; AA, acrylic acid; HMAMD, N-(hydroxymethyl)acrylamide; HEAMD, N-(2-hydroxyethyl)acrylamide; HPAMD, N-(2-hydroxypropyl)-acrylamide.
[b] Ratio of the two monomers charged to reaction.
[c] Using LVT No. 1 spindle, 60 rpm.
[d] GPC in 0.05 M Na_2SO_4 solution.

hydrolysis or degradation of the N-(hydroxyalkyl)-acrylamide was not significant or that the copolymerization was approximately random.

Results and Discussion

Evaluation of the Copolymers: The polymer solutions were evaluated for their deposit control and dispersant activities. The tests included calcium phosphate inhibition, calcium carbonate inhibition, iron oxide dispersion, and clay dispersion. The procedures for these tests have been previously reported (12). A commercially available polyacrylic acid was also tested for comparison. The results are shown in Tables II to V.

Calcium Phosphate Inhibition: Control of soluble phosphates is important to the success of phosphate/phosphonate treatment programs. The test procedure included mixing calcium chloride and sodium orthophosphate solutions, allowing the resulting solution to equilibrate for specified time, filtering the mixture, and measuring residual, soluble phosphate ion concentration. High soluble phosphate concentrations indicate good deposit control, since the treatment inhibited calcium phosphate crystallization. The percent inhibition was calculated according to the equation in Table II. As shown in Table II, the results indicated that when compared to the polyacrylic acid, copolymers of acrylic acid and N-(hydroxyalkyl)-acrylamide were quite effective in inhibiting calcium phosphate formation at the normal use dosage of 10 to 20 ppm. Among the three copolymers tested, acrylic acid/N-(2-hydroxyethyl)acrylamide appeared to be the most effective.

Calcium Carbonate Inhibition: The test procedure included mixing calcium chloride solution (with and without treatment) and calcium carbonate solution. After equilibrium, filtration, and pH adjustment, the residual calcium ion concentration was then titrated by EDTA solution. A higher residual calcium ion concentration indicates better inhibition activity and, therefore, more effectiveness in controlling calcium carbonate deposition in the treated water. As shown in Table III, at dosages of 1 to 5 ppm, the polyacrylic acid was more effective than the acrylic acid/N-(hydroxyalkyl)-acrylamide copolymers.

Dispersant Activity: Tests were conducted utilizing iron oxide and clay suspensions in order to establish the efficacy of the copolymers as dispersants for suspended particulate matter. Kaolin clay was used for the clay dispersion study. According to the procedure, separate 0.1% iron oxides and 0.1% clay suspensions in

Table II: Calcium Phosphate Inhibition Test

Testing Conditions	Solutions
Temperature 70°C	36.76 g of $CaCl_2 \cdot 2H_2O$/L of H_2O
pH 7.5	
17 hr equilibrium	0.4482 g of Na_2HPO_4/L of H_2O
Ca^{2+} : 250 ppm as $CaCO_3$	
PO_4^{3-} : 6 ppm	

$$\% \text{ Inhib} = \frac{\text{ppm } PO_4^{3-}(\text{treated}) - \text{ppm } PO_4^{3-}(\text{control})}{\text{ppm } PO_4^{3-}(\text{stock}) - \text{ppm } PO_4^{3-}(\text{control})} \times 100$$

% Inhibition

Sample[a]	Active Dosage 5 ppm	10 ppm	20 ppm
PA	34	42	67
AA/HMAMD	47	59	56
AA/HEAMD, Sample 2	55	59	65
PA	17	38	42
AA/HEAMD, Sample 3	28	67	83
AA/HPAMD	7.9	59	68

[a] See footnote a and sample # in Table I.

deionized water were prepared. The hardness of each slurry was adjusted to 200 ppm Ca^{2+} as $CaCO_3$, and the resultant media were each mixed until uniform suspensions resulted. The pH of each suspension was adjusted to about 7.5. In this test, higher values in the difference of measured transmittance (Δ%T) indicate better dispersing activity as more particles remain suspended in the aqueous medium.

As shown in Tables IV and V, copolymers were quite effective in dispersing iron oxide and clay as compared to polyacrylic acid. Among the polymers tested, acrylic acid/N-(2-hydroxyethyl)acrylamide seemed to be the most effective.

Mechanism: It has been suggested that the hydroxyl functionality is important in the adsorption of anionic polyelectrolytes from water onto metal oxides (13-14). The mechanism involves hydrogen bonding of the hydroxy groups to negatively charged surface oxide ions:

$$MO^- + HOA^{n-} \rightleftharpoons MO^- \ldots HOA^{n-}$$

where A^{n-} represents an anion or anionic polyelectrolyte, and MO^- is the metal oxide ion. Study also showed that polycarboxylates containing hydroxyl groups

Table III: Calcium Carbonate Inhibition Test

Testing Conditions	Solutions
Temperature 70°C	3.25 g of $CaCl_2 \cdot 2H_2O$/L of H_2O
pH 9.0	
5 hr equilibrium	2.48 g of Na_2CO_3/L of H_2O
Ca^{2+} : 442 ppm	
CO_3^{2-} : 702 ppm	

$$\% \text{ Inhib} = \frac{\text{mL EDTA titr.(treated)} - \text{mL EDTA titr.(control)}}{\text{mL EDTA titr. (stock)} - \text{mL EDTA titr.(control)}} \times 100$$

% Inhibition

Sample[a]	Active Dosage 1 ppm	3 ppm	5 ppm
PA	35	59	69
AA/HMAMD	2.8	53	63
AA/HEAMD, Sample 2	4.6	51	58
PA	15	29	71
AA/HEAMD, Sample 3	6.1	29	51
AA/HPAMD	0.9	19	49

[a] See footnote a and sample # in Table I.

Table IV: Iron Oxide Dispersion Test

Testing Conditions	Solutions
Temperature 25°C	0.1% Solution of Fe_2O_3 in H_2O
pH 7.5	3.68 g of $CaCl_2 \cdot 2H_2O$/100 mL of H_2O
Ca^{2+} : 200 ppm as $CaCO_3$	

$$\Delta \% \text{ Transmittance} = \% \text{ T (control)} - \% \text{ T (treated)}$$

Δ % Transmittance

Sample[a]	Active Dosage 5 ppm	10 ppm	20 ppm
PA	2.0	11	19
AA/HMAMD	18	30	35
AA/HEAMD, Sample 2	29	30	31
PA	5.7	12	23
AA/HEAMD, Sample 3	41	40	47
AA/HPAMD	26	43	45

[a] See footnote a and sample # in Table IV.

Table V: Clay Dispersion (Kaolin) Test

Testing Conditions	Solutions
Temperature 25°C	0.1% Solution of Hydrite UF in H_2O
pH 7.5	3.68 g of $CaCl_2 \cdot 2H_2O$/100 mL of H_2O
Ca^{2+} : 200 ppm as $CaCO_3$	

Δ % Transmittance = % T (control) - % T (treated)

Δ % Transmittance

Sample[a]	Active Dosage 5 ppm	10 ppm	20 ppm
PA	22	28	31
AA/HMAMD	31	35	30
AA/HEAMD, Sample 2	41	39	35
PA	36	43	55
AA/HEAMD, Sample 3	60	51	49
AA/HPAMD	49	59	49

[a] See footnote a and sample # in Table I.

adsorb at pHs significantly above the point of zero charge. Although there is no direct correlation between the extent of adsorption and activity differences as shown in this study, the possible contribution from the hydroxyl group cannot be overlooked.

The copolymers in this study were all in a 3:1 mole ratio of acrylic acid with N-(hydroxyalkyl)acrylamide. Different mole ratios and other molecular weight ranges may have different testing results. However, by continuing this kind of systematic study, we may be able to understand more about the relationship between polymer structure and its activity in water treatment. Hopefully, it can facilitate in choosing or synthesizing a specific polymer for the desired application.

Conclusion

The objective of this study was to correlate polymer structure versus its activity for corrosion control in water-treatment applications. Raw material and manufacturing costs were not considered. At the same active dosages, acrylic acid/N-(hydroxyalkyl)acrylamide (3:1 mole ratio) copolymers showed an overall improved performance in calcium phosphate inhibition and iron oxide and clay dispersion as compared to the polyacrylic acid. Among the three copolymers tested, acrylic acid/N-(2-hydroxyethyl)acrylamide appeared to be the most effective. Differences in activity were attributed to the additional functional group in the copolymer versus the homopolymer.

Acknowledgments

The author gratefully acknowledges the support of his colleagues at Betz and, in particular, W. R. Snyder for activity testing and L. D. Chadwick for ^{13}C NMR spectroscopy.

Literature Cited

1. "Handbook of Industrial Water Conditioning", 8th ed., 1980, Betz Laboratories, Inc., Trevose, PA; p. 169.

2. Nichols, J. D.; Clavin, J. S.; Blasdel, J. E. Hydrocarbon Process 1980, 59 (10), 75.

3. Vorchheimer, N. In "Polyelectrolytes for Water and Wastewater Treatment"; Schwoyer, W. L. K. Ed., CRC Press, Boca Raton, FL 1981; Chap. 1., p.1.

4. Hann, W. M.; Natoli, J. NACE Corrosion/84, New Orleans, LA 1984; Paper No. 315.

5. Godlewski, I. T.; Schuck, J. J.; Libutti, B. L. U.S. Patent 4,029,577, 1977.

6. May, R. C.; Geiger, G. E. U.S. Patent 4,303,568, 1981.

7. Woerner, I. E.; Boyer, D. R. NACE Corrosion/84, New Orleans, LA 1984; Paper No. 315.

8. Hagstrand, W. E. Proc. Int. Water Conf., 44th, Pittsburgh, PA 1983; p. 118.

9. Gunderson, L. O.; Grove, M; Keust, H. U.S. Patent 3,285,886, 1966.

10. Kawasaki Y.; Hanno, K. U.S. Patent 4,432,884, 1984.

11. Kopecek J.; Bazilova, H. Eur. Polym. J. 1973, 9,7.

12. Snyder W. R.; Feuerstein, D. U.S. Patent 4,427,568, 1984.

13. Morrison, Jr., W. H. J. Colloid Interface Sci., 1984, 100 (1), 121.

14. Davis, J. A.; Leckie, J. O. In "Chemical Modeling in Aqueous Systems," Jenne, E. A.; Ed.; ACS Symposium Series, 1979, Vol. 93, p. 299.

RECEIVED January 21, 1986

26

Performance Aspects of Plasma-Deposited Films

H. P. Schreiber[1], J. E. Klemberg-Sapieha[2], E. Sacher[2], and M. R. Wertheimer[2]

[1]Chemical Engineering Department, Ecole Polytechnique of Montreal, Montreal, Quebec, H3C 3A7, Canada

[2]Engineering Physics Department, Ecole Polytechnique of Montreal, Montreal, Quebec, H3C 3A7, Canada

Large volume microwave plasma apparatus (LMP) has been used to produce thin, integral films on a variety of substrates. Through control over such variables as substrate temperature and plasma power density, films have been produced from organo-silicones and inorganic (SiN) starting materials. Measurements of moisture permeation through such films indicate that the plasma deposited films are highly resistant to water transport, making them particularly attractive in applications calling for corrosion resistance. To illustrate the effect, the moisture transport characteristics of polyimide substrate with deposited films some 0.5 μm in thickness were measured and found to give up to two orders of magnitude reductions in permeation coefficients. Silicon nitride films were particularly effective in this regard. In another illustration plasma-deposited organo-silicone polymers strongly inhibited the corrosive attack of metal films immersed in aggressive media. Deposition kinetics in microwave plasmas are roughly an order of magnitude greater than in lower frequency (e.g. r.f.) plasmas. This, combined with the feasible scale-up of microwave plasma apparatus holds out the promise of larger-scale applications for plasma produced films in the corrosion protection area.

Glow discharge or "cold" plasmas are gaining increased currency for the deposition of novel and potentially valuable macromolecular coatings. The range of properties attainable by a plasma-polymer is wide, and depends critically on such variables of the plasma deposition process as choice of monomer, substrate temperature (T_s), power density (p), the excitation frequency (ν), and others including monomer flow rate, reactor geometry, etc... Control over these variables can produce crosslinked, dense deposits which adhere tenaciously to

0097–6156/86/0322–0291$06.00/0

substrates including metals, polymers, ceramics, etc. To the extent that thin plasma-deposited films function as effective barriers to the diffusion of water and other aggressive fluids, plasma-polymers may serve a valuable role as coatings for the protection of metals against corrosion.

The potential usefulness of plasmas as routes to corrosion protection has been recognised for some time, the work of Williams and coworkers (1,2) being an early example of using glow discharges to overcoat steel. Yasuda and coworkers (3) have made extensive use of plasma-deposited CH_4 structures as "primers" for the enhanced adhesion of conventional topcoats, thereby indirectly also improving the corrosion resistance of the composite coating. Earlier research from our laboratory (4) gave empirical evidence of the corrosion protection due to plasma-deposited organo-silicone polymers, and of the excellent high temperature stability of the plasma coatings. The present paper is directed mainly at the question of water permeability through plasma films, the presence of water at a surface being considered essential for the initiation and propagation of corrosion processes.

As in earlier reported work (4), the present study has used a large-volume microwave plasma (LMP) facility. The choice is based on the favorable deposition kinetics at microwave frequencies, and on the relative ease of scaling experimental LMP apparatus to industrially useful size.

Experimental

All plasma experiments were performed with an LMP reactor which has been described elsewhere in detail (5). The reactor is equipped with a heatable, rotating platen, 15 cm. in diameter, to which are fastened specimens for surface modification. The reactor operates at 2.45 GHz frequency and at power in the range 0.1 - 2.5 kW.

Primary interest was in the barrier properties obtained from plasma organo-silicones and from inorganic "SiN" coatings. Spectral grade HMDSO was used in the former case, while mixtures of SiH_4 and NH_3 were used to produce the SiN structures. The substrate in much of the work was DuPont Kapton type H polyimide film, 51 μm thick. Substrate temperatures extended to 450°C, as described earlier (6). The thickness of plasma-polymer deposits was about 0.5 μm. Moisture permeation was evaluated by the routine of ASTME-96-53 T (water vapor transmission of materials in sheet form). Additional, more precise data, were obtained with both a Dohrmann Envirotech Polymer Permeation Analyser, modified as previously described (6), and a Mocon "Permatran W" moisture permeation apparatus.

To test the corrosion protection conferred by LMP - produced films, glass microscope slides bearing 5000 Å - thick layers of aluminium (by vacuum evaporation) were overcoated with P-PHMDSO films. In this experimental series plasma deposits were maintained at thicknesses near 1000 Å, and were produced at T_s ranging from 100°C to about 300°C. Plasma-coated and control samples were placed in a bath of alkaline cleaning fluid (pH 8.5) and inspected periodically for loss of Al, as described in an earlier publication (5).

Finally, scanning electron microscopy (SEM) was used to study the morphology of plasma deposits obtained from HMDSO.

Results and Discussion

a) Moisture barrier properties

As stated, the capability of plasma deposits to reduce the access of water to corrosion-sensitive surfaces may be an important motivation for their application in corrosion protection. In order to study this property, Kapton polyimide film was selected as the substrate because of its high inherent permeability to water and its ability to resist elevated temperatures. The response of Kapton film overcoated by PPHMDSO to the permeation of water vapor is shown in Fig. 1. Clearly, the presence of the organo-silicone plasma film greatly reduces water permeation. The magnitude of the effect is much enhanced when plasma polymers are produced at high T_s and p. It was shown earlier (4) that as T_s and p increase, the plasma polymer becomes denser and increases considerably in the proportion of inorganic linkages. By way of illustration, at T_s in the 400°C range, the density of P-PHMDSO is about 1.8 $gm.cm^{-3}$ (4); these rigid, glassy structures therefore serve particularly well as moisture barriers and, by implication, as corrosion inhibiting coatings for metals. The results in Fig. 1 suggest that at low power, the permeation changes are related primarily to dielectric heating of the substrate, while at high power the permeation changes are principally

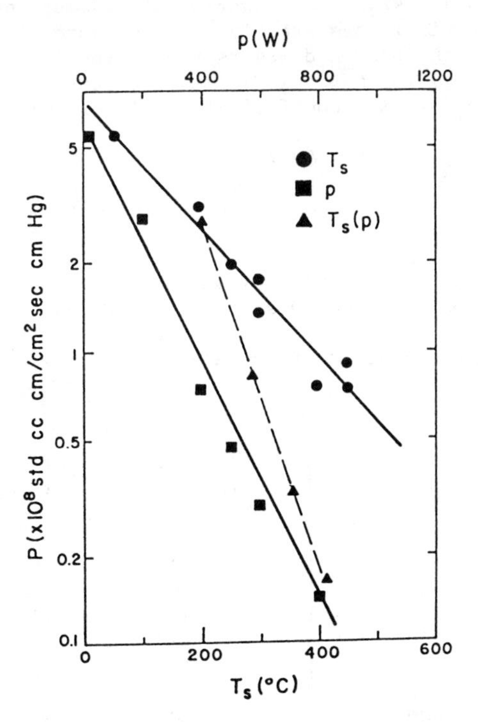

Figure 1. Permeation coefficient of PPHMDSO-Kapton, as function of T_s (●), power (■) and dielectric heating effects (▲) (from ref. 6).

to be associated with the changed chemistry of plasma deposits. The changes in deposit chemistry, in turn, may be associated with increased monomer fragmentation under these operating conditions.

The change in water vapor flux, ΔF, due to the presence of 0.5 μm plasma films may be expressed by

$$\Delta F = \left(\frac{F_o - F}{F_o}\right) \times 100 \quad (1)$$

Here F_o is the water vapor flux (mg.m^{-2} day^{-1}) through a control film, while F is the corresponding flux through an overcoated film. Values of ΔF are convenient for a comparison of the moisture barrier efficacy of SiN and organo-silicone plasma deposits. Pertinent data are given in Table I. It is evident that SiN deposits are particularly effective moisture barriers, being superior to organo-silicone films under any given set of plasma variables. Again, T_s and p are shown to be vital to the control of barrier film. properties. Thus, at constant total power (e.g. 600 watts) stronger reduction in water transport is produced at increasing T_s, the effect being equally important for HMDSO and SiN deposits, as shown in the table. The important effect of p at constant T_s is also documented in Table I (for T_s=250°C). At all LMP operating conditions used in this work, the superior performance of SiN as compared with HMDSO deposits is manifest. Indeed, when SiN films are produced at T_s > 250°C, and p in the range of 1.0 kW. the reduction of moisture permeability conferred to Kapton increases greatly. An estimate of the permeation coefficients of the plasma deposits themselves (i.e. of the P-PHMDSO and the SiN films) further illustrates the superiority of the SiN coating. The permeation coefficient, π of a film is given by

$$\pi = \frac{\Delta W}{t} \frac{\ell}{A} \frac{1}{C_{H_2O}} \quad (2)$$

Table 1. Comparison of water flux changes, ΔF, due to plasma deposits of HMDSO and SiN on Kapton.

In all cases plasma films are 0.5 μm thick

Plasma Operation T_s (°C)	p(W)	ΔF (%) for (PPHMDSO)	P(SiN)
100	600	11	40
200	200	10	-
200	600	18	70
200	800	38	-
200	1200	64	-
250	400	23	-
250	600	37	85
250	1000	65	93
250	1200	78	>99

where ΔW is the weight increment due to water sorption, t is the contact time, ℓ and A are the thickness and area of the film and C_{H_2O} is the vapor pressure of water under conditions chosen for analysis. Since $\Delta W/At$ is equal to the flux parameter F, in equation (1), it follows that

$$\pi = \frac{F.\ell}{C_{H_2O}} \qquad (3')$$

or

$$F = \pi.C_{H_2O} \times 1/\ell \qquad (3'')$$

The linearity between F and $1/\ell$ has been documented by experiments to be reported elsewhere. In the case of a composite film, consisting of materials, 1 and 2, (e.g. Kapton and SiN, or PPHMDSO), we can write

$$\frac{\ell}{\pi} = \frac{\ell_1}{\pi_1} + \frac{\ell_2}{\pi_2} \qquad (4)$$

The permeation coefficient of Kapton under test conditions similar to those used here has been reported as 4.47×10^{-11} (g.cm/cm^2.sec. cmHg), in close agreement with our own measurements. Thus, solving equation (4) for π_2, we find

$$\pi_{SiN} = 2.8 \times 10^{-14} \quad \text{g.cm/cm}^2\text{.s.cmHg}$$
$$\pi_{PPHMDSO} = 2.5 \times 10^{-13}$$

The permeation coefficients of the two thin film materials differ by an order of magnitude. To the extent then, that water initiates and propagates the corrosion of metal substrates, plasma films of SiH_4/NH_3, produced under the stated conditions seem to be particularly promising corrosion inhibitors.

While, in the case of PPHMDSO, the higher inorganic content produced at high T_s and p enhances film barrier properties, an equally important factor may be the dependence of plasma film morphology on these deposition variables. SEM well documents these effects (5), as illustrated in Figure 2(a) and (b). At low T_s, for example near room temperature as in Fig. 2(a), PPHMDSO films consist of agglomerates of spheroidal particles with clearly distinguishable interparticle boundaries. This type of plasma product morphology has been previously discussed in the literature (8,9). At high T, such as the 400°C used to produce the specimens shown in Fig. 2(b), much smoother films are formed, the few remaining discrete particles now being imbedded in a continuous matrix of plasma-polymer. Logically enough, the latter films would be much more suitable as vapor or fluid barriers than the discontinuous structures exemplified in Fig. 2(b).

b) Corrosion protection

Qualitative evaluation of the specimens immersed in alkaline cleaning fluid has demonstrated (5) the capability of PPHMDSO topcoats to protect the Al films supported on glass slides. The intrusion of

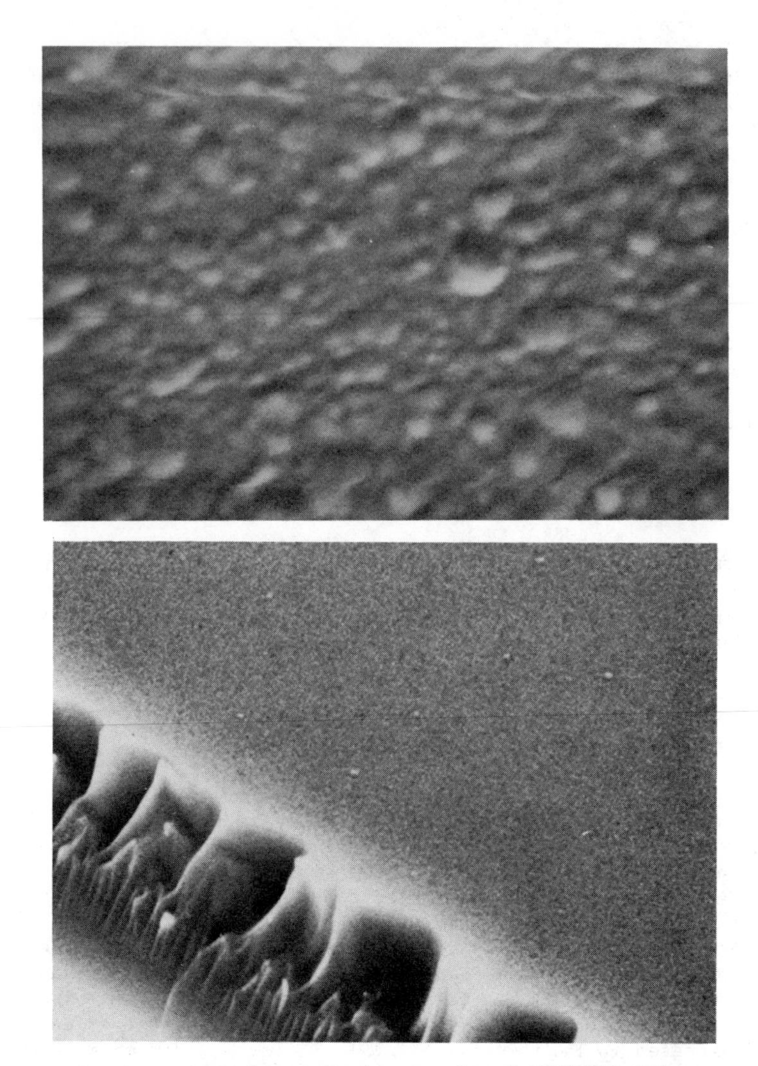

Figure 2. Scanning electron micrographs of PPHMDSO films at substrate T = near ambient (top), and at T_s = 400 °C (bottom). Reproduced with permission from ref. 5.

fluid to the Al-glass interface in uncoated, control samples was rapid, with significant loss of Al evident after about 15 min. of immersion. In these samples total loss of Al occured within the first 60-75 min. of contact. P-PHMDSO deposited at T_s < 100°C was ineffective in this test, but at T_s > 100°C increasing degrees of protection were given by the plasma films. In samples coated at T_s = 150°C the Al layers remained intact for about 4h. immersion, with gradual deterioration noted thereafter. Samples coated at T_s > 250°C showed no sign of metal loss even after several days of immersion. Evidently, in these systems, a shift from marginal to excellent corrosion protection takes place in the critical T_s range of 150-250°C. We attribute this in part to the denser, more inorganic structure of the plasma deposit and to stronger bonding at the plasma polymer/metal interface. The increased contiguity of plasma films, discussed in terms of SEM results above, would, of course, also contribute greatly to the improved performance.

The implied capability of these plasma deposits to inhibit corrosion at metal surfaces may be of practical as well as of basic importance. An important consideration in this respect is the rapid rate of deposition for such protective coatings attainable at microwave frequencies. Since plasma technology is still in a process of evolution, optimum deposition kinetics cannot yet be stated; however, the marked effect of excitation frequency on the deposition of organo-silicones can be documented (10), as in Fig. 3. Here, using terminology and comparative data due to Yasuda et al. (3), it is shown that deposition rates in microwave plasmas exceed those at lower (e.g. radio) frequencies by about an order of magnitude. Coupled with the relative ease of scaling microwave plasma reactors to a size pertinent to industrial coating operations (11), these features place microwave plasma processes into the domain of industrial relevance.

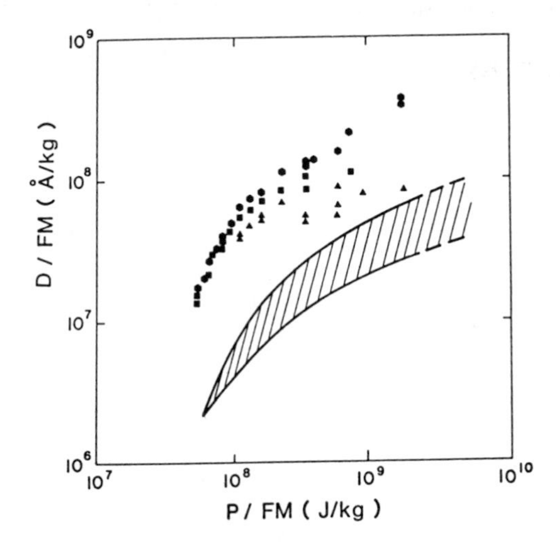

Figure 3. Deposition rates (D) for organo-silicones, normalized to flow rate (F) and monomer molecular weight (M), as function of plasma power (P). Shaded develope: 13.56 MH_z plasmas; data points are for 2.45 GH_z plasmas at monomer pressure 0.1 Torr (▲), 0.2 Torr (●) and 0.4 Torr (■).

Acknowledgment

Parts of this work were supported by the Natural Sciences and Engineering Research Council, Canada, by the National Research Council, Canada and by the Ministry of Education, Québec.

Literature cited

1. T.Williams and M.W. Hayes, Nature 209, 769 (1966) and 261, 614 (1967).
2. T. Williams and W.H. Edwards, Trans. Inst. Met. Finish. 44, 119 (1966).
3. Y. Matsuda and H. Yasuda, This Solid Films, 118, 211 (1984).
4. H.P. Schreiber, M.R. Wertheimer and A.M. Wrobel, Thin Solid Films, 72, 487 (1980).
5. M.R. Wertheimer, J.E. Klemberg-Sapieha and H.P. Schreiber, Thin Solid Films 115, 109 (1984).
6. E. Sacher, J.E. Klemberg-Sapieha, H.P. Schreiber and M.R. Wertheimer, J. Appl. Polym. Sci., Polym-Symp, 38, 163 (1984).
7. E. Sacher and J.R. Susko, J. Appl. Polym. Sci., 23, 2355 (1979).
8. L.F. Thompson and G. Smolinsky, J. Appl. Polym. Sci., 16, 1179 (1972).
9. J. Grebowicz, T. Pakula, A.M. Wrobel and M. Kryszewski, Thin Solid Films, 65, 351 (1980).
10. M.R. Wertheimer and M. Moisan, J. Vac. Sci. Technol. A3, 2643 (1985).
11. J. Kieser and M. Neusch, Thin Solid Films, 118, 203 (1984).

RECEIVED January 27, 1986

Humidity Testing of Silicone Polymers for Corrosion Control of Implanted Medical Electronic Prostheses

Philip R. Troyk, Michael J. Watson, and James J. Poyezdala

Pritzer Institute of Medical Engineering, Illinois Institute of Technology, IIT Center, Chicago, IL 60616

Adhesion tests are insufficient to qualify polysiloxanes for the corrosion control of electronic assemblies, and parameters such as material hardness may be of equal importance as the bond strength of the polymer to the protected assembly. Silane coupling agents in the form of commercially available surface primers do not seem to enhance the corrosion control of silicone encapsulated electronic devices despite higher measured bond strengths for primed samples as opposed to unprimed ones. This is probably due to water soluable contaminants in the primers. Accelerated temperature-humidity tests upon 72 interdigitated test substrates encapsulated with various silicone elastomers treated with and without surface primers show a strong dependence upon the material hardness in controlling corrosion. Electrical leakage current measurments combined with visual examination were used as performance criteria for the encapsulants.

Silicone elastomers have been used for the protection of integrated circuits, epoxy glass printed wiring boards, high voltage assemblies and other electronic assemblies exposed to harsh environmental conditions. Among the most severe conditions is exposure to high humidity with resultant corrosion of the electrical conductors and components. The corrosion takes place not only as the result of electrochemical couples, due to the variety of metals used in a typical electronic assembly, but also due to forced voltage potentials present within the assembly during normal operation. The latter may present a higher stress than the former since the potential differences during operation may exceed typical electrochemical couples by orders of magnitude. This is particularly the case in high density assemblies where the spacing between printed-wiring-board (PWB)conductors may be as small as 0.003" to 0.0010". With the advent of technology such as small-outline packaging (with pin spacings of 0.03") the corrosion protection of PWBs has become a major industry problem.

Another application for which polymers, and in particular sili-

0097-6156/86/0322-0299$06.00/0

cones, have been used for corrosion control is that of medical implanted electronic devices. In the past decade, hermetic cans have emerged as the preferred method of corrosion control for implanted medical devices. Although hermetic cans with glass-on-metal electrical feedthroughs have been suitable for devices such as pacemakers and simple monopolar neurostimulators, they are not easily miniaturized. Miniaturization is important because the effectiveness of sensors and stimulators is often significantly enhanced by reducing their invasive side effects. The miniaturization of electronic circuitry is possible using modern thin-film techniques, and such technology has already resulted in the design and fabrication of numerous recording and stimulation micro-electrode arrays, biosensors, and integrated circuits applicable to the development of subminiature medical electronic implants (1,2,3,4,5,6). However, the successful design and implementation of complete and functional implants depends not only upon the fabrication of these thin-film devices, but also upon the ability to protect them during use from the hostile physiological environment. Unless significant advances in hermetic packaging are achieved, the relatively large size of hermetic cans will continue to prelude their use when miniaturization is important and presently no universally accepted method of encapsulation exists for moisture protection of non-hermetic implants.

Even when size might allow hermetic packaging such as sealed welded cans, the need for multiple feed-throughs in devices such as neuromuscular stimulators presents new packaging problems which have not formerly been present in other medical implants such as pacemakers. These feed-throughs must be protected from corrosion and electrical leakage, and this is usually accomplished by overcoating the hermetic package with an encapsulant (7,8). The resulting insulated feed-through and wire connector can occupy as much as one-third the volume of a modern pacemaker. Electrical leakage, which significantly affects the impedance at these sites, is permissible only for low impedance circuits. The increased use of sensors with high impedance outputs in implanted systems places greater demands upon the encapsulant's insulating capabilities, and the use of highly corrosion-resistant metals as compensation for encapsulant inadequacies in preventing implant degradation is no longer a viable design philosophy.

Encapsulants are an alternative to the welded can. A properly chosen encapsulation system can provide adequate corrosion protection for electronic assemblies and a number of organic barrier coatings and encapsulants have been used to protect electronic systems from deleterious effects of water vapor including products based upon epoxy, imide, and silicone polymers. The latter, especially those polymerized by addition mechanisms have gained wide acceptance for military and space assemblies which must be protected from humidity and altitude effects, since polysiloxanes are easily processed, their performance has been found satisfactory, and they provide versatility in package designs. However, in medical electronic implants the failure incidence of encapsulants (especially silicone polymers) for electronic devices such as neuromuscular stimulators is much greater than in nonmedical devices such as plastic-pak integrated circuits, even when the latter are subjected to accelerated stress tests (9). It is not clear whether this poor success rate is due to improper application and processing of the encapsulants or to material inadequacies. Silicone polymers are often misused in the sense that they regarded as vapor barriers. However for any system, the high failure rate, undoubtedly, is a result of the severe exposure conditions that exist in a physiological environment. Not only is the implant exposed to an aqueous system, but the electrolytic environment and ionic mobility present a milieu conducive to current leakage. In this regard, it should be noted that chlorides, as well as sodium ions, have been implicated in the degradation of silicone nitride and silicone rubber encapsulated circuits (10).

Mechanisms of Implant Failure

Body fluids, contain a high concentration of ions and can cause implant failure in the following two ways:

1) Migration of ions (especially sodium) into unprotected semiconductors results in the degradation of junction characteristics such as reverse bias leakage currents so that the chip no longer performs to specifications and eventually fails.
2) Leakage paths form between normally insulated conductors due to fluid invasion, so that areas of the circuit which must be insulated from each other do not remain so, degrading circuit operation. These leakage currents may cause redox reactions resulting in the corrosion of conductors such as chip metallizations. Similarly, electrochemical couples in an aqueous environment can produce corrosion. The entry of fluid may be direct, through mechanical defects in the package, or by diffusion of water through the polymer encapsulant.

Migration of Ions. In recent years, the first type of failure has been fairly easy to avoid. Chips are protected from contamination during normal handling by a process known as "passivation." In this procedure, several protective layers of ceramic material are deposited or thermally grown over the chip. The passivating layer serves as a barrier to migrating ions. Silicon dioxide offers partial protection against some contaminants. Unfortunately, it is permeable to sodium ions and does not offer complete protection (11,12). Although used for internal passivation layers, almost no commercially available chips have final passivation layers of silicon dioxide. In the past few years, silicon nitride has emerged as another passivation material. Sodium penetration of a properly applied nitride layer is very low--even undetectable (12,13,14). Many semiconductor manufacturers now offer nitride passivated chips. However, the application of silicon nitride can have variable results and the presence of impurities will reduce the effectiveness of the layer as an ionic barrier. Trace amounts of oxygen within the deposition equipment can result in an oxy-nitride layer which is less effective against ionic penetration. In addition, many nitride-passivated devices have exposed underlying thermal oxide at each edge of the die. Lateral migration of ions is possible and has been shown to adversely affect device reliability (15). At the present time the most commonly used final passivation layers are the phosphosilicate glasses (PSG). PSG is particularly effective as a final passivation coating, and is characterized by simplicity of application. Layers of PSG have been widely used because they are in less tensile stress than layers of chemically-vapor-deposited silicon dioxide (16,17,18), and have the ability to trap ions (16). The lower stress prevents cracking of the coating.

Leakage. By far the majority of implanted electronic devices fail due to mechanisms of the second type (i.e. leakage). In a poorly encapsulated implant, this failure mode can cause rapid deterioration in a matter of days, hours, or even minutes. As the size of the implant designs are reduced, the risk of this type of failure is greatly increased because due to the small dimensions involved, the time for fluid entry into the package becomes very short.

1. Mechanical Defects. Direct entry of highly ionic fluids such as saline into the package results in the formation of conductive paths between electrochemical couples or conductors of differing potentials. The site of entrance can vary, depending upon the package design, but most frequently involves the interface between the encapsulant and a dissimilar material. Failure due to direct saline entry can be one of two general types: those caused by defects in the bulk material used for encapsulation and those caused by the mechanism of creep.

The nature of bulk defects is straightforward. Pinholes, cracks, or other surface imperfections result in rapid diffusion of saline into the implant resulting in corrosion of electrical conductors. This corrosion lowers the resistance of the original leakage path until normal operation of the electronic circuit is impared and the implant fails.

Creep failures are not often as easy to predict and may not show up immediately. Although the device may not fail as rapidly as in the case of a bulk defect, the final result is identical: corrosion of electrical conductors. In a creep failure, the saline diffuses between two adjoining and usually differing materials. The presence of the fluid tends to separate the two surfaces, thus permitting more fluid to leak into the device. A creep failure is primarily caused by the lack of, or poor adhesion between the two materials. A common site for such leakage is the interface between an exiting lead wire and the encapsulation material. Often, an inappropriate choice for the wire insulation may accelerate the failure. Materials such as Teflon are difficult to bond to, (unless chemically etched) however are frequently used for implanted wires. The output leads from implants such as neurostimulators are usually of a low impedance nature and many implants are designed on the premise that some electrical leakage is tolerable at this site. Noble metal connections can reduce the corrosion to an "acceptable" level. This assumption is not true for leads which enter the package from sensors such as micro-electrodes which are characterized by relatively high impedances. The trend for neuroprosthetic devices is towards closed-loop control in which the use of high impedance bioelectric sensors will be common. In addition, differing potentials within multi-circuit cables can result in corrosion even when the conductors are fabricated from highly corrosion resistant materials such as MP35N.

Another common location for creep failures of encapsulated assemblies is at sharp corners or edges. Many encapsulants such as polyimides must be applied in thin coats, and coverage of points, edges, or corners is difficult or impossible. Sharp corners, characteristic of most thin-film devices provide ideal conditions for the initiation of creep failures due to the resulting irregularity of the encapsulant coverage.

2. Diffusion of Water. Although the protection against moisture afforded to encapsulated implants by some high molecular-weight organic materials depends on the barrier effectiveness of the encapsulant, for silicones, other factors, notably bond conditions at the encapsulant/substrate interface (adhesion), thermal stability, mechanical integrity, and electrical compatibility play a more important role. In fact, water vapor permeability may not be the most critical property that will determine the acceptability of a polymer as an encapsulant for electronic assemblies. This appears to be the case for industrial and military electronic assemblies which use the generally preferred polysiloxanes, which are by no means impermeable to water vapor, a consequence of their helical chain molecular structure. It has been previously well accepted (19), although not widely realized, that the dominant property of silicone polymers for use against moisture penetration of electronic assemblies is their ability to adhere (with appropriate primers) to the protected assembly.

Interface Properties of Silicone Polymers

Hydrophilic surface sites permit the adsorption and condensation of water. Effective blocking of hydrophylic sites on encapsulated surfaces would require chemical reaction of these sites with functional groups of appropriate coatings, or deposition of strongly adherent films of organic compounds on the protected surface. Inactivation of hydrophylic surface sites in the corrosion control of electronic devices had been proposed as early as 1969 by White of Bell Telephone Labs (19). He demonstrated the effectiveness of many different types of coatings in protecting integrated circuit chips from moisture damage. Silicone and imide materials showed excellent protection against moisture penetration in this study. In subsequent high pressure steam tests, superior performance was seen for the silicone polymers, in spite of their relatively high water vapor permeability. Both classes of materials contain groups which can react with surface hydroxyls present on almost all hydrophilic surfaces. From this study emerged the theory that moisture protection of normally hydrophilic surfaces by the highly semi-permeable silicones was a consequence of reactions with surface hydroxyl characterized by strong adhesion of the polymer to the protected surface. Various theories of adhesive bonding are well documented (20,21) and will not be repeated here. Similarly, the need for rigid cleaning protocols to enhance adhesion has been previously well described (22).

Most commercially available silicone elastomers, and especially those which polymerize by addition-cure, do not, or weakly, adhere to most surfaces. Adhesion promoters known as surface primers are available and are often used to adhere the silicone to the protected surface. These primers are generally formulated from silane coupling agents which contain functional groups that react with the protected surface as well as the polymer (23). There is a trend in the chemical industry towards the blending of coupling agents and base resins to form "self-priming" materials.

Adherent films would not necessarily require formation of covalent bonds at the interface, since localized intermolecular dispersion forces that are operative in the adsorption of coatings (with good wetting properties) should provide stable interfacial bond conditions. Among candidate materials which could fulfill the requirements of good adhesion and substrate protection from moisture are epoxy-modified polyurethanes and epoxy-siloxane polymers.

Silicones for Electronic Encapsulation

Since the initial work of White (19), the Bell System and other major semiconductor users have extensively used silicones in the protection of numerous thin-film and thick-film devices (24,25). The materials primarily have been condensation-cure silicones in xylene dispersion. The performance of encapsulated semiconductors used within the Bell System is well documented (26), and studies continually in process support the use of silicones for the prevention of electronic component deterioration under conditions similar to the *in-vivo* environment.

More recently Wong (27), has reported the results of tests designed to characterize parameters of electronic silicones by Thermogravimetric, Fourier Transform-Infrared, and Gas Chromatography/Mass Spectrometry. This work has concentrated upon condensation-cure materials used within the Bell System, rather than the end-blocked free radical (addition-cure) silicones.

Based upon the theory that the moisture protection of electronic assemblies by encapsulants is directly related to the adhesive bond between the encapsulant and the underlying surface, adhesion testing has emerged has the primary method of encapsulant qualification for implant use. Pacemaker manufacturers have performed lap shear and butt tensile testing of selected substrate/encapsulant combinations after high humidity exposure and these tests relate the bond strengths to the corrosion protective ability of the encapsulant (28). Donaldson (29) has also investigated the effects parameters such as electric field strength (30), and tensile forces within actual implants. Corrosion is an electrochemical process whose prevention depends upon the ability of the encapsulant to prevent the flow of electrical leakage currents. Electrical leakage current tests combined with bond strength tests would provide not only a measure of the polymer's ability to remain in contact with the protected surface, but also to act as an insulator to electrical leakage currents which inevitablity will result in corrosion.

Other investigators have noted that assemblies in which the silicone elastomer was poorly bonded to the protected surface have survived implantation longer than expected if considering adhesion as the primary protective mechanism. In this regard, Donaldson (31) has advanced the theory of osmotic pumping as an alternate mechanism of corrosion protection by silicones. For submersion in physiological saline, diffusion of water through the silicone into the device, is opposed by an osmotic gradient out of the device.

Methods

We tested 6 different silicone casting elastomers in conjunction with 3 surface primers. The materials and their designations are as follows:

Surface Primers

Designation	Name	Manufacturer	Formulation
Type A	Primecoat 1200	Dow Corning	Silane-Silicate Blend
Type B	Chemlok 607	Lord Chemical	Blended Silanes
Type C	SS 4155	General Electric	Ethylorthosilicate

Silicone Elastomers

Designation	Name	Manufacturer	Cured Hardness Durometer (Shore A)
Type 1	RTV-615	General Electric	45
Type 2	Sylgard 184	Dow Corning	35
Type 3	MDX-4-4210*	Dow Corning	25
Type 4	96-083	Dow Corning	63
Type 5	Sylgard 567	Dow Corning	38
Type 6	RTV-3140	Dow Corning	22

*MDX-4-4210 was blended with 10% Sylgard 527, by weight, to decrease viscosity for casting purposes.

The first 5 silicone elastomers are addition-cure products. The 6th material is a condensation-cure product which has been previously used for medical implant protection (7). As verified by lap shear tests, (described below), the first 3 elastomers provide minimal adhesion in the absence of a surface primer. The last 3 are self-priming materials. To permit visual examination, we tried to select clear materials. All of the elastomers are clear except for types 5 and 6.

Types 1, 2, and 3 were selected for test with each of the primers, types A, B, and C. In addition, type 4 elastomer was used as a primer for types 1, 2, and 3. All elastomers types 1-6 were also tested without primers.

Testing was performed by encapsulation of 4 PWB, G-10, epoxy-glass interdigitated test pattern substrates per test group. The patterns selected were taken from the IPC B-25 standard test board, patterns A and B. These patterns use interdigitated line widths and spacings of 0.006" and 0.012" respectively. Two single-sided A patterns (0.006" lines and spacings) and two single-sided B patterns (0.012" lines and spacings) were prepared for each elastomer-primer combination tested. Following encapsulation, each of the patterns was continuously submersed in 85° C water and placed under a continual voltage stress of 20VDC. Electrical leakage current (under 9VDC bias) was measured for each of the test substrates on a daily basis.

Preparation of each test substrate was performed as follows: A 10" long, Kynar (polyvinylidene fluoride) insulated wire (insulation thickness, 0.005") was soldered to each electrical half of the substrate. The substrate was then completely grit-blasted (40 PSI) using 240 mesh (50um) aluminum oxide grit. Following grit-blasting, the substrate was cleaned in a series of solvents as follows: Methylene Chloride-1 min. wash, Isopropyl Alcohol rinse, Acetone-1 min. wash, Isopropyl Alcohol rinse, Freon TMS-2 min. ultrasonic clean in a Branson B-220 ultrasonic cleaner, Freon TF running rinse. During the cleaning procedure the substrate was handled, with gloves, only by the ends of the connecting wires. After cleaning, a vacuum-bakeout was performed for 12hrs at 120°C in a Precision model 10 vacuum oven at 1-2 torr. The vacuum system was filtered by an activated alumina filled vacuum filter. Upon removal from the vacuum oven, designated substrates were primed with the appropriate primer according to the manufacturer's instructions. Each of the manufacturers was contacted directly in order to verify the procedure as outlined in the data sheet. A substrate ready for encapsulation was placed in a half-open mold which supported the substrate horizontally by its wires. The appropriate encapsulating elastomer was mixed and de-aired and poured into the mold under vacuum casting. This procedure produced encapsulated substrates with 0.100"±0.015" encapsulant thickness over and under the substrate. The diameter of the mold was such that a thickness of at least 0.3" of encapsulant covered the substrate's edge. As a result of the vacuum casting no bubbles were visible in the encapsulated substrates. Curing of the addition-cure elastomers was by forced-air convection at a prescribed temperature of 120°-165°C depending upon the particular elastomer cured. Curing of the type 6 material was done in a humidity chamber of >90%rh for 72hrs. The specific combinations of primers and elastomers tested are listed in table 1. Type 4 elastomer was used as a primer due its previously demonstrated high bond strength in experiments in our laboratory. It was applied in a very thin (<0.005") coat over the substrate and cured before elastomer casting.

A Keithley 610A electrometer with a shielded box and a 9VDC battery comprised the leakage current measurement apparatus. Resolution of the system was at least 0.1pA, however measurements less than 0.5pA were not recorded. Due to the low levels of measured leakage current, a shielded box was essential to the measurement system. The volume resistivity of most silicones shows a strong temperature dependence. Therefore all leakage current measurements were made at room temperature. The substrates were removed from the water bath for a period of 1hr prior to measurement. Visual examination was done on a periodic basis, documenting visual evidence of discoloration, debonding, or corrosion. A dry lap shear bond test, as described in ASTM-D905 (modified), to G-10 for each of the tested materials was performed on an Instron Universal Testing Instrument. Double lap shear samples were prepared in the same manner as the test substrates.

Results

The measured values of leakage current for each of the 72 test substrates over the initial 2 month test period are listed in table I. In addition, visual test results are reported at days 30, 45 and 60. For each group of 4 samples, samples 1 and 2 are the A pattern (0.006" spacing), and samples 3 and 4 are the B pattern (0.012" spacing). The initial baseline leakage currents for each of the samples was in the low pA region, and for many of the samples, the measurement error below 0.5pA was probably greater than the actual leakage current. The Kynar insulated wire for connection to the substrates was a poor choice. Submicroscopic defects in the insulation resulted in discoloration and corrosion in the wires of a number of the samples. Some of the samples failed due to wire breakage. However, subsequent testing on specially prepared samples indicated that the measured values of leakage current were not affected by the wire discoloration; these samples continued to give consistent reading up to the day of wire breakage.

All samples primed with type B primer showed visible corrosion at a very early date. Within one week of water submersion, all of these samples showed some signs of deterioration. In each case, the side of the pattern maintained at positive potential was the first to discolor and then corrode. In contrast none of the samples primed with type 4 material showed visible deterioration at day 60. With the exception of the type B samples, all of the other samples failed to show any noticable deterioration until one month into the test. At this time, most of the types A, B, and C primed samples showed discoloration and debonding. Correspondingly their leakage currents, at day 30, were an average of 35 times greater than those samples primed and encapsulated with the self-priming elastomer types 4, 5, and 6.

The samples on which no primer was used, and the encapsulating material was nonpriming are perhaps the most interesting. Their leakage currents were consistently lower than the types A, B, and C primed samples in spite of the fact that adhesion as tested in lap-shear tests was considerably less than the primed samples. Results from the lap shear bond tests are presented in table II. Note that when used without surface primers, elastomer types 1, 2, and 3 demonstrated the lowest relative bond strengths when compared to the primed and self-priming samples. However, the leakage currents from the unprimed samples were an average of 50 times lower (day 45) than those primed with primer types A, B, and C, and their overall visual appearance was superior to that of the primed samples. The best performance was seen in types 4, 5, and 6 samples. None of these samples showed any signs of significant deterioration-visual or electrical, at day 60. Type 6 material consistently showed overall superior performance.

Discussion

The rapid initial rise of leakage to steady-state values characteristic of most of the test samples was not surprising in light of the relatively high water vapor pemeability of the silicones. Within 6hrs of water submersion most of the samples demonstrated currents of about the same order of magnitude as those which persisted throughout the test.

In many regards, however, this study has raised more questions than it has answered. The lack of performance of all of the primed samples and the corresponding superior performance of the unprimed samples using types 1, 2, and 3 elastomers raises serious questions concerning the validity of adhesion testing as the primary criteria in the selection of silicones for corrosion control of electronic assemblies. The poor performance of the primers A, B, and C, is

Table I. Electrical leakage current measurements for interdigitated G-10 silicone encapsulated substrates over a 60-day period. Visual examinations made at day 30, 45, and 60 are coded as follows: d = discoloration of the conductors, db = debonding of the encapsulant, c = corrosion, na = not applicable.

PRIMER	ENCAPSULANT	SAMPLE#	ELECTRICAL LEAKAGE IN AMPERES FOR ELAPSED TIME					
			0	6 hrs	day 15	day 30	day 45	day 60
TYPE A	TYPE 1	1	<0.5p	640n	700n	200n d	42n d,db	41n d,db
		2	3.0p	140n	360n	74n d,db	42n d,db	41n d,db
		3	3.0p	150n	120n	66n d,db	12n d,db	9.8n d,db
		4	2.0p	74n	100n	82n	9.4n	9.4n
	TYPE 2	1	<0.5p	20n	400n	260n d	62n d	39n d,db
		2	<0.5p	20n	540n	400n d	83n d	59n d
		3	<0.5p	5.8n	160n	86n c,db	24n c,db	18n c,db
		4	1.0p	20n	220n	82n	22n	22n db
	TYPE 3	1	1.0p	320n	3.2u	620n d	86n d,db	90n c,db
		2	64p	1.8u	2.0n	440n d	26n d	40n d
		3	3.0p	32n	620n	300n	86n	80n
		4	1.5p	120n	320n	510n	43n db	36n db
TYPE B	TYPE 1	1	2.9p	160n	8.8n	74n c,db	3.0n c,db	1.6n c,db
		2	1.0p	96n	32n	68n c,db	9.4n c,db	1.7n c,db
(note 1)		3	1.5p	6.4n	2.8n	2.2n c,db	960p c,db	290p c,db
		4	2.0p	700n	failed @ day 10, current >100uA			
	TYPE 2	1	<0.5p	7.4n	16n	9.4n c,db	1.8n c,db	620p c,db
		2	1.0p	7.2n	8.0n	8.0n c,db	2.8n c,db	1.8n c,db
		3	<0.5p	2.6n	2.0n	3.8n c,db	440p c,db	420p c,db
		4	2.5p	540p	3.0n	3.4n c,db	16n c,db	18n c,db
	TYPE 3	1	<0.5p	3.5n	5.4n	2.8n c,db	2.5n c,db	2.5n c,db
		2	<0.5p	140n	14n	4.0n c,db	540p c,db	420p c,db
		3	1.5p	4.0n	2.0n	420p c,db	150p c,db	180p c,db
		4	<0.5p	5.4n	2.2n	3.0n c,db	280p c,db	210p c,db
TYPE C	TYPE 1	1	7.0p	38n	140n	wire failure		
		2	5.0p	32n	220n	40n d	34n d	30n d
		3	<0.5p	14n	68n	24n d	8.8n d	8.6n d
		4	2.0p	7.8n	48n	28n d	7.6n d	6.2n d
	TYPE 2	1	<0.5p	28n	320n	44n	370n	wire failure
		2	<0.5p	540n	2.2u	550n c,db	360n c,db	200n c,db
		3	<0.5p	8.4n	120n	68n db	14n db	14n d,db
		4	2.5p	50n	220n	52n db	26n db	22n db
	TYPE 3	1	65p	1.2u	660n	140n db	18n db	22n db
		2	440p	190n	1.4u	160n d	180n d	160n d
		3	110p	840n	140n	24n	18n	16n
		4	40p	600n	900n	72n	160n c,db	140n c,db

TYPE 4	TYPE 1	1	<0.5p	50p	5.4n	3.4n	720p	520p
		2	<0.5p	60p	8.2n	1.4n	950p	280p
		3	<0.5p	4.0p	16p	50p	wire failure	
		4	1.0p	150p	120p	14n	120p	58p
	TYPE 2	1	<0.5p	80p	12n	6.0n	3.6n	560p
		2	<0.5p	20p	36p	8.8p	54p	60p
		3	<0.5p	220p	10n	440p	620p	2.9n
		4	<0.5p	64p	30p	5.4n	26p	38p
	TYPE 3	1	<0.5p	14p	260p	120p	39p	35p
		2	5.0p	46p	1.2n	1.6n	220p	200p
		3	8.5p	7.0p	32p	30p	22p	32p
		4	<0.5p	14p	30p	68p	18p	26p
NONE	TYPE 1	1	<0.5p	50p	5.6n	100n db	500p db	360p db
		2	<0.5p	220p	32n	5.4n db	2.9n db	2.7n db
		3	<0.5p	3.0p	40p	320p db	28p db	72p db
		4	1.0p	40p	32n	5.0n c,db	2.2n c,db	2.8n c,db
	TYPE 2	1	1.0p	1.2n	32n	400n db	2.1n db	7.2n db
		2	1.5p	1.2n	8.8n	84n db	320p db	600p db
		3	1.2p	4.2p	56p	66p	21p	47p
		4	<0.5p	3.0p	10p	30p	22p	19p
	TYPE 3	1	5.0p	8.0p	5.0n	24n	720p	840p
		2	1.0p	12p	300p	660p	380p	170p
		3	3.5p	14p	60p	40p	30p note 2	1.2u
		4	1.5p	4.5p	22p	60p	56p	50p
	TYPE 4	1	<0.5p	1.4u	7.2n	1.8n	710p note 2	4.0n
		2	<0.5p	56p	84n	12n	12n	wire failure
		3	<0.5p	3.0n	10n	840p	180p	200p
		4	<0.5p	1.8p	22p	25p	20p	23p
	TYPE 5	1	<0.5p	160p	3.0n	500p na	420p na	340p na
		2	<0.5p	52p	320p	500p na	600p na	580p na
		3	<0.5p	56p	2.2n	180p na	35p na	76p na
		4	<0.5p	14p	48p	200p na	45p na	46p na
	TYPE 6	1	<0.5p	22p	50p	40p na	35p na	38p na
		2	<0.5p	60p	120p	70p na	15p na	32p na
		3	<0.5p	6.0p	16p	40p na	18p na	38p na
		4	<0.5p	6.0p	28p	46p na	23p na	28p na

TABLE 1

[1] All B-primed substrates showed corrosion at day 10.

[2] Foil-pattern defect caused local metal bridge.

Table II. Lap shear (PSI) test results for G-10 substrates with various primer-elastomer coatings. (ASTM-D905 modified)

PRIMER	ENCAPSULANT	LAP SHEAR	PRIMER	ENCAPSULANT	LAP SHEAR
TYPE A	TYPE 1	451	TYPE C	TYPE 2	339
	TYPE 2	389		TYPE 3	163
	TYPE 3	149			
			NONE	TYPE 1	41
TYPE B	TYPE 1	512		TYPE 2	46
	TYPE 2	276		TYPE 3	80
	TYPE 3	111		TYPE 4	575
				TYPE 5	201
TYPE C	TYPE 1	292		TYPE 6	291

perhaps easier to explain than the good performance of the unprimed samples. The manufacturers' original purpose of these primers was for bonding of silicones in non-electrical applications. The manufacturers presently provide no data concerning their electrical properties. One might postulate that these primers contain watersoluable contaminants, and that although they may remain bonded, their bulk and surface resistivities drop by orders of magnitude in the presence of moisture. There would seem to be no other reasonable explanation as to how groups such as type A-type 3 could sustain high levels of leakage current without significant visible deterioration. Condensation of water at a debonded polymer-substrate interface sufficient to produce leakage currents in the near uAmpere region would have to result in rapid conductor corrosion. Although we have not, as of yet, performed tests to determine the degree of possible contamination for each primer, the results of this study and others seem to support the concept that the primers degrade the performance of the elastomers from a corrosion control, and an electrical insulation standpoint. The use of primers such as type C in military high-voltage assemblies has been associated with drift of high resistance circuits such as high-voltage voltage dividers (32). The degraded performance of materials such as types 1, 2, and 3 with the use of a primer cannot be more clearly seen than with type B primer. However, in spite of the rapid deterioration of the type B primed samples, they did not have the highest leakage currents. Although low electrical leakage is a necessary requirement for prolonged substrate survival, a weakness of leakage current measurements as sufficient criteria of encapsulant acceptance is the difficulty in determining the current density across the substrate. Local sites of high current flow can result in local corrosion without causing correspondingly high values of total leakage. Low total current as well as a uniform current density is necessary for acceptance of an encapsulant. Note that even for those samples which did not demonstrate early visible signs of corrosion, but did show elevated levels of leakage currents, the protecting primer-elastomer combination may not be acceptable for many high-impedance assemblies. An upper level of 60nA was chosen in this study as an acceptable total leakage current level. This represents a 5% error in the current which might flow in a 10Mohm resistor with a voltage differential of 12VDC. None of the primers A, B, or C could meet this limit, whereas the samples with types 4, 5, and 6 materials consistently met or exceeded this limit.

Perhaps the most surprising samples were the unprimed substrates encapsulated with types 1, 2, or 3 elastomers. The low bond strengths as seen in table II for these materials did not seem to be significant in their corrosion-protective ability. It may be that low substrate surface tension, with a correspondingly high degree of wetting by the protecting encapsulant is more important than the actual interfacial bond strength. Maintaining this surface wetting is obviously enhanced by strong adhesion, especially in assemblies with complex geometries and high local mechanical stresses. However, in certain assemblies adhesion is often lost due to the presence of sharp corners and close component spacings, characteristic of high-density PWBs. It may be that surface wetting could also be maintained by using a low durometer elastomer (not necessarily self-priming) as a "priming layer" with an overcoat of a more rigid elastomer. It is interesting to note that the materials which performed the best in this study were not only the self-priming ones, but also the ones characterized by low durometer readings. Both types 5 and 6 materials are self-priming, and of relatively low durometer (38 and 22) as compared with type 4 having a durometer of 63. Although the manufacturer's published value for the hardness of type 2 material is 35, our measurements found it to be in the range of 45-50. Interestingly, the type 2 samples demonstrated higher currents and more rapid corrosion than the type 1 samples of durometer 45 (measured at 40). Type 5 and 6 materials performed better than type 4, with type 6 (of lowest durometer) currents lower by approximately an order of magnitude. Type 6, however is a condensation-cure material and cannot be completely cured in closed molds or in thick sections. Most of our implant devices are cast, and addition-cure materials are preferred. The specification and quality control of material hardness may have to be tightly controlled for silicones used to protect electronic assemblies.

Rigid cleaning protocols are essential to low surface tension. The presence of water soluable contaminants upon the encapsulated surface provides the ideal conditions for the initiation of corrosion. This is especially true when using the semi-permeable silicones. As a demonstration of this, we encapsulated two substrates (B pattern) using type 1 elastomer without any cleaning performed. Solder fluxes, finger prints, and other contaminants normally present during assembly were not cleaned off. These samples showed visible corrosion within 5hrs of water submersion. In contrast, the unprimed type 1 samples in this study survived over 60 days of exposure with comparatively little degradation.

Dust control is a major problem in corrosion control by silicone polymers. Dust particles which may bridge, or meerly lay upon conductors will absorb moisture diffusing through the silicone and result in a localized site of corrosion. We found this to be the case in a number of our early samples. Following the results of this study, all of our encapsulated assemblies will be cleaned and encapsulated in a clean room which is serviced by HEPA filters.

The degree of life acceleration that this test represents is uncertain. Fitting data to a temperature dependent first-order exponential (Arrhenius) failure curve is often inaccurate due to initial model assumptions. Depending upon the estimate of activation energies used, and the criteria used to define failure, an acceleration factor in a range of 17-50 may result (33). Based upon a frequently sited rule of thumb, that reaction rates double for every 10°C, an acceleration factor of 32 would be used. We have no evidence at this time that a first-order exponential model is valid, nor that the 85°C accelerated test does not, in itself, induce stresses which would normally not be possible during the service life of the device. This is a criticism which is often sited against accelerated life tests. We do, however, have preliminary data from other humidity studies in our laboratory in which the time to corrosion for the type B primed substrates at 65°C was measured. Based upon these studies an acceleration factor of 25 was calculated for the 85°C test.

Conclusions

As part of our continuing program of evaluation of appropriate encapsulants for medical implants as well as non-medical electronic high-density assemblies, we shall pursue the question of surface wetting vs. strong adhesion as a criteria for selecting encapsulants suitable for corrosion control. We are investigating the use of silicone gels in place of surface primers in applications where a low-durometer, non-adhering material, with high wetting properties, may provide good humidity protection in the presence of high local mechanical stresses.

Low leakage current for the qualification of an encapsulant as a corrosion resistant coating is a necessary, but not sufficient, condition. The use of adhesion tests as a sole qualification of an encasulant, used essentially as an insulator, is at minimum incomplete, and sometimes invalid. The bond strengths seen for types A, B, and C primer with their poor corrosion resistant performance demonstrate the inappropriate nature of classical bond strength tests. High bond strength accompanied by low leakage current is probably an indication of a high degree of surface wetting, good cleaning protocols, and low encapsulant contamination. All of these factors seem necessary for good corrosion control using silicone encapsulants.

Acknowledgments

We thank Robert Swendsen, and John Bartels of Northrop Corp. D.S.D. for assistance in experimental design and the supply of the test materials. We thank Amy Yang and Tracy Larkins for recording the data. We also thank Jane Adams and Jerry Jeka of IITRI for assistance in the bond strength tests.

Literature Cited

1. Wise, K. D. 9th Quarterly Report, NIH-NINCDS-N01-NS-1-2384 1984.
2. Wise, K.D.; Angell, J.B.; Starr, A. IEEE Trans. Biomed. Eng. 1970, BME-17, 238-246.
3. Knutti, J. W.; Allen, H. V.; Meindl, J. D. Biotelemetry Patient Monitoring 1979, 6, 95-106.
4. Esashi, M.; Matsuo, T. IEEE Trans. Biomed. Eng. 1978, BME-25, 184-191.

5. Moss, S. D.; Johnson, C. C.; Janata, J. IEEE Trans. Biomed. Eng. 1978, BME-25, 49-54.
6. Mercer, H. D.; White, R. L. IEEE Trans. Biomed. Eng. 1978, BME-25, 494-500.
7. Donaldson, P. E. K. IEEE Trans. Biomed. Eng. 1976, BME-23, 281-285.
8. Donaldson, P. E. K.; Sayer, E. Med. Biol. Eng. Comput. 1977, 15, 712-715.
9. Ko, W. H.; Spear, T.M. Eng. Med. Biol. 1983, 2, 24-38.
10. Ianuzzi, M. IEEE Trans. Comp. Hybrids, Manufac. Technol. 1983, 6, 2, 191.
11. Schnable, G. L.; Kern, W.; Comizzoli, R. B. J. Electrochem: Solid State Science and Tech. 1975, 7, 1092-1103.
12. Mcmillan, R. E.; Misra, R. P. IEEE Trans. Elec. Insul. 1970, EI-5, 1, 10-17.
13. Maguire, C. F.; Jarret, Q.T.; Bartholomew, C. Y. Solid State Tech. 1972, 15, 4, 46-52.
14. Kerr, D. R.; Logan, J. S.; Burkhardt, P. J.; Pliskin, W. A. Res. Dev. 1964, 8, 4, 376-384.
15. McDonald, B. A. Solid-State Electron. 1971, 14, 17-28.
16. Schlacter, M. M.; Schlegel, E. S.; Keen, R. S.; Lathlaen, R. A.; Schnable, G. L. IEEE Trans. Elec. Dev. 1970, ED-17, 1077.
17. Kern, W.; Schnable G. L.; Fisher, A. W. RCA Rev. 1976, 37, 3-54.
18. Paulson, W. M.; Kirk, R. W. 12th Ann. Proc. Reliab. Phys. 1974, 172-179.
19. White, M. L. Proc. IEEE 1969, 57, 9, 1610-1615.
20. Cagle, C.V. "Handbook of Adhesive Bonding"; McGraw Hill: New York, N.Y., 1973.
21. Wake, W. C. Polymer 1978, 19, 291-309.
22. Rantz, L. E. Proc. Soc. Manu. Eng. Los Angeles, 1978.
23. Plueddemann, E. P. "Silane Coupling Agents"; Plenum, N.Y., 1982.
24. Jaffe, D. IEEE Trans. Parts, Hybrids, & Packaging 1976, 3, PHP-12 , 182-187.
25. Soos, N.A.; Jaffe, D. Proc. 28th Electron. Comp. Conf. 1978.
26. Peck, D. S.; Zierdt, C. H. Proc. IEEE, 1974, 62, 2, 185-211.
27. Wong, C. P. Polymers in Electronics, Ed. T. Davidson, ACS Symposium Series, Washington D.C., 1984.
28. Cobian, K. E.; et. al. Adhesives Age 1984, 17-20.
29. Donaldson, P. E. K. Proc. Int. Conf. Biomed. Polymers 1982, p. 143-150.
30. Donaldson, P. E. K.; Sayer, E. Med. & Biol. Eng. & Comput. 1977, 15, 712-715.
31. Donaldson, P. E. K.; Sayer, E. Med. Biol. Eng. Comput. 1981, 19, 483-485.
32. R. Swendsen, Northrop Corp. D.S.D., Rolling Meadows, IL., Personal Communication.
33. Donaldson, P. E. K.; Sayer, E. Med. Biol. Eng. Comput. 1981, 19, 403-405.

28

Corrosion Behavior of Epoxy and Unsaturated Polyester Resins in Alkaline Solution

H. Hojo, T. Tsuda, K. Ogasawara, and T. Takizawa

Department of Chemical Engineering, Tokyo Institute of Technology, Ookayama, Meguro-ku, Tokyo, 152, Japan

The effect of temperature and concentration on corrosion behavior and corrosion mechanism of epoxy and polyester resins in NaOH solution were studied, and were discussed by considering their structures.

Resins used were two types of epoxy resins cured with anhydride and amine and iso-phthalic type polyester resin.

Different behaviors and mechanisms were clearly recognized between these resins. Epoxy resin cured with amine showed no degradation during immersion because of its stable crosslinks. Epoxy resin cured with anhydride showed the uniform corrosion with the softening and dissolution of the surface and also behaved similar to the oxidation corrosion of the metal at high temperature obeying linear law.

Iso-phthalic polyester resin was corroded with the formation of the color changed surface layer and corrosion rate of the resin were controlled by diffusion process of the solution through the layer. Thus similar behavior was observed to oxidation corrosion of metal obeying Wagner's parabolic law. The difference of behaviors of these resins were mainly due to the position of ester bonds in the structures.

Method to predict the retention of the strength of resins after immersion was also proposed by applying the concept of corrosion mechanism in metal.

Fiber reinforced plastics have seen much service in industry because of their excellent mechanical and chemical properties and also economical point of view. At present corrosion resistant fiber reinforced plastics are in use as large tanks, vessels, reactors and pipes.

Corrosion resistant FRP structures and also resin linings have resin or resin-rich surface layer to protect the structures from corrosive attack. Therefore, the study of corrosion behavior of resin is essentially important.

0097–6156/86/0322–0314$06.00/0

In alkaline and acid solutions, thermosets containing ester groups are degrated mainly due to the hydrolysis of the esters (1-6) But in case of the resin crosslinked by ester bonds such as some epoxy resins (7-9), corrosion behavior is though to be different from that of the resin which has ester bonds in the main chain.

In this paper, the effect of temperature and concentration on corrosion behavior and corrosion mechanism of epoxy and unsaturated polyester resins in NaOH solution were studied, and were discussed by considering their chemical structures. Corrosion rate studies were also made by applying the concept of metallic corrosion.

EXPERIMENTAL

Resins used were two types of epoxy resins (EP) and an unsaturated polyester resin (UP) as shown in Figure 1. EP is the bisphenol-A type resin cured with methyl-tetrahydrophthalic anhydride (MTHPA) or 1,8-p-menthandiamine (MDA). UP is the iso-phthalic type resin which has ester bonds in the main chain and is crosslinked by styrene (10).

Flexural test specimens were made according to ASTM D790 from casted resin sheet of 2mm thickness.

Test environments used were NaOH solution with various concentrations of 10 to 40wt.%. Immersion tests were carried out at temperatures of 20 to 104°C for up to 3000 hrs, and after immersion weight measurements and flexural tests were performed at room temperature (Testing speed: 2mm/min, Span: 40mm). Optical and scanning electron microscopes and infrared spectroscope (IR) were further used to study the degradation mechanism of the resins.

CORROSION BEHAVIOR OF EPOXY RESINS

(a) Epoxy Resin Cured with MTHPA

Figures 2 and 3 show the weight change of epoxy resin cured with MTHPA (MTHPA-EP) at various concentrations and temperatures. Typical weight change is shown in the curve of 20wt.% solution in Figure 2.(also see Figure 10). At first the weight increases and secondly decreases and thirdly again increases remarkably with an increase of immersion time. The last weight gain, however, can be recognized only at high temperature and concentration range. At the first stage, penetration rate of the solution into the resin is higher than dissolution rate of corrosion products, thus causing to increase in weight. At the second stage, dissolution rate is higher than penetration rate, as a result weight of the resin decreases. Thirdly severe penetration of the liquid into and through the corroded layer causes the marked increase in weight.

Figure 4 shows the retention of flexural strength of immersed specimens. Strength retention was defined as the ratio of strength after immersion to original one. Strength decreases markedly with increasing temperature and concentration because of the severe corrosion.

At the specimen surface, soft corroded layer was formed during immersion and the layer could be easily removed by wiping lightly with acetone-soaked paper and in some conditions this layer was dissolved spontaneously. The thickness of this layer was defined as corrosion depth x, and measured at various test conditions. As shown in Figure 5, corrosion depth increases linearly with time, and

(a) Epoxy resin cured with MTHPA

(b) Epoxy resin cured with MDA

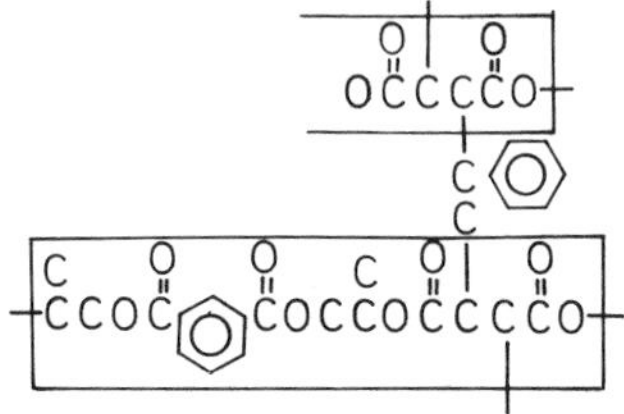

(c) Iso-phthalic type unsaturated polyester resin

Figure 1. Chemical structures of epoxy and unsaturated polyester resins.

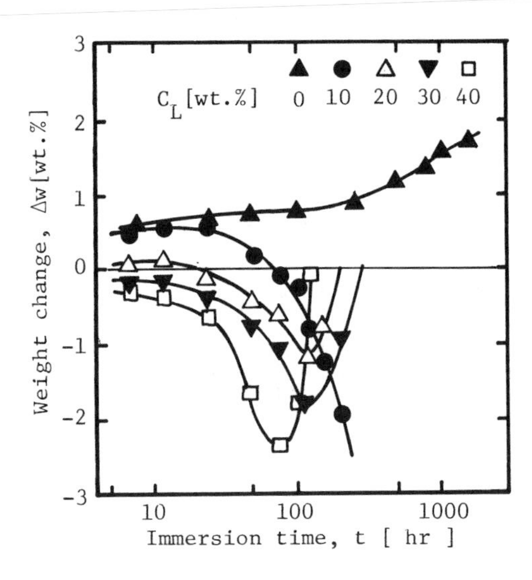

Figure 2. Effect of concentration of NaOH solution on weight change of MTHPA-EP.

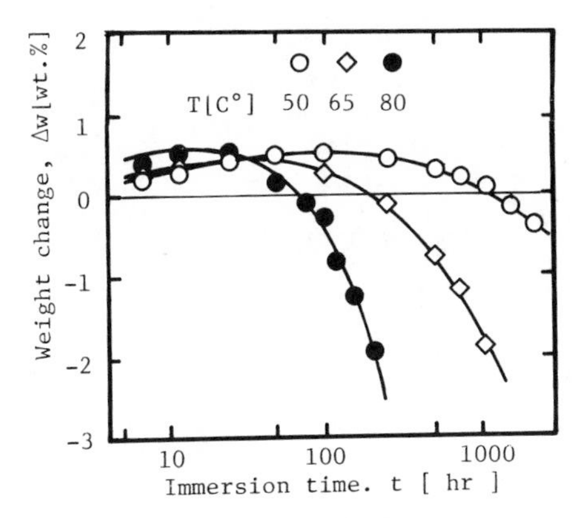

Figure 3. Effect of temperature of 10wt.% NaOH solution on weight change of MTHPA-EP.

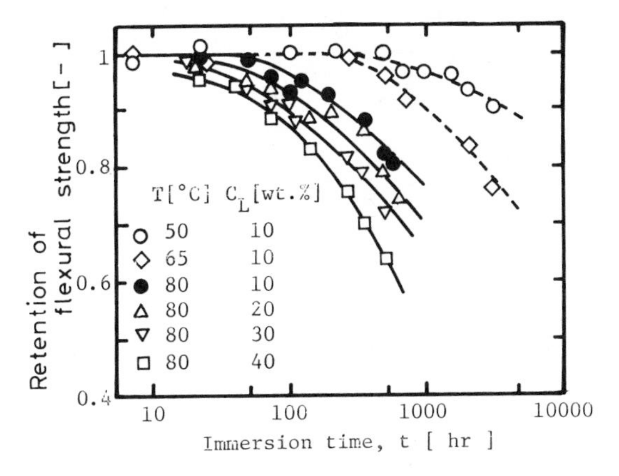

Figure 4. Retention of flexural strength of MTHPA-EP in various conditions.

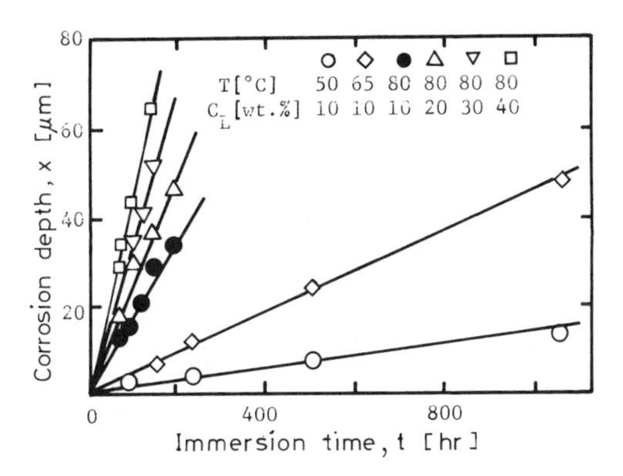

Figure 5. Variation of corrosion depth of MTHPA-EP with immersion time in various conditions.

this behavior is similar to that of an oxidation of metal at high temperature with the formation of rough, porous oxide scale. This has shown that the concept of corrosion rate in metals can be applied even in plastic materials.

Figure 6 shows the IR analysis of the soft corroded layer

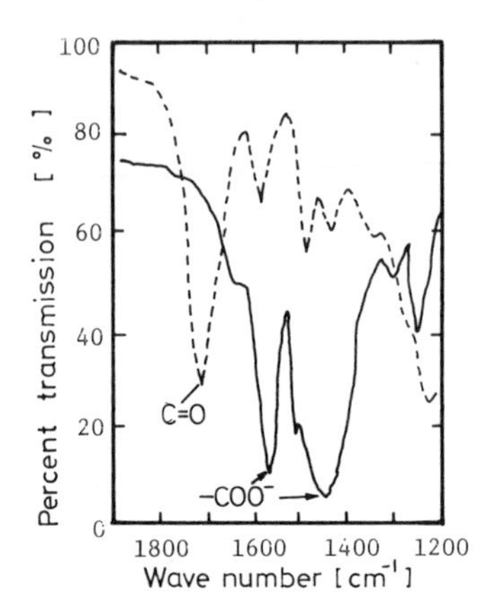

Figure 6. IR spectrum of MTHPA-EP. ----- before immersion, —— after 360 hrs immersion in 40wt.% NaOH at 80°C.

compared with noncorroded resin. In case of the corroded resin, ester peak around $1730cm^{-1}$ disappears and carboxylate peak appears near $1570cm^{-1}$ and $1440cm^{-1}$. This has been proved by the hydrolysis of the esters as shown below,

$$R-\overset{O}{\overset{\|}{C}}-OR' + OH^- \longrightarrow R-\overset{O}{\overset{\|}{C}}-O^- + R'OH \quad (1)$$

Thus, severe corroded layer is formed because MTHPA-EP has relative short main chain, and these main chains are crosslinked by the ester bonds.

IR analysis of the resin specimen below the corroded layer showed no sign of corrosion, which implies that the chemical attack progresses gradually from the surface and also corrosion behavior depends strongly on higher reaction rate shown in Equation 1 than penetration rate of NaOH solution into the resin.

(b) Epoxy Resin Cured with MDA

The effect of concentration of NaOH solution on weight change of epoxy resin cured with MDA(MDA-EP) is shown in Figure 7. Only a small amount of weight gain was observed although after 3000 hrs immersion. The weight gain, however, decreases with an increase of concentration. The main reason for this behavior is the increase of wetability as shown in Figure 8. Figure 9 presents the change of flexural strength after immersion, and it holds same strength as that of before testing even longer immersion at 80 °C. Therefore, MDA-EP is not attacked by NaOH because of its stable crosslinks of C-N bonds.

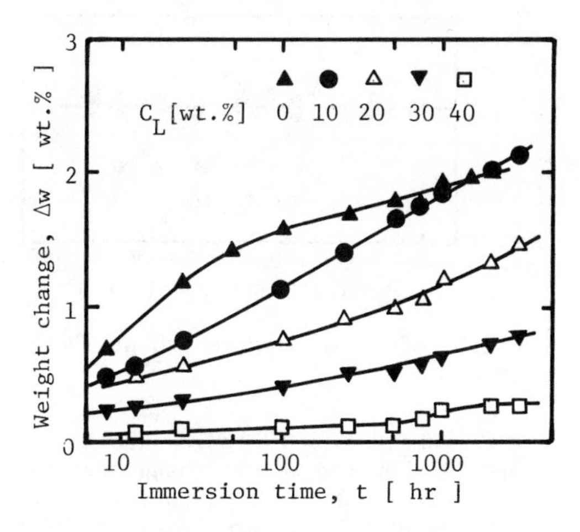

Figure 7. Effect of concentration of NaOH solution on weight change of MDA-EP at 80°C.

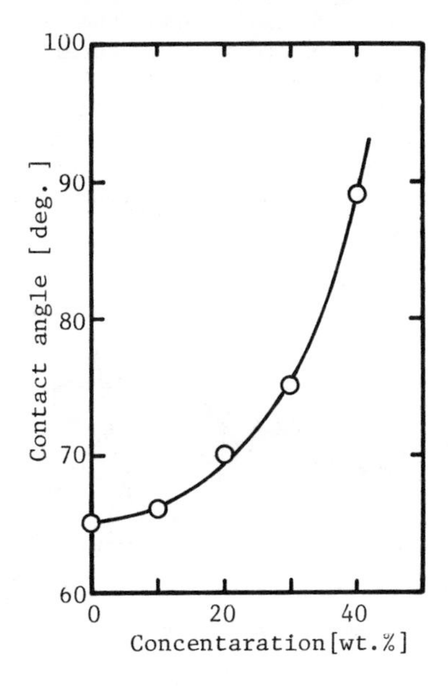

Figure 8. Effect of concentration of NaOH solution on the wetability to MDA-EP at 80°C.

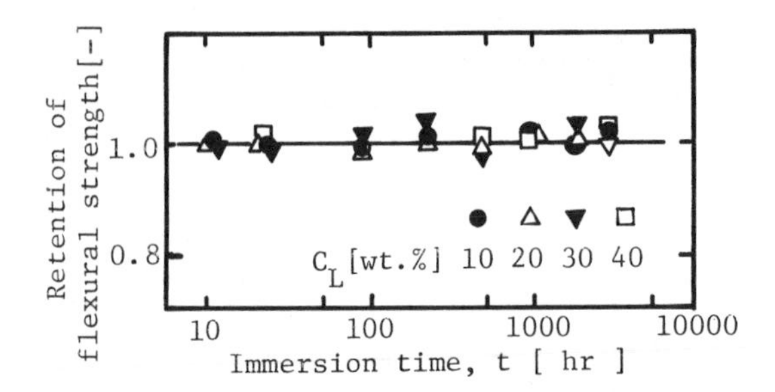

Figure 9. Retention of flexural strength of MDA-EP in various concentrations at 80°C.

CORROSION BEHAVIOR OF UNSATURATED POLYESTER RESIN

Figures 10 and 11 show the weight change and the retention of strength for iso-phthalic unsaturated polyester resin (iso-UP). These behaviors show almost the same tendency as MTHPA-EP, however, as shown in Figure 12 the concentration influences flexural strength and the strength becomes minimum at the concentration of 30wt%. This behavior is thought to depend on contradictory tendency of the wetability and the reactivity with the concentration.

In iso-UP, color changed rubber-like layer was clearly observed as shown in Figure 13. IR analysis of the color changed layer showed the same results of hydrolysis of esters as Figure 6. The thickness of hydrolyzed area was measured by IR, and the results agreed well with that of the color changed layer measured by optical microscope as shown in Figure 14. The thickness of the layer was defined as corrosion depth x. Figure 15 shows the variation of depth x with immersion time t, and the following relation holds,

$$x^2 = k_1 t \quad \text{or} \quad x = k_2 t^{1/2} \tag{2}$$

where, k_1 and k_2 are constants.

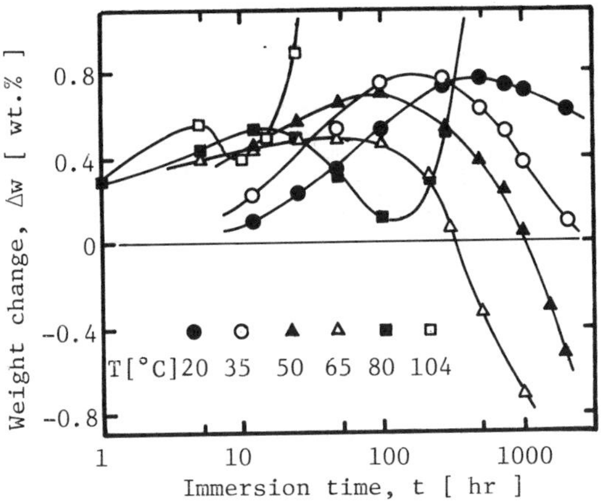

Figure 10. Effect of temperature of 10wt.% NaOH solution on weight change of iso-UP.

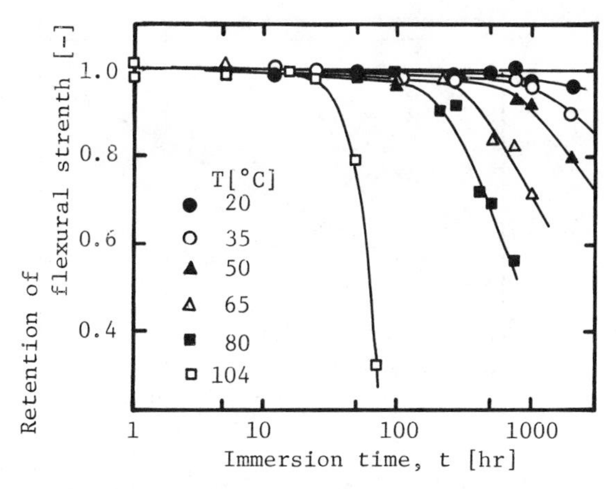

Figure 11. Effect of temperature of 10wt.% NaOH solution on retention of flexural strength of iso-UP.

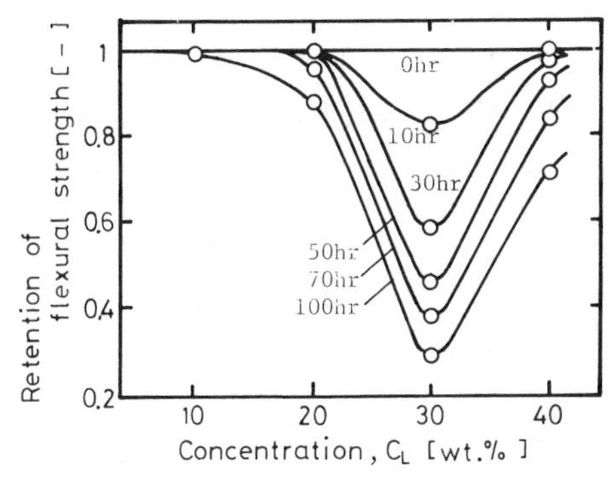

Figure 12. Retention of flexural strength of iso-UP vs. concentration of NaOH solution at 80°C.

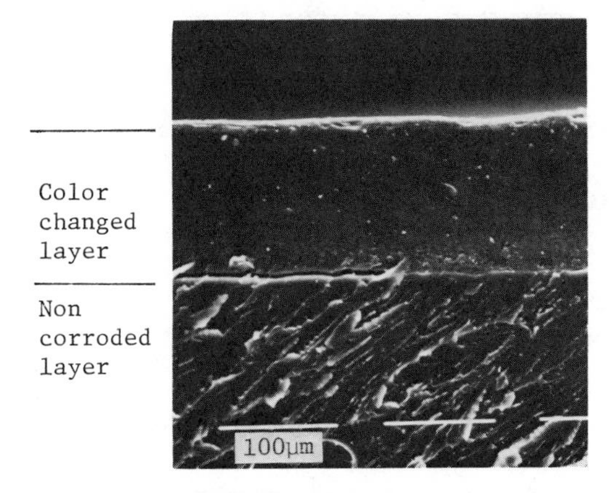

Figure 13. Scanning electron micrograph of fractured surface of iso-UP after flexural test. (504 hr, 80°C, 10wt.% NaOH solution)

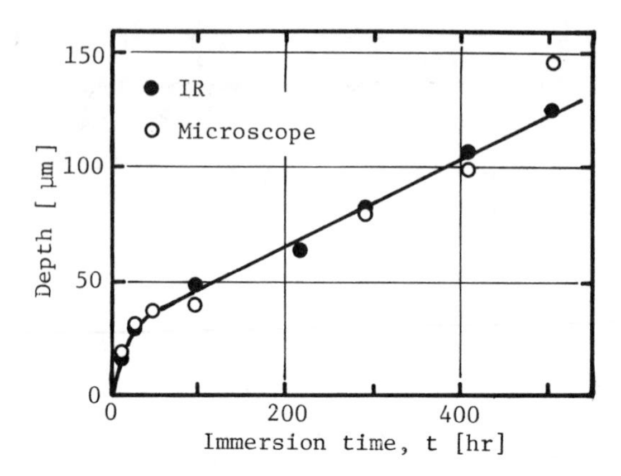

Figure 14. Comparison of the thickness of hydrolyzed area measured by IR and optical microscope.

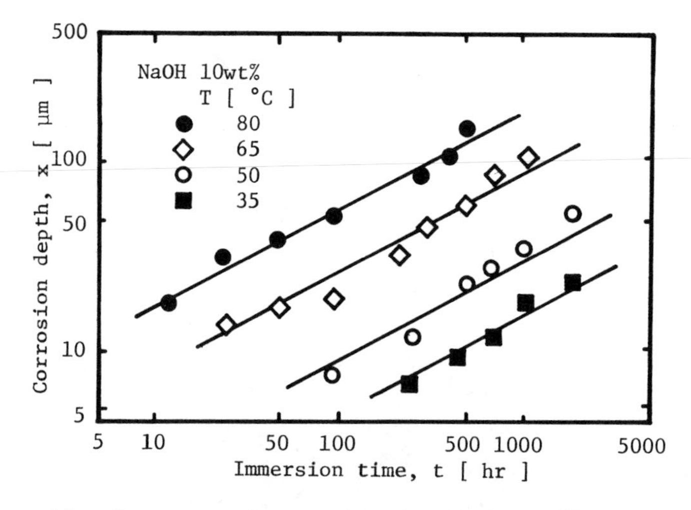

Figure 15. Variation of corrosion depth of iso-UP with immersion time in various conditions.

Iso-UP has ester bonds only in the main chain where hydrolysis occurs, so a part of reaction products from the main chain dissolves into the solution. While the crosslink formed by styrene remains unaffected because of its stable C-C bonding. As a result, the corroded surface layer resists the diffusion of NaOH solution. This mechanism is just like an oxidation of the metal at high temperature with formation of thick, cohered oxide scale, and can be expressed by similar relation of Wagner's parabolic law as shown in Equation 2. The concept of corrosion in metals can be applied in this case too.

PREDICTION OF STRENGTH OF RESINS IN CORROSIVE ENVIRONMENT

In MTHPA-EP and iso-UP resins, the strength of the resin below the corroded layer of specimen was same as that of before immersion. Assuming that the corroded layer has no strength, the prediction of strength after immersion was made by estimating the corrosion depth in the following way.

(a) Epoxy Resin Cured with MTHPA

Rate of ester hydrolysis depends on both concentrations of ester bonds in the resin and solution at constant temperature, then

$$C_E dx/dt = k_3 C_L^{\alpha} C_E^{\beta} \quad (3)$$

where C_E is concentration of ester bonds per unit volume of the resin, C_L is concentration of the solution, α and β are order of reaction, and k_3 is reaction rate constant.
From Arrhenius' equation, k_3 is given as a function of absolute temperature T,

$$k_3 = A_1 \exp(-E/RT) \quad (4)$$

From Equations 3 and 4, the corrosion rate becomes,

$$dx/dt = A_2 \exp(-E/RT) C_L^{\alpha} \quad (5)$$

Equation 5 is confirmed by Figures 16 and 17, and from these results x is given as

$$x = A_2 \exp(-18.8 \times 10^3/RT) C_L^{0.56} \cdot t \quad (6)$$

(b) Iso-phthalic Unsaturated Polyester Resin

k_2 in Equation 2 was obtained from Figure 15 and plotted against reciprocal temperature as shown in Figure 16. The same relation with temperature as that of Equation 4 holds, thus

$$x = A_3 \exp(-12.8 \times 10^3/RT) t^{1/2} \quad (7)$$

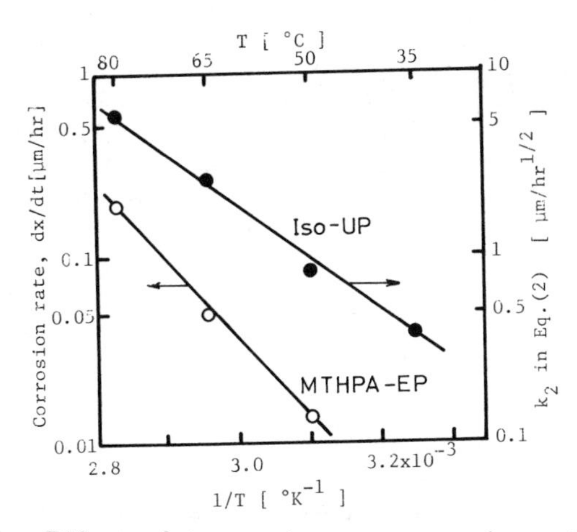

Figure 16. Effect of temperature on corrosion rate and k_2 in Equation 2.

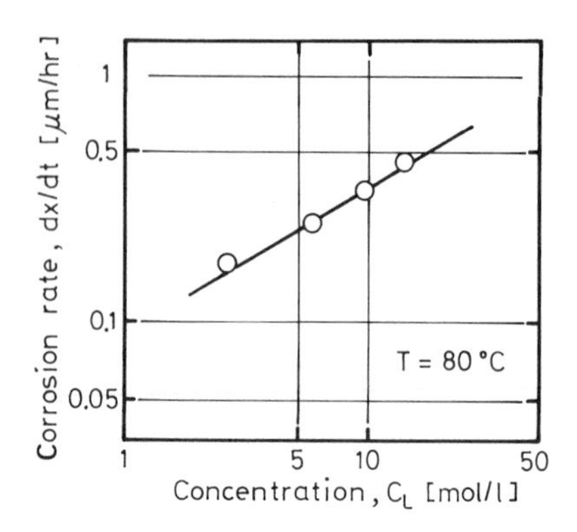

Figure 17. Effect of concentration on corrosion rate of MTHPA-EP.

(c) Prediction of Strength after Immersion

The apparent flexural strength σ_B' after immersion is expressed as

$$\sigma_B' = \frac{3Pl}{2bh^2} \tag{8}$$

where h, b, and l are thickness, width and span of the specimen before immersion respectively.

The true flexural strength σ_B is equal to that of the original resin σ_{B_o}, thus

$$\sigma_B = \sigma_{B_o} = \frac{3Pl}{2(b-2x)(h-2x)^2} \tag{9}$$

Equations 8 and 9 give the retention of flexural strength S,

$$S = \frac{\sigma_B'}{\sigma_{B_o}} = \frac{(b-2x)(h-2x)^2}{bh^2} \tag{10}$$

Calculated values of retention of strength by Equation 10 for MTHPA-EP were compared with experimental values as shown in Figure 18. Calculated values were well coincided with experimental values.

As the retention of strength S varies only with x, a master curve can be obtained by plotting the terms $\exp(-18.8\times10^3/RT)C_L^{0.56}\cdot t$ or $\exp(-12.8\times10^3/RT)t^{1/2}$ as shown in Figure 19. By using these master curves, the retention of strength after long term immersion at any temperature and concentration can be predicted.

CONCLUSIONS

Different behaviors and mechanisms of corrosion were clearly recognized in alkaline solution between two types of epoxy resins and a polyester resin depending on their different chemical structures. Epoxy resin cured with MDA showed no degradation during immersion because of its stable crosslinks. Epoxy resin cured with MTHPA showed the uniform corrosion with the dissolution of the surface, and the resin behaved similar to the oxidation corrosion of metals at high temperature obeying linear law. On the other hand, iso-phthalic unsaturated polyester resin was corroded with the formation of the color changed rubber-like surface layer. Thus corrosion rate of this resin was controlled by the diffusion of the

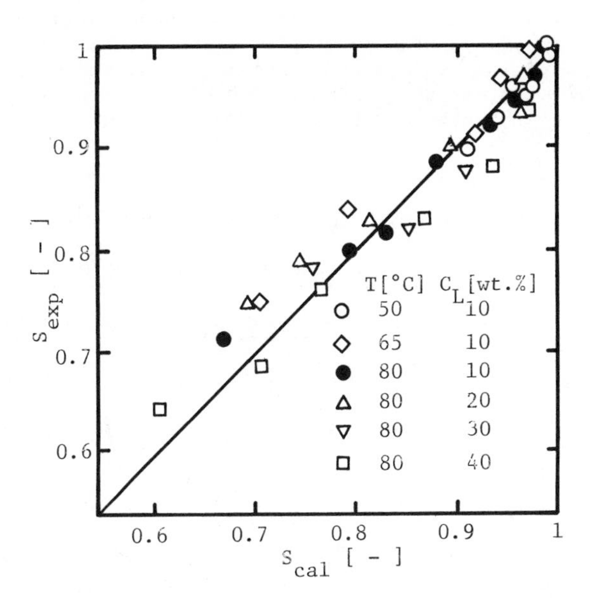

Figure 18. Comparison of experimented results with calculated results for retention of flexural strength of MTHPA-EP.

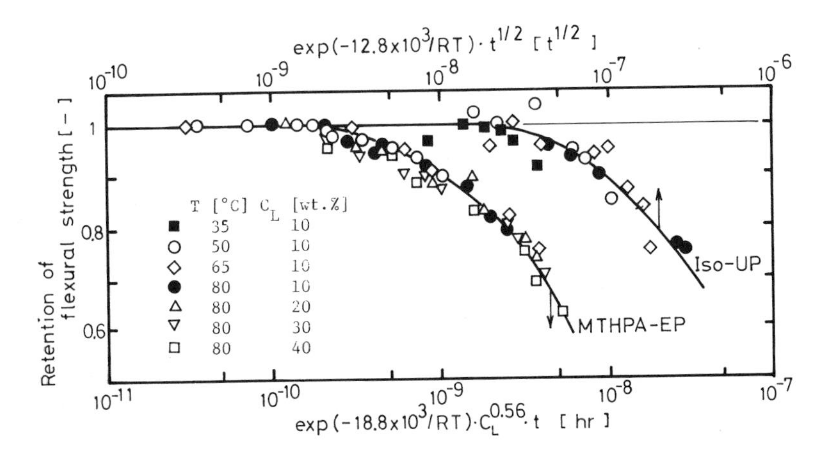

Figure 19. Master curves to predict retention of flexural strength of MTHPA-EP and iso-UP.

solution through the layer, and was similar to the oxidation corrosion of metals at high temperature obeying Wagner's parabolic law. Therefore, the concept of corrosion mechanism in metals can be applied also to the plastic materials. Finally, in order to predict the strength of the resin after immersion, master curve including effects of temperature and concentration of solution was proposed by applying corrosion rate concept as in metals.

Literature Cited

1. K.H.G. Ashbee, F.C. Frank, F.R.S. and R.C. Wyatt, Proc. Roy. Soc. 1967, A300, 415.
2. C.M. Vetters, Proc. 25th Ann. Tech. Conf. RP/Comp. Div., SPI, 1970, 4-B.
3. R.C. Allen, Proc. 33rd Ann. Tech. Conf. RP/Comp. Inst., SPI, 1978, 6-D.
4. H. Hojo, K. Tsuda & M. Koyama, Proc. 3rd Int. Conf. in Coating Sci. Tech. (Athens), 1977, p.221.
5. H. Hojo, K. Tsuda, Proc. 34th Ann. Tech. Conf. RP/Comp.Inst., SPI, 1979, 13-B.
6. H. Hojo, K. Tsuda, K. Ogasawara & K. Mishima, Proc. 4th Int. Conf. on Composite Materials (Tokyo), 1982, p.1017.
7. R.F. Fisher, J. Polym. Sci., 1960, 44, 155.
8. Y. Tanaka & H. Kakiuchi, J. Polym. Sci., 1964, A-2, 3405.
9. M. Jinbo, K. Ochi & M. Yamada, Kobunshi Ronbunshu, 1980, 37, 57.
10. M.B. Launikitis, "Managing Corrosion Problems with Plastics", NACE, 1977; Vol.1, p.190.

RECEIVED January 21, 1986

Structure–Property Relationships in Tin-Based Antifouling Paints

D. Ibbitson[1], A. F. Johnson[1], N. J. Morley[2], and A. K. Penman[2]

[1]School of Polymer Science, University of Bradford, Bradford, BD7 1DP, England
[2]International Paint PLC, Stoneygate Lane, Felling, NE10 0JY, England

Compositionally homogeneous copolymers from tributyltin methacrylate and methylmethacrylate and compositionally homogeneous and heterogeneous terpolymers of tributyltin methacrylate, methyl methacrylate and 2-ethyl hexyl acrylate have been synthesised and the thermal and mechanical properties of the polymers, and paints made from these polymers, measured in relation to polymer composition and composition distribution. The self polishing and anti-fouling characteristics of the materials have also been examined. It has been shown that there are marginal advantages to be obtained in the use of compositionally controlled multicomponent polymers for anti-fouling paints.

The protection against corrosion of ships and other marine equipment makes demands on the protective coatings over and above those which might be expected for land based metal structures. Whatever the environment, mechanical damage to the protective coating through impact is probably the major cause of their failure and consequential corrosion at the site of damage. With ships there are unique problems with those parts of the structure which reside below the water-line in that the metal substrates in this situation not only have to withstand impact damage but must also resist fouling by marine organisms. The ideal coating should provide both corrosion and anti-fouling resistance. The most common practical solution to the problem is to apply two coatings, the first which is specifically aimed at corrosion protection and the second which provides protection against anti-fouling. Our concern here is with anti-fouling coatings only.

Fouling of hulls leads to a significant increase in the surface roughness which in turn introduces an economic penalty for ship owners. The importance of surface roughness has been reviewed by Christie (1), who also describes the development of self polishing

0097–6156/86/0322–0327$06.00/0

copolymer (SPC) anti-fouling paints which prevent fouling over extended periods and which also decrease the surface roughness of hulls in service. These paints have evolved from the work of Montermoso et al (2) and were developed into commercial self-polishing paint systems by Milne and Hails (3). The benefits of self polishing paints systems are now widely recognised and SPC paints hold a major share of the anti-foulant market.

Generally, antifouling coatings containing organotin compounds may be divided into two types: polishing and non-polishing. In conventional non-polishing paints the tin compounds are physically trapped into a polymer matrix and the anti-fouling behaviour is dependent on the leaching of the biocide from the matrix. With self-polishing paints the anti-fouling mechanism is more complex. The biocide is attached to the polymer substrate and has to be released by a hydrolysis mechanism. The hydrolysis reaction modifies the surface of polymer making it hydrophilic and therefore more susceptible to removal by the frictional forces developed at the hull-water interface when a ship moves through water.

Recently, concern has been expressed about possible environmental problems which might result from the release of those paints in which the tin compounds are not chemically bonded to the base polymer. Obviously it is desirable to reduce the amount of biocidally active material from anti-fouling paints to the lowest practicable level. An understanding of the structure property-relationships in the copolymer or multicomponent polymers which are used in SPC paints is essential in order to make the most effective use of the biocide. To fully evaluate the structure-property relationships, it is necessary to decouple a very large number of interacting parameters, for example, molecular features such as chemical composition, chemical composition distribution or molecular weight and their influence on polymer hydrophilicity, film forming character or mechanical properties. It is almost inevitable with such a complex set of interacting factors that there will be no single ideal polymer structure for all purposes but rather that there might be some optimum structure which satisfies a number of the major criteria for any given end use.

An early study of the influence of composition heterogeneity on the physical properties of copolymers was undertaken by Nielson (4), but to the knowledge of the authors, there have been no similar investigations involving tin-based polymers. In this work a range of different acrylic co- and terpolymers has been prepared and the effects of composition and composition distribution on the physical and performance-related properties of the polymers in their native state and in paints have been examined.

As a necessary preliminary to the study of how compositional heterogeneity affects the properties of the polymers, compositionally heterogeneous and homogeneous co- and terpolymers had to be synthesised. It is common in copolymerisations for the relative reactivity of the co-monomers to be different (5) so that during polymerisations carried out to high conversion in a free-running batch reactor, the initially formed polymer is richer in the more reactive monomer, whereas, at the end of the reaction the polymer produced contains a greater proportion of the less reactive monomer. In such circumstances, compositional heterogeneity is inevitable in the polymer which is finally isolated from such a process.

Compositionally uniform copolymers of tributyltin methacrylate (TBTM) and methyl methacrylate (MMA) are produced in a free running batch process by virtue of the monomer reactivity ratios for this combination of monomers (r (TBTM) = 0.96, r (MMA) = 1.0 at 80°C). Compositionally homogeneous terpolymers were synthesised by keeping constant the instantaneous ratio of the three monomers in the reactor through the addition of the more reactive monomer (or monomers) at an appropriate rate. This procedure has been used by Guyot et al (6) in the preparation of butadiene-acrylonitrile emulsion copolymers and by Johnson et al (7) in the solution copolymerisation of styrene with methyl acrylate.

EXPERIMENTAL

MODELLING AND SIMULATION

The modelling techniques which are commonly used for polymerisation reactions have been reviewed (8). Deterministic analytical models based on the detailed chemistry of co- and terpolymerisations have been used to assist with the design of reactor conditions for the synthesis of polymers with specific composition and structure. Typical data used with these models are shown in Table 1. Simpler mass balance models have been used for the design of reactor control strategies for the production of compositionally homogeneous polymers. For control purposes, use has been made of the observation that for many terpolymerisations (and copolymerisations, although control was unnecessary for the monomer combination used in this work) individual monomers are consumed by an apparent first order process. When this situation pertains (and if the reactor is assumed to be isothermal and the initiator has a long half-life) then it may be shown (9) that the following equations are true for terpolymerisations carried out in a semi-batch reactor.

$$dA/dt = -k1*A + F(1) * A(F) + [F(1) + F(2)] * A/V \qquad (1)$$

$$dB/dt = -k2*B + [F(1) + F(2)] * B/V \qquad (2)$$

$$dC/dt = -k3*C + F(2) + C(F) + [F(1) + F(2)] * C/V \qquad (3)$$

where the monomer feeds are F(1) = [A/A(F)] * (k1 - k2) and F(2) = [C/C(F)] * (k3 - k2) and k1, k2 and k3 are apparent first order rate constants for individual monomers but at a specific monomer composition. In these equations the three monomer concentrations (moles) are designated by A B and C and it is assumed that B reacts more slowly than A and C. The concentration of the more reactive monomers in the feeds are A(F) and C(F) (moles/1) and F(1) and F(2) are the feed rates (1/min). These equations define the feed profile for the production of compositionally homogeneous products. Precise kinetic constants are necessary in order to execute effective experimental control of the polymersiation reactors and the methods used to obtain these data and some typical constants for one ternary system are reported overleaf.

Table 1. Kinetic parameters used in simulation studies of the copolymerisation of MMA with TBTM.

kd	=	1.50.10 exp(-30800/RT)	s
k11	=	3.20.10 exp(-7000/RT)	1/s/mol
k12	=	1.75.10 exp(-9480/RT)	1/s/mol
k21	=	4.20.10 exp(-7520/RT)	1/s/mol
k22	=	6.60.10 exp(-6300/RT)	1/s/mol
kt11	=	1.23.10 exp(-3000/RT)	1/s/mol
kt22	=	1.77.10 exp(-2840/RT)	1/s/mol
kt12	=	(kt11*kt22)	

Monomer 1 = MMA and monomer 2 = TBTM

k11 = homopolymerisation constant for monomer 1

k22 = homopolymerisation constant for monomer 2

k12 and k21 are the cross propagation constants

kt12 an average termination constant

A full description of the modelling and control of multicomponent polymerisations is beyond the scope of this presentation since there are many exceptions to the above simplistic model hence details will be described elsewhere (9).

POLYMER SYNTHESIS

All polymerisations were carried out in nitrogen purged xylene solutions in a thermostatically controlled one litre glass reactor. Semi-batch processes were carried out in a similar reactor which was provided with calibrated peristaltic pumps (computer controlled when necessary) for delivering the monomer feeds. Typically, experiments were carried out at 80°C with monomer concentrations which gave solids contents in the range 10 - 60% at 100% conversion.

The control strategies for determining the feed policies were decided on the basis of a numerical solution of the terpolymerisations described by equations 1 - 3 using a microcomputer and a general purpose simulation package, BEEBSOC (10). Where necessary, these data were acquired in the course of this study, otherwise literature values were used. The apparent first order rate constants in terpolymerisations have been shown to be composition dependent. The variation in rate constants with composition at 80°C can be described by the following equations:

$$k(MMA) = -0.025 * X + 0.0346 \quad (4)$$

$$k(TBTM) = 0.036 * X + 0.0065 \quad (5)$$

$$k(2EHA) = 0.056 * X + 0.0062 \quad (6)$$

where k( ) is the apparent rate constant and X the initial weight fraction of the respective monomer in the reactor feed.

The following materials were used as supplied: tributyltin methacrylate (International Paint plc), 2-ethylhexyl acrylate (Aldrich Chemical Co. Ltd.), methyl methacrylate (ICI plc) sulphur-free xylene and chloroform (May and Baker Ltd.).

POLYMER CHARACTERISATION

A number of methods were explored for monitoring the progress of polymerisations. In each case samples were removed from the reactor at appropriate time intervals and analysed off-line. Gas-liquid chromatography proved to be unreliable for analysis of residual monomer concentrations because of monomer decomposition on the columns and the relatively low volatility of the tin-containing monomer. Gravimetric analysis of the polymer produced with time by precipitation was also shown to be inaccurate, particularly at low conversion, because of incomplete isolation of lower molecular weight material and the retention of residual solvent and monomer by the precipitate. The polymerisations were successfully followed using gel permeation chromatography (GPC) to monitor residual monomers.

The chromatograph (Waters Associates) was fitted with PL Gel columns (Polymer Laboratories Ltd.) and two infrared detectors in series. Chloroform was used as the eluant. Infrared detectors were used because the tin containing acrylic monomer has a characteristic carbonyl stretching frequency at 1620 cm^{-1} which is well removed from that of other acrylic monomers which have carbonyl absorptions at the more characteristic wavenumber 1720 cm^{-1}. Both peaks obeyed the Beer Lambert law. A typical chromatogram showing the separation of residual monomers and polymer is shown in Figure 1.

POLYMER TESTING

Polymer films of approximately 1000 microns wet film thickness were laid down with a bar applicator on PTFE coated glass panels and the solvent allowed to evaporate at ambient temperature for a standard period of seven days. A typical plot of solvent weight loss with time is shown in Figure 2. The thickness of the wet film was dictated by the need to have adequate mechanical strength in the dry films in order that they might be suitable for subsequent mechanical test procedures. Dry film thicknesses were approximately 300 microns as measured by micrometer. The dried polymer films were examined by dynamic mechanical thermal analysis (DMTA) (Polymer Laboratories Ltd.). Typical DMTA data for a polymer and paint are shown in Figure 3. Tensometry (Instron Model 1026) was used to obtain mechanical performance data on both polymer and paint films.

PAINT PREPARATION AND TESTING

Paints were prepared from polymers of different composition and composition distribution using a standard copper thiocyanate based formulation similar to that which has been described by Hails and Symonds (11). A rotating disc technique (3) was used to measure the polishing rate (which is a measure of hydrolysis rate) of polymer and paint films. Standard coated panels were attached to a disc (Figure 4) in a radial display and this disc then rotated at a constant speed (1400 rpm) in a thermostatically controlled tank (25°C) of replenished sea water. They hydrolytic stability of the films was assessed by the rate of change of film thickness as measured by a surface profiling technique (Ferranti Surfcom).

Anti-fouling tests were carried out on brush coated plastic laminate panels which had been given a primary coating of anti-corrosion paint. Performance was measured by visual observations of the panels after prolonged immersion (4 - 12 months) in a known high-fouling estuarine environment.

RESULTS AND DISCUSSION

The relative reactivity of TBTM and MMA is such that compositionally homogeneous copolymers are produced to complete conversion of monomers in a free-running batch reactor. The reactivity of 2EHA is significantly less than that of either of the other two monomers in ternary polymerizations and control action is required during polymerization in order to produce homogeneous products. The influence of controlled monomer feed on the instantaneous ratio of coreactants can be seen in Figure 5. The ratio of MMA to 2EHA remains constant throughout the reaction with controlled MMA feed to the reactor and the ratio of TBTM to 2EHA is constant up to approximately 97% conversion of monomers with a controlled TBTM feed. The small amount of uncontrolled material which is introduced into the product beyond 97% conversion has been considered insignificant. The error in making absolute measures of residual monomer concentrations by GPC increases as the concentrations of the monomers decreases and the ratio of two inaccurate small numbers can be misleading and probably is responsible for the large deviation shown in the TBTM/2EHA ratio at conversions >97%.

Glass transition data for copolymers and terpolymers of controlled and uncontrolled composition are shown in Figures 6 and 7. The Tg's calculated using the equations 7 and 8 of Fox (12) and Woods (13) have been used with the following hompolymer Tg's; methyl methacrylate, 108°C; tributyltin methacrylate, 0°C; 2-ethylhexyl acrylate, -50°C (14-16) are also shown.

$$Tg = w(1).\ Tg(1) + w(2).\ Tg(2) + w(3).\ Tg(3) \quad (7)$$

$$Tg = [A(1).\ w(1)/Tg(1)] + [A(2).\ w(2)/Tg(2)] + [A(3).\ w(3)/Tg(3)] \quad (8)$$

In equations 7 and 8 w(i), Tg(i) and A(i) are the weight fraction and glass transitions of monomer where i = 1, 2 or 3 and A(i) is an adjustable parameters.

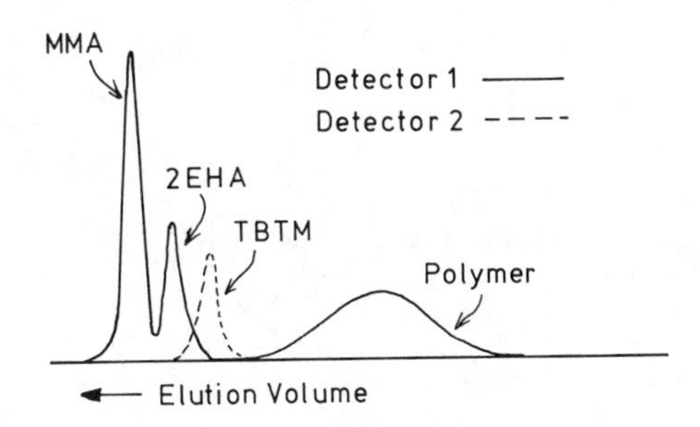

FIGURE 1. Typical GPC chromatograms of residual monomers with detector 1 set at 1720 cm(-1) and detector 2 at 1620 cm(-1). Columns; 10, 100, 1,000 and 10,000 nm., each 30cm long and packed with 10μ gel particles.

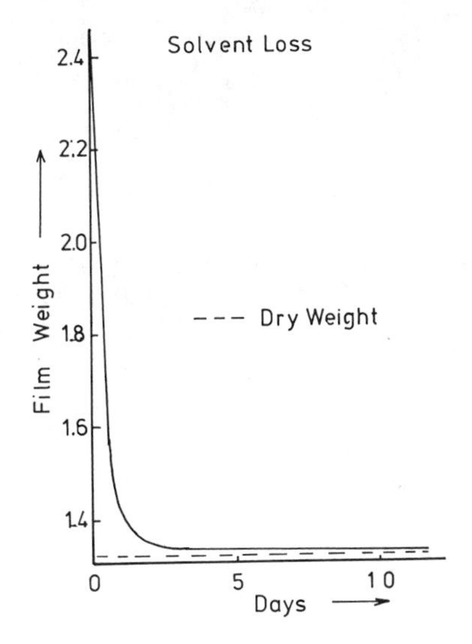

FIGURE 2. Typical solvent loss (by weight) from a thin film at ambient temperature. Weight of dry material.

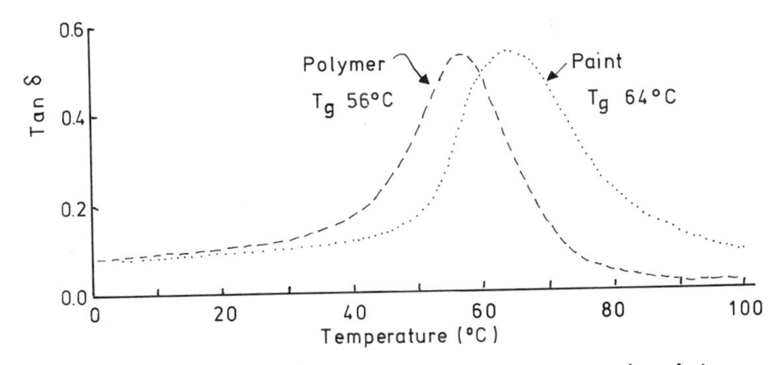

FIGURE 3. Typical DMTA data for terpolymer and paint.

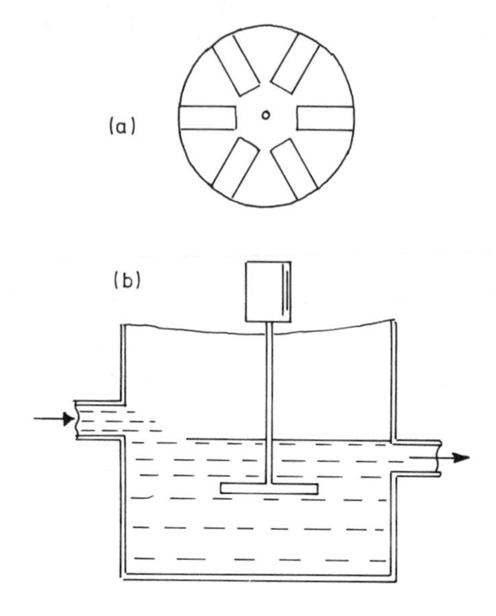

FIGURE 4. Schematic diagram of rotating disc polishing equipment. Rotor speed 1400 rpm, temperature 25°C. (a) plan view of disc and samples (b) side view of disc mounted in water tank.

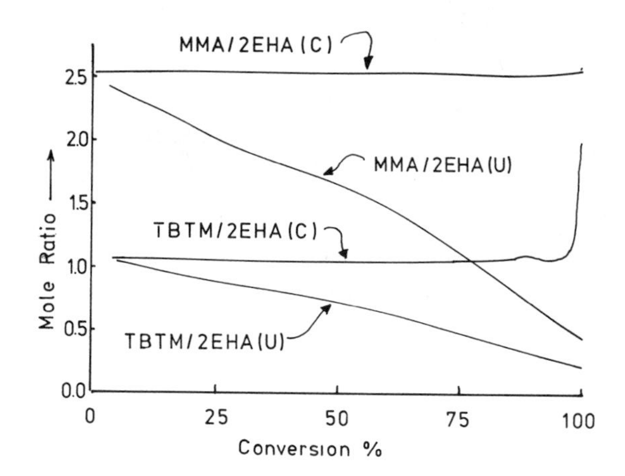

FIGURE 5. Monomer ratio in controlled (C) and uncontrolled (U) typical terpolymerisation taken to high conversion. Control was achieved by feeding both TBTM and MMA to a semi-batch reactor at 80°C. Monomer ratios measured by GPC.

There appear to be no reported values for the Tg of poly(TBTM) and it proved difficult to measure any meaningful transition by differential scanning calorimetry. The value of 0°C was selected as it gave a reasonable fit to experimental data when using equations 7 and 8.

For both the binary and ternary systems the trends in the Tg's are as might be expected in a qualitative sense. Neither equation 7 nor 8 give good fits to the experimental data over the complete composition range for either the binary or ternary copolymer cases but the general trend in Tg's with composition is as might be predicted. These equations are sensitive to the values of the homopolymer Tg's used. There is no obvious reason why there should be such a sharp change in the observed Tg's at about 0.5 mole percent of methyl methacrylate and for the binary polymer our observations can be attributed to the small amounts of residual solvent which remains in the polymers using the methods we have adopted to produce the polymer (and paint) films used for the analyses. In the commercial context, paint films are assumed to be dry in a relatively short time after application (24 hours or less). Ships may enter service well within the timescale of the seven day drying period used for our laboratory prepared films. Although the solvent removal rate is very rapid initially, the diffusion rate of solvent from the film soon becomes very slow (see Figure 2). After seven days the films might retain between 2-5% solvent by weight depending on the composition of the polymer from which the solvent has had to escape. In this work the data have been obtained with materials containing 4.0 + 0.5% residual solvent. Precise studies of solvent evaporation under controlled conditions have not been carried out but it is evident that the lower than predicted Tg Values for high methyl methacrylate content copolymers (Figure 6) results from the plasticizing effect of residual solvent. It is appropriate that measurements are made on polymer and paint films containing residual solvent as these measurements are more realistic in relation to the end use of the materials. For scientific purposes measurements on the fully dried films are the only ones of relevance. Some typical Tg data for fully dried films are shown in Table 2. On average the observed Tg's are 10°C higher than for films containing small amounts of residual solvent. What is not known is the contribution to the properties of slow solvent removal from paint films in service. In the case of non-self polishing paints the consequences must be a trend to coatings which are more brittle in character. In the case of self-polishing paints the situation is more complex in that the solubilizing effect of the sea water at the coating-water interface is likely to counteract the embrittlement which might otherwise be observed.

What is apparent is that Tg's are insensitive to compositional heterogeneity of the polymers. There is some evidence that the Tg's of compositionally homogeneous polymers are higher than for the heterogeneous materials but the difference is small. For the comonomer systems reported this observation is not too surprising since the reactivity ratios of TBTM and MMA are very similar. It is the reactivity of the 2EHA which is significantly different but this monomer is only incorporated to a relatively small extent, <12% in the terpolymer, hence the overall impact of compositional heteogeneity on the thermal and mechanical properties is not large.

Table 2. Values for Xylene-free Binary and Ternary Copolymers

Polymer	Tg + 4% Xylene	Tg Xylene Free
Std. 1 TBTM/MMA 52/48	57	65
MMA/TBTM 50/50	49	58
MMA/TBTM/2EHA (U) 50/37.5/12.5	42	57
MMA/TBTM/2EHA (C)	46	60

U = Uncontrolled polymerisation

C = Controlled polymerisation

The paints made from both compositionally controlled polymers and heterogeneous polymers show little difference in their thermal behaviour, but the Tg's for the former are again slightly higher than for the latter (Figure 6 and 7). The paints, as might be expected, have higher Tg's than the base polymers since they contain approximately 60 volume % inorganic material.

Typical tensile test data are shown in Table 3 for both binary and ternary polymers. the most significant features of these data are that binary copolymers have the greater tensile strength, whereas the ternary materials have greater elongation at break for a given TBTM content. The compositionally controlled terpolymers have increased tensile strength and elongation at break when compared with the uncontrolled polymers. The inorganic particulate materials reinforce the paint films and the elongation to break is less but the tensile strength is greater than the pure polymer.
The major factor influencing the thermal and mechanical properties is the composition differences between the polymers. Gel permeation chromatography measurements have shown that the molecular weight averages and molecular weight distributions are not significantly different for the samples which have been studied and are therefore not seen as important as far as Tg and mechanical measurements are concerned.

The long term fouling and self-polishing are still in progress but it is already apparent from the data from polishing experiments summarised in Table 4 that the TBTM concentration in copolymers has to be in excess of 25 mole % in order to achieve reasonable polishing rate (although the correlation of the accelerated disc test data and in service performance of the paints is not simple). Above 25 mole % tin monomer the rate of polishing increases in

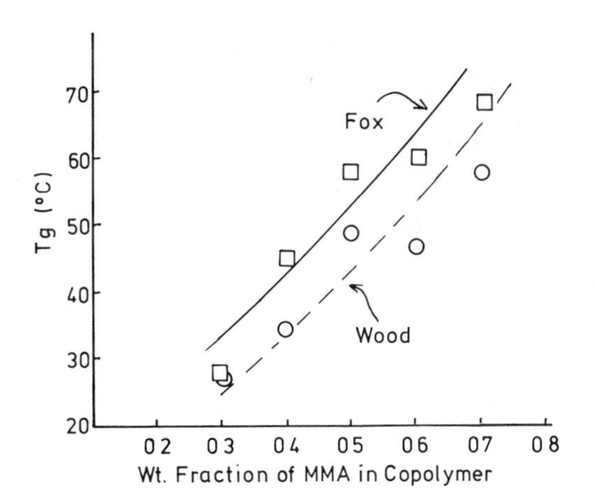

FIGURE 6. Glass transition data for homogeneous TBTM-MMA Copolymers at different compositions.

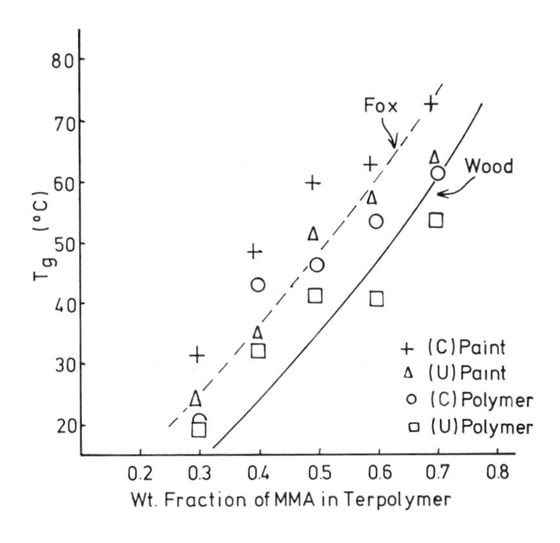

FIGURE 7. Glass transition data for TBTM-MMA-MMA Terpolymers at different compositions.

proportion to the amount of TBTM in the copolymer but above 30 mole % the polishing rate is greatly increased. In the case of the terpolymers, the ratio of MMA to 2EHA is also a significant factor in controlling polishing rate. the hydrolysis rate can be controlled through the use of a hydrophobic monomer such as 2EHA.

Table 3. Typical tensile test data

Polymer Composition	Tensile$_2$Strength Nm		Cross Head Speed mm/min		Elongation To Failure %	
Copolymer TBTM/MMA Mole Ratio	Polymer	Paint	Polymer	Paint	Polymer	Paint
Std.2	2.3	3.5	5	5	2.5	1.1
12.5/87.5	-	-	5	5	-	-
18.3/81.7	2.1	3.2	5	5	0.7	0.4
15.9/74.1	1.5	2.4	5	5	1.8	1.0
33.5/66.5	1.3	3.8	50	50	268	22.2
44.0/56.0	0.8	1.8	50	50	513	100
Uncontrolled Terpolymers TBTM/MMA/2EHA						
7.5/87.5/5.0	2.5	-	5	5	1.2	-
10.9/81.7/7.4	1.4	1.7	5	5	2.5	2.0
15.5/74.1/10.4	1.1	2.2	5	5	5.6	4.8
20.0/66.5/13.5	0.6	2.7	50	50	300	21
26.2/66.5/17.8	0.1	0.3	50	50		287
Controlled Terpolymers						
7.5/87.5/5.0	5.1	3.2	5	5	3.3	1.6
10.9/81.7/7.4	1.8	3.0	5	5	3.4	1.0
15.74.1/10.4	1.4	2.1	5	5	3.6	1.3
20/66.5/13/5	1.1	3.4	50	5	154	16
26.2/56/17.8	0.1	0.1	100	100	454	330

Std. 2 = 27.4 (TBTM)/72.6 (MMA)

Table 4. Data from polishing experiments

Paint	Average depletion of paint film μ	Comparison with Std. 1 = 1.0	Comparison with Std. 2 = 1.0
Rotor Speed 1400 rpm 120 Days			
Std. 1	15.2	1.0	11.7
Std. 2	1.3	0.09	1.0
TBTM/MMA			
12.5/87.5	0.9	0.06	0.7
18.3/81.7	0.8	0.05	0.6
33.5/66.5		Polished away	
44.0/56.0		Polished away	
Rotor Speed 1400 rpm 120 days			
Std. 1	17.1	1.0	13.15
Std. 2	1.3	0.08	1.0
Uncontrolled TBTM/MMA/2EHA			
7.5/87.5/5.0	1.3	0.08	1.0
10.9/81.7/7.4	2.1	0.12	1.62
20.0/66.5/13.5	1.3	0.08	1.08
26.2/56.0/17.8	1.4	0.06	0.08
Controlled TBTM/MMA/2EHA			
7.5/87.5/5.0	1.1	0.06	0.8
26.2/56.0/17.8	3.1	0.18	2.4

Std. 1 = Industrial Standard

Std. 2 = 27.4 (TBTM)/72.6 (MMA)

CONCLUSIONS

1. The polymerization control strategy which is based on the fact that each of the monomers in a co- or ter- polymerization is lost in a first order manner has been shown to be satisfactory for the synthesis of homogeneous multi-component polymers. The MMA/TBTM/2EHA is not a demanding system in that only a small amount of the relatively unreactive 2EHA have been used.
2. It has been demonstrated that relative to the compositionally heterogeneous polymers the homogeneous terpolymers show:
   (i) a marginal increase in the mechanical properties,
   (ii) a marginal increase in the Tg's in both the pure polymer and corresponding paint, and
   (iii) enhanced self-polishing characteristics with lower tin content.
3. A hydrophobic comonomer such as 2EHA considerably retards the self polishing rate of TBTM/MMA/2EHA terpolymers; the terpolymers have a significantly lower self-polishing rate than TBTM/MMA copolymers with the same TBTM content. The third monomer also acts as an internal plasticiser reducing the Tg and increasing the elongation to break when similar tin content binary and ternary polymers are compared.

Acknowledgments

The authors gratefully acknowledge the financial support of the Science and Engineering Research Council and Mr. R. Gosden for assistance with the computer simulations.

Literature Cited

1. A.O. Christie, ACS Preprints, Org. Coatings and Plast. Chem. 1978, 39, 585-9.
2. J.C. Montermoso, T.M. Andrews and L.P. Marinelli, J. Polym. Sci., 1958, 32,523.
3. A. Milne and G. Hails, Brit. Pat. 1,457,590.
4. L.E. Nielson, J. Am. Chem. Soc., 1953, 75,1435-9.
5. G.E. Ham, (Ed) "Copolymerization", Interscience, NY, 1966.
6. A. Guyot, J. Guillot, C. Grailiat and M.F. Llauro, J. Macromol. Sci-Chem., 1984, A21 (6 and 7), 683-99.
7. A.F. Johnson, B. Khaligh and J. Ramsay, ACS Symp. Ser., ACS Annual Mtg., New York, 1981.
8. W.H. Ray, J. Macromol Sci. Rev. Macromol. Chem., Chem., 1972, C8(1),1,
9. D. Ibbitson, A.F. Johnson, N.J. Morley and A.K. Penman, to be published.
10. R. Gosden and A.F. Johnson, to be published.
11. G. Hails and J.D. Symonds, US Pat. 4,191,579.
12. T.G. Fox, Bull. Am. Soc. Phys., 1956, 1(3), 123.
13. L.A. Woods, J. Polym. Sci., 1958, 28, 319-330.
14. J. Brandrup and E.H. Immergut, Polymer Handbook, 2nd Ed., Wiley-Interscience, 1975.
15. M.S. Matheson, E.E. Auer, E.B. Bevilacqua and E.J. Hart, J. Am. Chem. Soc., 1949, 71, 497-504.
16. N.A. Ghanem, N.N. Messiha, N.E. Ikladious and A.F. Shaaban, Eur. Polym. J., 1979, 16, 339-342.

RECEIVED March 5, 1986

Polyurethane Foam Component Lifetimes

K. B. Wischmann

Sandia National Laboratories, Albuquerque, NM 87185

Access deterrent foams are generated by mixing two separately stored and pressurized components upon demand. Investigations have been conducted concerning the aging of both components of three separate polyurethane foam formulations. The polyol component of the first formulation, a propylene oxide adduct of phosphoric acid, hydrolyzes rapidly to give phosphoric acid. Since phosphoric acid can corrode the container as well as adversely affect the final foam properties, a second formulation not containing acid adduct was investigated. This second polyol was isothermally aged at room temperature, 60°C and 71°C and reactions followed by acid number determination. A reaction between the polyol and the blowing agent, Freon 11, was found to give high acid content. Attempts to add inhibitors to lengthen this initiation period failed. Finally, a third formulation was designed which placed the Freon 11 in the isocyanate component thereby precluding the incompatibility of the blowing agent with the polyol. Subsequent aging studies indicate a long term (6-8 years) storage foam system could be achieved. *This work was performed at Sandia National Laboratories supported by the U. S. Department of Energy under Contract Number DE-AC04-76DP00789.

Stored prepacked polyurethane foam components, e.g., isocyanate, polyols, are prone to chemical aging, thereby jeopardizing their intended function. In fact, many vendors of these Freon blown materials will not guarantee their product for more than 90 days. This is for a variety of reasons such as moisture attack on the isocyanate, blowing agent separation and general material instability, e.g., thermal degradation, incompatibility. Because of the expense to change-out foam components, we would like our systems to last as long as possible.

Although generally considered chemically inert, Freon 11 (tri-

0097–6156/86/0322–0341$06.00/0

chlorofluoromethane, CCl_3F), the blowing agent used in these foams, is unstable under certain conditions. For example, this popular industrial refrigerant and aerosol, in the presence of moisture, will react with steel or aluminum forming free hydrochloric acid (1,2). As a result, efforts are made to maintain anhydrous conditions or provide acid scavengers. Although not widely known, Freon 11 will react with primary and secondary alcohols including polyols to liberate aldehydes, ketones and hydrochloric acid (3,4). All the above mentioned reactions can be accelerated with temperature. At higher temperatures thermal decomposition of Freon 11 may produce hydrochloric acid (5). Efforts have been made to find suitable stabilizers for this fluorocarbon; unfortunately these efforts have met with limited success (6-8). One of the better stabilizers, α-methylstyrene was reported to be effective but for only 3 months (6). Because of long term requirements, we became concerned with the reliability of these foam systems.

In this work three different polyurethane foam formulations were investigated. The first was a deployed foam system in which a corrosion study was performed. Due to incompatibilities, a second and a third system had to be formulated and aging studies performed to insure a foam system that provided adequate aging characteristics. To determine the latter two systems' longevity, we started an accelerated aging program to simulate the individual foam component lifetimes. Specifically, the amine equivalent in the isocyanate component and hydroxyl equivalent and acid number in the polyol component were followed at various temperatures (ambient, 60°, 71°C). Lifetime estimates were made by Arrhenius modeling of the data. The following results and discussion describe the efforts made to evaluate the aging characteristics of these foam systems.

Experimental

The formulated isocyanate and polyol components supplied by a vendor (Coplanar Corp.) were aged in one gallon steel vessels. The vessels were rated at 250 psi and equipped with Jenkins ball valves and pressure relief diaphragms (set for 170 psi). The vapor pressure of Freon 11 at the highest aging temperature (71°C) was 60 psig giving a 4/1 safety margin. The actual isothermal aging was carried out at ambient, 60° and 71°C.

Analysis of the isocyanate was accomplished by performing an amine equivalent determination (per ASTM D1638). The polyol component was analyzed for hydroxol equivalent by an acetylation procedure developed at Sandia National Laboratories. An 0.8 gram sample is acetylated with a 1/9 acetic anhydride-pyridine mixture for 2 hours at reflux temperature and then the free acetic acid is back titrated with base and compared to a blank. From this information a hydroxol equivalent can be calculated. The polyol acid number was determined by ASTM D2849.

Background - Corrosion Study of Formulation 1

A two component polyurethane formulation was stored in the field in separate 208 liter (55 gal) vessels of 0.95 cm wall thickness and under a 0.35 to 0.9 MN/m^2 (50-100 psi) over pressure. Upon demand the components are mixed and discharged to form a rigid foam; requirements necessitate a foam density of 0.016 to 0.032 g/cc (1-2 lb/ft^3) and 1-3 minute tack time. The polyol component consists of a polypropylene oxide adduct of phosphoric acid. Upon aging in the presence of moisture, this phosphate ester can hydrolyze to phosphoric acid which could lead to corrosion of the metal container as well as change the resultant foaming characteristics. To determine whether a potentially hazardous condition prevailed a corrosion study was conducted.

The polyol component was known to contain 26.8% by weight of the phosphate ester. From a known hydroxyl number, a molecular weight of approximately 382 was calculated. Assuming complete hydrolysis, about 7% by weight of phosphoric acid would be formed. Actually complete hydrolysis is unlikely, however, a worst case situation was desired for this accelerated study. There are two commonly used methods to determine corrosion rates: 1) a weight loss technique and 2) Tafel extrapolation (electrochemical method). A weight loss experiment was performed in: 1) 7% phosphoric acid in water (dissociated), again a worst case situation and 2) with concentrated acid (essentially undissociated) as it might appear in an organic medium. The initial rate of 0.188 and 0.043 mm/year in 7% and concentrated phosphoric acid respectively decreased with time, most likely due to corrosion product buildup and insolubility in concentrated acid. A value of approximately 0.102 mm/year appears to be a reasonable estimate for long term exposure in the 7% acid and about 0.003 mm/year for the concentrated phosphoric acid. From this data it was concluded that the tested steel shows better resistance in concentrated rather than 7% phosphoric acid. The initial corrosion rate for mild steel in 7% acid (pH=1) was verified by the Tafel extrapolation method. Results indicate an initial rate of 0.18 mm/year which was in close agreement to the value determined by the weight loss method.

The above results must be tempered with the following considerations that were omitted from this study: 1) the vessel was under a constant pressure of 0.35 to 0.7 MN/m^2 (50-100 psi) and may reach 2.0 MN/m^2 (300 psi) when the material is dispensed, this would affect the corrosion rate; 2) the test is very sensitive to environmental changes, e.g., temperature, solution homogeneity; 3) the tests were not conducted on actual container materials, data in the literature show that some steels corrode at a much higher rate in phosphoric acid; and 4) there may be surface defects in the vessel that could lead to accelerated local attack and premature failure. With these qualifications, assuming complete hydrolysis (which is unlikely) and incomplete disassociation, excessive corrosion would not be expected. However, these chemical changes will affect the foaming characteristics, thereby yielding a product that does not meet design specifications.

Since the phosphate ester's only purpose was fire retardation, a new nonphosphate system was recommended for future applications. The following discussion addresses the results of an accelerated aging study on this formulation.

Discussion - Accelerated Aging Study of Formulation 2

A second new dispensable rigid polyurethane foam formulation was acquired that did not contain the phosphate adducts. Critical design requirements were the same as in the first formulation. The respective isocyanate and polyol components were placed in ovens for isothermal aging. We have previously found that about a 10% change in amine and hydroxyl equivalents would alter foaming, e.g., tack time, sufficiently to deviate from design requirements. Thus, lifetime estimates are based on a 10% change in the above analytical parameters.

After 180 days aging, the isocyanate aged at a slow controlled rate (see Figure 1) whereas the polyol showed a dramatic change at 71°C and 180 days aging (see Table I). Before discussing the reasons for the polyol's unusual behavior a description of the isocyanate aging will follow. First, it was believed that component A would show the most pronounced effects of aging, since isocyanates are particularly susceptible to moisture attack. In a 14 month study, reaction with the water was assumed to be the primary aging reaction determining isocyanate lifetime. If one uses the amine equivalent as a damage parameter, an Arrhenius plot can be constructed (see Figure 2). From this data an activation energy (ΔE) of ~9.3 kcal/mole was calculated. This ΔE corresponds quite nicely with literature values (9) for other similar isocyanate reactions with water. Employing the Arrhenius plot, lifetime estimates can be made with various aging temperatures. For example, if the isocyanate component A was aged continuously at 23°C, it would take 6-8 years for a 10% change in amine equivalent to take place. If aged at 49°C (120°F), it would take only 2 years for the same amount of aging (see Figure 3). Based on these projections, a material change-out would be recommended in 6-7 years.

Table I. Foam Long-Term Stability for Formulation II - Chemical Analysis

	0	30 days	90 days	180 days
Component A (Isocyanate)				
Amine Equivalent, Ambient	142	142	142	142
Amine Equivalent, 60°C		142	144	145
Amine Equivalent, 71°C		144	147	150
Component B (Polyol)				
Hydroxyl Equivalent, Ambient	104	105	104	106
Hydroxyl Equivalent, 60°C		107	104	108
Hydroxyl Equivalent, 71°C		107	107	191
Acid Number, Ambient	0.2	0.8	0.6	1.0
Acid Number, 60°C		2.0	1.5	2.0
Acid Number, 71°C		2.0	3.0	111.0

The polyol component B experienced catastrophic change in both hydroxol equivalent and acid number at 71°C and 180 days aging. Because of the large increase in acid number, it was initially thought that Freon 11 (trichlorofluoromethane) was generating free hydrochloric acid by reaction with other components. Free hydrochloric acid was verified by adding $AgNO_3$ to an aged polyol sample and obtaining a precipitate of AgCl. The literature certainly confirms the instability of Freon 11 (1), therefore a superior stabilizer was

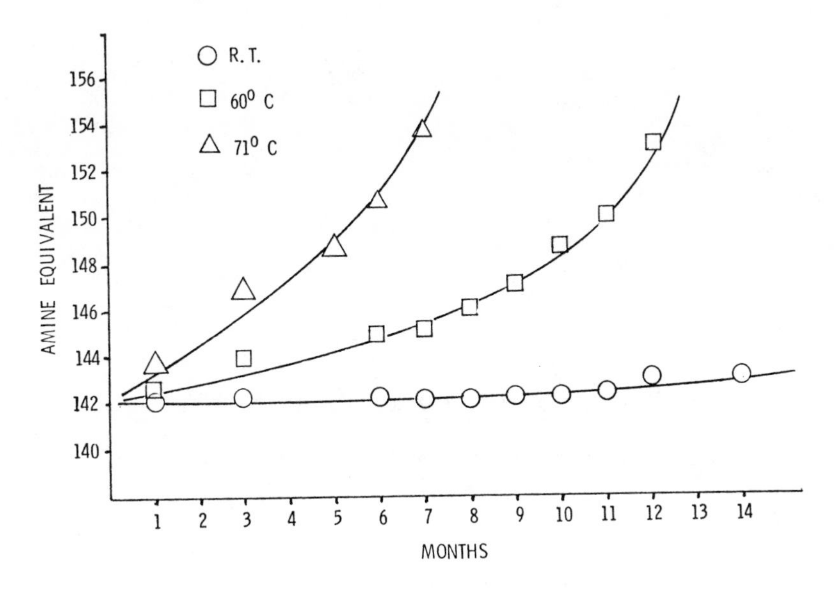

FIGURE 1. COMPONENT A - AMINE EQUIVALENT.

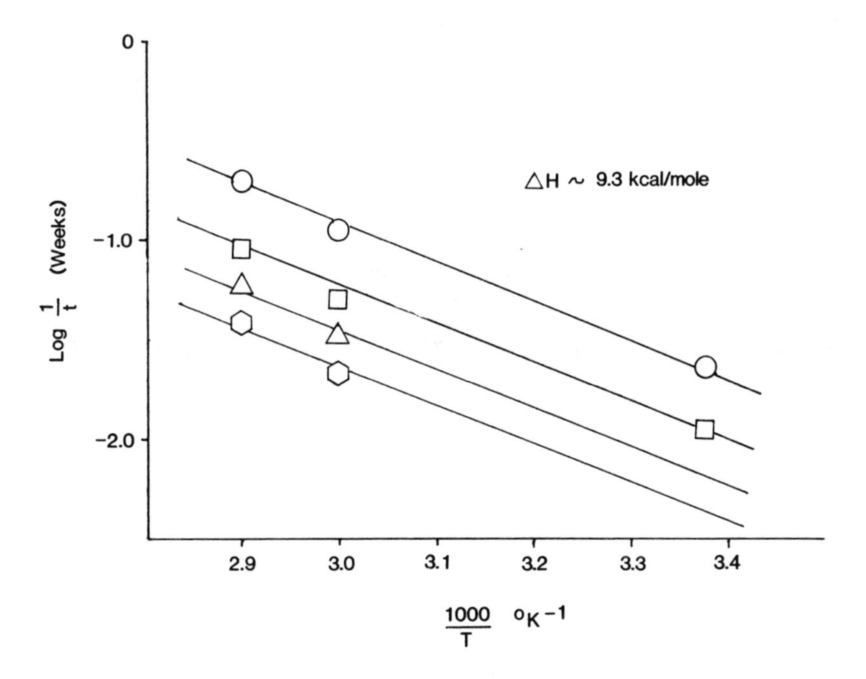

FIGURE 2. ARRHENIUS PLOT OF AMINE EQUIVALENT.

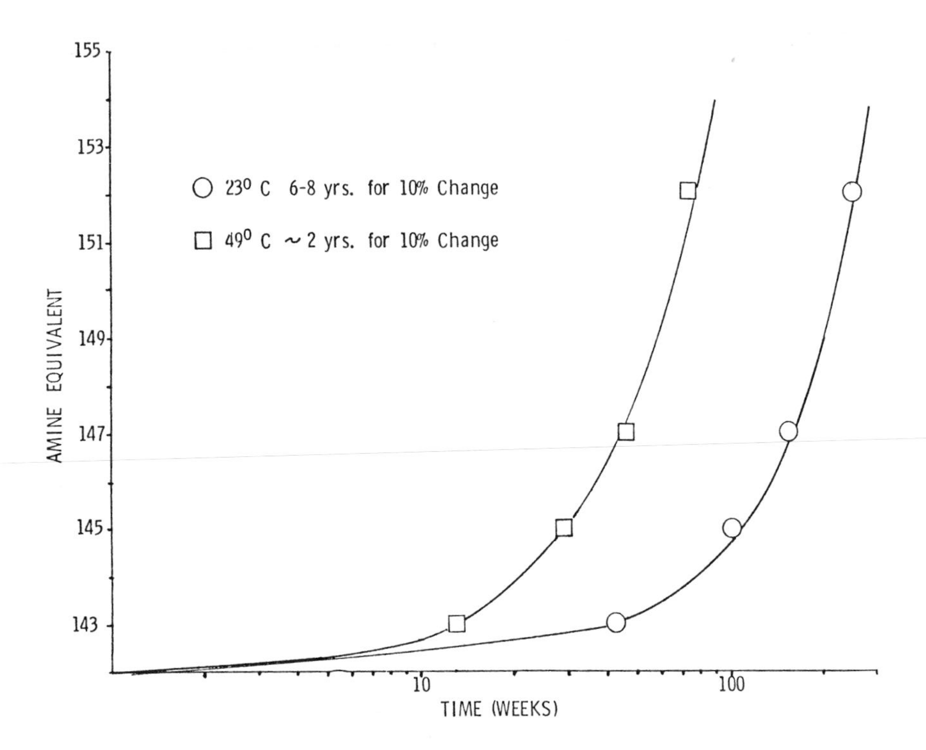

FIGURE 3. LIFETIME PREDICTION OF ISOCYANATE COMPONENT.

sought. Communication with the manufacturer, revealed that the stabilizer used in Freon 11 was a material called alloocimene (2,6-dimethyl 2,4,6-octatriene). The manufacturer suggested employing 1% by weight α-methylstyrene as a stabilizer. The polyol aging study was repeated using the newly stabilized trichlorofluoromethane, termed Freon 11A. At exactly the same aging station (180 days), again large changes in hydroxyl equivalent and acid number occurred.

$$CCl_3F + CH_3CH_2OH \longrightarrow CHCl_2F + CH_3CHO + HCl$$

Why the reaction takes place so dramatically at about 180 days rather than gradually is not understood; perhaps it is simply due to an induction period or perhaps the stabilizer is expended at that time.

This reaction is not a widely known text book reaction, in fact, it was only found in the patent literature (4). However, this reaction explains the evidence of free hydrochloric acid. Since a carbonyl group, e.g., aldehyde, was generated by the above reaction, this group should be observable in the IR. Subsequent IR scans of the aged polyol revealed formation of a carbonyl at about 5.8 microns. Unaged polyol shows negligible carbonyl formation. According to the patent (4), this reaction is peculiar to any chlorofluoroalkane containing three or more chlorines, i.e., trichlorotrifluoroethanes, $C_2F_3Cl_3$; tetrachlorodifluoroethanes, $C_2F_3Cl_4$. Dichlorofluoroalkanes, i.e., Freon 12, dichlorodifluoromethane, are apparently free from such reactions. The reaction of Freon 11 with a polyol appears inescapable; consequently long term storage of foam systems of this composition are not advisable.

Formulation 3

At this juncture, it was decided to make a radical formulation change. First, because of the incompatibility of the Freon 11 with the polyol, the Freon 11 would be removed from the polyol and placed in the isocyanate component. Second, since 1/1 component ratios are necessary to accommodate the mixing machine, different isocyanate and polyol components were formulated to establish appropriate viscosities. The final formulation is shown below.

Isocyanate Component A		Polyol Component B	
PAPI-580	117 pbw	PEP-550	117 pbw
Freon 11	33 pbw	DC-197	1 pbw
		DABCO	5 pbw
		Freon 12	25 pbw

In subsequent aging studies the polyol component showed virtually no change in acid number at any of the aging temperatures (ambient, 60°, 71°C) over 13 months. The isocyanate was shown to age similarly to the isocyanate in Formulation 2. As a result, the above formulation is being employed in the field.

Summary And Conclusions

Two different dispensable rigid polyurethane foam formulations were found inadequate over a long period of time. The first system contained phosphoric acid adducts which can hydrolyze in the presence of moisture, thereby changing their foaming characteristics. A reformulated system excluding the phosphate esters also exhibited poor aging, simply because the Freon 11 blowing agent reacts with the polyol liberating free hydrochloric acid, an unacceptable situation. The latter reaction was not common knowledge, yet, it must be recognized in view of the wide-spread popularity of this refrigerant and blowing agent.

Finally, a third formulation was devised which excluded the use of Freon 11 in the polyol component. Freon 11 was placed in the isocyanate component and both isocyanate and polyol components were changed to meet viscosity considerations. Subsequent aging studies showed the isocyanate to age similarly to the previously aged (Formulation 2) isocyanate. The polyol showed virtually no increase in acid number at any aging temperature over 13 months. Thus, at ambient temperature we would expect a 6-8 year system lifetime on the isocyanate before a 10% change in analytical properties would dictate a material change-out. The polyol appears to have a greater lifetime, but would probably be replaced at the same time.

Acknowledgments

The author gratefully acknowledges S. L. Pohlman for conducting the corrosion study, S. L. Erickson for the analytical data, and C. Arnold for helpful discussions.

Literature Cited

1. Church, J. M.; Mayer, J. H. J. Chem. Eng. 1961, 6, 449.
2. Parmelee, H. M.: Downing, R. C. Soap Sanit. Chem. 1950, 26, 114.
3. Parmelee, H. M.; Downing, R. C. Pro. Chem. Specialties Manufacturers Assoc. 1950, 45, 47.
4. Bauer, A. W. U.S. Patent 3 183 192, 1965.
5. Eiseman, B. J. Progr. Refrig. Sci. Technol. 1973, 2, 643.
6. Degginer, E. R.; Knapp, W. A.; Zuem, H. E. U.S. Patent 3 352 789.
7. Blodgett, F. W. U.S. Patent 3 361 833.
8. DuPont, Belgium Patent 621 364.
9. Wright, P; Cumming, A. P. C. "Solid Polyurethane Elastomers"; McClearen and Sons, 1969.

RECEIVED January 22, 1986

Rubber Coatings for Fiberglass Protection in an Alkaline Environment

P. Dreyfuss[1], R. D. Vargo[2], R. S. Miller[3], and R. Bright

CEMCOM Research Associates, Inc., 9901 George Palmer Highway, Lanham, MD 20706

Glass is known to be readily attacked by strong alkali. This was reaffirmed in the present study when samples of style 3701 E-glass fabric lost 90% of their original tensile strength after immersion in aqueous alkali at pH 13 and 80°C for 7 days. The purposes of this study were to gain some understanding of the relative importance of the factors that influence the degradation of glass and coated glass in alkaline environments and to use that knowledge to develop a protective coating. Factors evaluated included the composition, surface area and and nature of the surface of the glass. The effects of the nature, uniformity, thickness and degree of bonding of glass coatings as well as different methods of applying coatings were considered. It was shown that well-bonded rubber coatings can lead to good protection of glass in alkaline environments.

Glass is known to be readily attacked by strong alkali (1). This was reaffirmed in the present study when samples of style 3701 E-glass fabric from Burlington Glass Fabrics Co. lost 90% of their original tensile strength after immersion in aqueous alkali at pH 13 and 80°C for 7 days. The purposes of this study were to gain some understanding of the relative importance of the factors that influence the degradation of glass and coated glass in alkaline environments and to use the knowledge gained to develop a coating that would protect glass in alkaline environments.

[1]Current address: Michigan Molecular Institute, 1910 W. St. Andrews Road, Midland, MI 48640
[2]Current address: Institute of Polymer Science, The University of Akron, Akron, OH 44325
[3]Current address: 1749 Dana Street, Crofton, MD 21114

0097-6156/86/0322-0349$06.00/0

Experimental

Materials. Table I lists the kinds and sources of glass fibers, bars and slides used in this study. E-glass fabrics were from Burlington Glass Fabrics Co. Glass beads were from Petrarch Systems, Inc. or Potters Industries, Inc. 3-Aminopropyltriethoxysilane as was obtained from Petrarch Systems, Inc. The bulk polybutadiene was Firestone's Diene 35 NFA, a noncrystallizying anionic polybutadiene of $\bar{M}_n$=150,000 and *cis:trans:* vinyl (%) = 36:54:10. Natural rubber latexes (45 and 55% solids) were from Killian Latex Inc. and SBR latex Polysar XE-404 was from Polysar Resins Inc. Dicumyl peroxide (DICUP R) was from Hercules, Inc. The aqueous alkali solution was prepared using 0.008g (0.022 moles)/l sodium hydroxide, 3.45g (0.062 moles)/l potassium hydroxide and 0.48g (0.006 moles)/l calcium hydroxide. Acetate splicing solution was obtained from Burlington Glass Fabrics Co. Alumina acid (80-200) mesh and calcium oxide were from Fisher Scientific Co. Boron oxide (99%) was from Alfa Products, Thiokol/Venton Division.

Weight Loss Studies. Glass samples were rinsed successively with toluene and acetone and dried in an oven at 80°C for 10 min. Glass samples for testing in water were placed in a large Soxhlet extractor (from a 120 mm joint) and extracted with distilled water for 24 hrs. before drying and reweighing. The weight loss was the difference between the initial weight and the final weight. Since none of the samples lost appreciable weight, this method was also used for washing samples for weight loss studies after immersion in alkali. After cooling, weighed samples for testing in alkali were placed in precleaned 1000 ml polypropylene bottles fitted with dividers from porous polypropylene when appropriate. Beads, chopped fibers and powders were placed in Teflon extraction thimbles before putting into the bottles. Long fibers were wrapped around special porous polypropylene holders as long as the bottles and then put into the bottles. The bottles were then filled with the aqueous alkali solution, closed with Teflon lined caps and placed for 7 days in a circulating water bath at 80°C. The samples were removed, thoroughly rinsed with water, dried overnight in an air oven at 80°C, cooled in a dessicator and reweighed. Dumbbells from cured rubber samples were similarly tested.

Alkali Durability Tests of Glass Fibers and Glass Fabrics. These tests were carried out using a modification of Burlington Test Procedure FP-017. For durability tests glass fibers were treated in the same way as for weight loss studies. Care was taken to run tensile tests only on those portions of the fiber, which had not been bent around the edges of the holder. Samples of glass fabric 27.9 cm. (11") long (in the warp direction) and 15.2 cm. (6") wide (in the fill direction) were cut from rolls of fabric and coated as desired. The prepared cloth specimens were placed in heavy-duty 4.5 - mil thick 24.13 x 40.64 cm Kapak heat-sealable pouches. Enough alkali solution to fill the bag was added, the pouch was heat-sealed using a Scotch Pak Pouch Sealer and placed in the circulating water bath at 80°C for 7 days. The fabric was removed from the alkali solution, rinsed with water, dried overnight in a vacuum oven at 60°C, and prepared for tensile testing.

Table I. Weight Loss Comparison for Different Glasses[1]

Glass	Form	Source	% Wt. Loss	Appearance
E	Fibers	Burlington Glass Fabrics Co.	water: ---	---
			alkali: 18.3	---
	Bars[2]	PPG Industries, Inc.	water: 0.07	No change
			alkali: 0.14	Slight Fogging
AR[3]	Fibers	Cem-FIL Corp.	water: 1.4	Stiff, not much change
			alkali: 1.6	Soft and wooly
Pyrex	Bars[2]	Five Points Glass Co.	water: 0.035	No change
			alkali: 0.04	No change
Quartz	Bars[2]	Five Points Glass Co.	water: 0.033	No change
			alkali: 0.10	Uniform fogging
Soda Lime	Slides[4]	Kimble	water: 0.07	Fogged, extremely discolored
			alkali: 0.35	Uniform fogging

1 After immersion in the medium shown in column 4 according to the procedure given in the experimental section.
2 E-glass were random sizes unsuitable for peel tests. Pyrex and quartz bars were 127 x 25 x 6.5mm.
3 Alkali-resistant glass.
4 Pre-cleaned microscope slides.

Tensile Testing. Tensile tests were run on an Instron 1000 using a 5.08 cm (2 inch) guage length and a speed of 12.7 cm (5 inches)/min. The procedures in Burlington Test Procedure FP-015 (ASTM-D-579) were followed. At least 5 specimens of each sample were tested. Fabric tests results are quoted for the warp direction only. The percent retention of tensile strength was calculated from the equation: % retention = (F_A/F_{NA}) x 100 where F_A and F_{NA} are the strength after and before alkali immersion, respectively.

Silane Treatment. Glass bars, fibers or fabric were soaked 5 min in stirred solutions (0.5, 1.0, 1.5, or 2.0%) of 3-aminopropyltriethoxysilane in 95% ethanol, then for 10 min. in 95% ethanol, and finally for 1 min. in absolute ethanol. The glass was heated overnight in a vacuum oven at 110°C to form a polysiloxane layer bonded to the glass (2).

Latex Treatment. A typical procedure for latex treatment is shown schematically in Figure 1. Details of the procedure are given in U.S. Patent Application Serial No. 06/701,747, filed 2/14/85. Blowers were used to dislodge any latex which might exist as "window pane" like films between bundles of fibres in the fabric. The temperature in the first two dryers was maintained at 75-85°C, too low a temperature to cure the latex during the drying time. The temperature in the third dryer was typically 120°C and curing of the rubber was continued at that temperature for 30 minutes. Microscopy revealed that the first dip results in rubber penetration of the fiber bundle but does not coat the bundle. The second and third dips produce coatings of about 10 micrometers each. Figure 2 is a schematic representation of the manner in which the coatings deposit on the fibers in the fiber bundle.

Peel Tests. E-glass plates (0.635 x 20.32 x 20.32 cm, 1/4 x 8 x 8") were cut to ~2.54 x 7.62cm (1 x 3") using a glass saw. The bars were cleaned by boiling in 2% Micro solution (a laboratory cleaner from International Products Corp.), washed for 48 hrs. with distilled water in a Soxhlet extractor, dried for 2 hrs. in an air oven at 150°C and stored in a dessicator over P_2O_5 until used. Peel test specimens for testing without immersion in the aqueous alkali were prepared and tested according to procedures previously described (3-5). Thus, dicumyl peroxide was mixed into polybutadiene on an open mill. Before bonding, the elastomer was pressed into a thin layer (~0.2 mm) by premolding for 1 hr. at 65°C, then pressed into a sheet of cotton cloth and again premolded for 1 hr. at 65°C. The cloth backed layer was then pressed against the cleaned glass slides for 2 hr. at 150°C in a PHI press at a pressure of ~6 psi/5 in ram in order to cure the elastomer. The thickness of the elastomer interlayer in the resulting cloth-elastomer-glass sandwich was ~0.2mm. 180° Peeling tests were carried out on strips of cloth backed elastomer layer after trimming them to a uniform width on the glass of 2 cm. The elastomer was peeled off the glass at a constant rate, 0.5 cm/min (0.0083 cm/sec) using a table model Instron. The work of adhesion, W_A, was calculated from the equation $W_A = 2P/w$, where P is the time average peel force and w is the width of the detaching layer.

Samples for testing in alkali were cured between a cloth backed rubber sheet and an unbacked rubber sheet using the same curing conditions and then were trimmed in two stages. In the first stage the specimens were cut apart so that the glass bars were completely

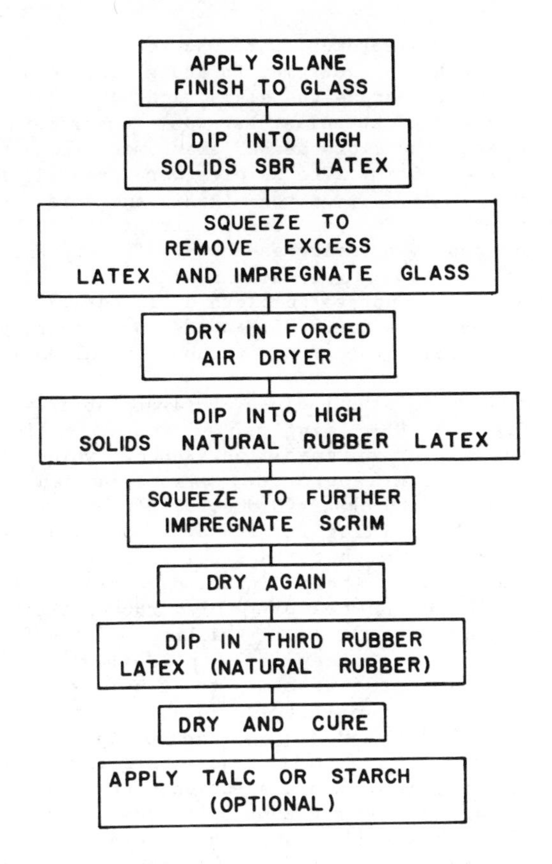

Figure 1. Flow sheet showing steps in coating process.

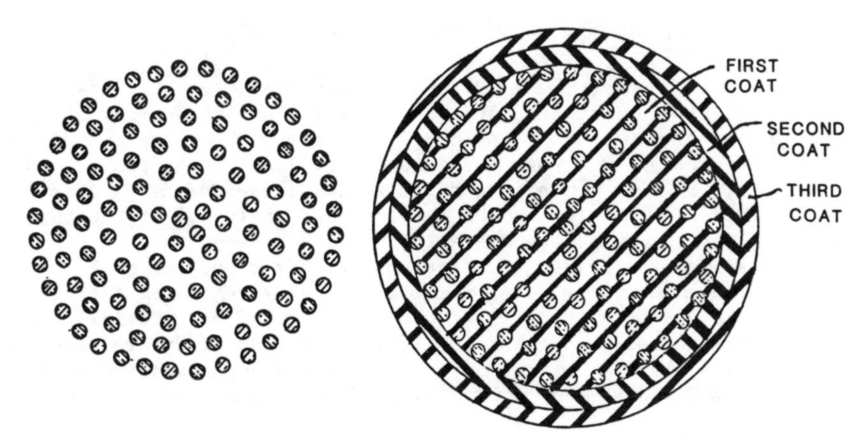

Figure 2. Schmatic view of glass fiber bundles.

embedded in the cured elastomer. Care was taken to insure that there were no holes through the elastomer to the glass, because alkali can seep through holes, degrade the glass, and invalidate the measurements. The thickness of the elastomer was 1 mm or more on all sides except the one with the cloth backed elastomer, where the elastomer thickness was 0.4-0.6 mm. The samples were treated with alkali as described above, removed from the alkali, and dried overnight in a vacuum oven at 60°C. For peel tests, strips of cloth backed elastomer were obtained by trimming as above. The strength of the rubber impregnated cotton cloth backing was retained after alkali treatment, even though untreated cloth disintegrated under the same conditions. Firestone's Diene 35 NFA was the rubber used for these studies. 0.05% parts per hundred of rubber of dicumyl peroxide was used for curing.

Coating with Bulk Polybutadiene. E-glass fabric was embedded in Firestone's Diene 35 NFA using procedures very similar to those used to prepare peel test specimens. Rubber, which had been mill-mixed with 0.05% dicumyl peroxide, was premolded between Mylar sheets to the desired thickness(0.308, 0.151, or 0.100 cm) and size (~ 30.5 x 18 cm) by molding for 1 hour at 60°C. and 40,000 lbs/5" ram. Fabric was cut so that the final size was at least one inch smaller than the rubber sheets in all directions. A sandwich was made from the fabric and two premolded rubber sheets of the same thickness and about half the total thickness of the final sandwich. The sandwich was cured in a press for 2 hours at 150°C and 5000 lbs/5" ram. In the cured specimen the fabric was embedded in the center of the molded specimen (0.15 -0.40 in thick). Samples were immensed in alkali before cutting to size for tensile tests.

Thinner coatings on fabrics and fibers were prepared by dipping the fabric or fibers into a solution containing 500 ml hexane, 50 g polybutadiene and 0.02 g dicumyl peroxide. Excess solution was squeezed off by passing the fabric or fibers between rollers, the hexane was removed by evaporation, and the rubber was cured for 2 hours in a vacuum oven at 150°C.

Microscopy. A Leitz Orthoplan microscope fitted with a Polaroid Land Camera was used to examine samples at low magnification. Scanning electronmicrographs were taken with an ISI-SS40.

Results and Discussion

Many of the factors that influence the degradation of glass in an alkaline environment are obvious. They include the composition of the glass, the surface area of the glass, and the nature of the surface of the glass. When coatings are applied to the glass, additional factors need to be considered also. These additional factors include the uniformity of the coating, the influence of the thickness of the coating, the degree of bonding between the coating and the glass, and the effect of different methods of applying the coating. A comprehensive consideration of each of these factors is beyond the scope of this paper but each of them was examined in at least a preliminary way. The conclusions of this study are based on weight loss studies, microscopy, peel test measurements and tensile properties before and after immersion in the aqueous alkali solution.

Studies on Glass Alone. Weight loss comparisons for different kinds of glass after immersion in water and alkali are given in Table I. The data showed that all the glasses listed were reasonably stable in water, that fibers were more severely degraded then bars, especially in alkaline solutions and that of the glasses tested, E-glass and soda lime glass were the most severely corroded by the alkali. The difference among the kinds of glass can be partially explained in terms of the relative stability in water and alkali of the various materials comprising the glasses. Some typical formulations are given in Table II (6-8).

Table II. Typical Weight % Composition of Glasses (6-8)[1]

Oxide	Glass				
	E	AR[2]	Pyrex	Quartz	Soda Lime
SiO_2	54.5	70.3	80.5	100	72
Na_2O	1±	11.8	3.8	---	14
CaO	17.0	---	---	---	13
Al_2O_3	14.5	---	2.2	---	1
B_2O_3	8.5	---	12.9	---	---
MgO	4.5	---	---	---	---
K_2O	4.0	---	0.4	---	---
ZrO_2	---	16.1	---	---	---
Li_2O	---	1	---	---	---

[1] Only the major oxides (1% or greater) are listed. Most glasses also have small amounts of other metal oxides such as Fe_2O_3, TiO_2, Mn_2O_3, etc.
[2]Alkali-resistant glass.

A comparison of the results in Table I with the compositions in Table II suggests that high concentrations of CaO and Al_2O_3 lead to decreased durability in alkali. This is not surprising when the results of % weight loss studies on these metal oxides alone and their known chemistry are considered. Al_2O_3, when exposed to the standard alkaline conditions, lost 8.3% by weight. Al_2O_3, when hydrated, is amphoteric, is soluble in strong alkalis, and forms compounds like $NaAlO_2$ or $Ca(AlO_2)_2$. Thus the high concentration of Al_2O_3 in E-glass can at least partially account for the observed weight loss. (B_2O_3 was totally soluble under the usual alkaline conditions and when hydrated, in acidic. However, the stability of Pyrex glass suggests that boron is present in glass not as B_2O_3 but in some form that is not readily attacked by alkali.) CaO, when exposed to the standard alkaline conditions, *gained* 13.3% by weight. This is understandable because CaO can react with water to form $Ca(OH)_2$ and with dissolved CO_2 to form $CaCO_3$. Both products have higher molecular weights than CaO. In order to account for the observed weight loss with E-glass and even more so with soda lime glass, it must be assumed that either these products slowly dissolve in the alkali or else they are washed off the surface of the glass. (Again the Na_2O present in significant quantities in soda lime cannot account for the severe weight loss observed with soda lime glass, because AR (alkali-resistant) glass also contains much Na_2O and is especially formulated to be and indeed is observed to be more resistant to alkali).

Tensile measurements of the E-glass and AR-glass fibers showed that although the E-glass fibers were stronger initally than the AR-glass fibers, the AR-glass retained its strength better in the alkali. E-glass fibers were too weak to test after immersion in the alkali, whereas AR-glass fibers showed 19% retention of initial strength.

The much greater weight loss observed with E-glass fibers relative to E-glass bars suggests that exposed surface area plays an important role in durability in alkaline solutions. The effect of surface area was examined by determining the percent weight loss of soda lime beads of different but known diameter after immersion in the alkali. As shown in Figure 3 the weight loss increased logarithmically with the square of the beads' diameters, i.e. of their surface area.

We concluded that the lack of durability in alkali of E-glass is related to its chemical composition and that a coating was needed to protect the glass. A suitable coating should both reduce the surface area exposed to the alkali and prevent contact between the degradable components in the glass and the alkali.

Studies on Coated Glass. Glass beads, fibers, and fabrics are commercially available with various polysiloxane coatings, which are chemically bonded to the glass through the silanol groups (2). Usually the polysiloxane coatings contain other functional groups, which can serve as bonding sites for other coatings. It was of interest to compare the degree to which these polysiloxanes rendered the glass corrosion resistant to alkaline solution. Weight loss data given in Table III, showed significant differences. Polysiloxanes with a long alkyl group like octadecyl gave coatings that resulted in a significant increase in the durability of the coated glass in alkali. Shorter alkyl substituents like ethyl, dimethyl and vinyl were deleterious and most polar substituents tested (glycidoxy-, methacryloxy-) seemed to encourage weight loss. The basic 3-aminopropyl group gave some protection. These data suggest that continuous nonpolar hydrocarbon coatings might protect the glass and that a polysiloxane coating with 3-aminopropyl-groups to promote bonding with the hydrocarbon layer (2,3) should not be disadvantageous.

Studies with Rubber Coatings. The above suggestions were verified in studies with rubber coatings on glass plates, fibers and fabrics. 180° peel tests on E-glass plates showed that with or without prior surface coating of the plates with the aminopolysiloxane, there was no decrease in the adhesion between the glass and peroxide-cured polybutadiene coatings. The results are summarized in Figure 4. Most samples failed cohesively in the rubber layer. The strength of the samples that were not exposed to alkali increased steadily with the percent of 3-aminospropytriethoxysilane (AS) in the solution used to treat the glass. The strengths of samples exposed to alkali were essentially constant at concentrations of 1% AS and above. Bars that were not coated with the polysiloxane failed adhesively and the scatter in the results ($\pm \sim 35\%$) reflected the fact that the surface of the bars was not as smooth as the surface of microscope slides and other bars used in previous studies (3-5).

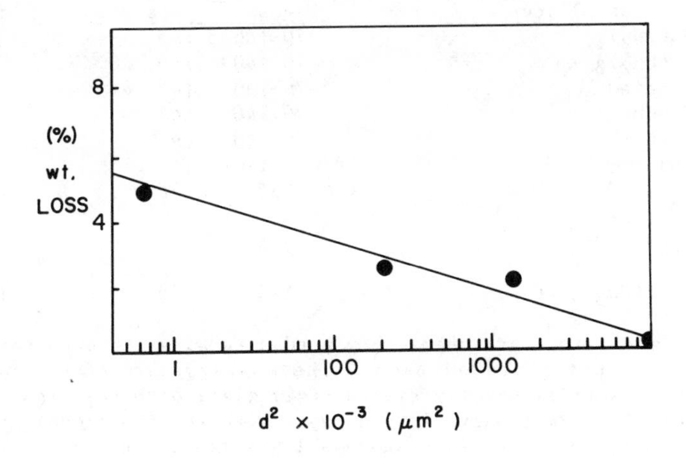

Figure 3. Effect of soda lime bead size on percent weight loss after immersion in alkali for 7 days at 80°C.

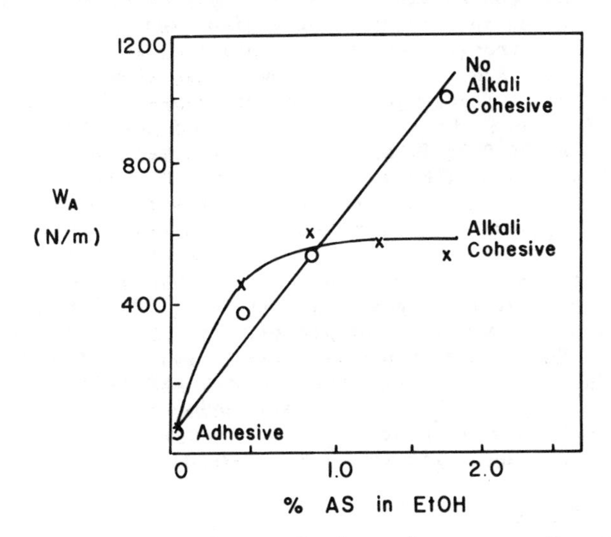

Figure 4. Peel test results on E-glass plates as a function of percent of 3-aminopropyltriethoxysilane in solution used to treat plates. x: results after immersion in alkali for 7 days at 80°C. o: initial results.

Table III. Studies with Silane Coated Soda Lime Glass Beads

Functional Group[1] on Silicon	Size Mesh	~µm	% Wt. Loss in Alkali
$CH_3(CH_2)_{17}$ -	70-140	145	0.86
$NH_2(CH_2)_3$ -	70-140	145	2.08
Uncoated	70-140	145	2.27
CH_2=CH-	70-140	145	3.18
CH_3CH_2-	70-140	145	3.58
$CH_3)_2$-	70-140	145	4.88
Uncoated	325	25	4.50
O $CH_2CHCH_2O(CH_2)_3$-	325	25	5.87
O CH_2=C(CH_3)CO(CH_2)$_3$-	325	25	10.89

[1]Silane coatings are chemically bonded to glass through the silanol groups on the glass surface. The coatings are often prepared by reacting a trialkoxyalkylsilane, $(RO)_3SiR'$, with the glass and then heating to form a polysiloxane layer with the functional group from R', if any, in the surface and available for further reaction (2). A dialkoxydialkylsilane, $(RO)_2SiR'_2$, can be used in the same way. If R'is CH_3 and the silane is $(RO)_2SiR'_2$, e.g., the polysiloxane layer will have two CH_3 groups and fewer bonds to the glass surface.

AR-fiber coated with the same polysiloxane and passed through a hexane solution containing polybutadiene and peroxide before curing, showed 100% strength retention in alkaline solution. E-glass fabric embedded in peroxide-cured polybutadiene showed nearly 100% retention of original strength for polybutadiene thicknesses of 1 mm or more. The strength typically observed for #3701 greige goods as received was 26.8-29.1 kN/m (153-166 lbs. per inch width). Strength retention in the alkali was at most 10%. Polybutadiene coated specimens had initial strengths of 16.6-19.8 kN/m (95-113 lbs per inch width), and strength retention in the alkali was 80-100% for coatings 1 mm thick or greater. (The strength of the composite varied with the thickness of the polybutadiene. Thicker coatings gave weaker overall composite strengths.) From latexes, coatings as thin as 0.03 mm gave good protection. Typical data is shown in Figure 5. Multiple dippings were necessary, since the first dip gave only penetration of the fiber bundle and no coating. The first latex dip is preferably into a latex other than natural rubber (e.g. SBR latex) because good adhesion between the glass and a coating from natural rubber latex is not obtained.

Acknowledgments

Samples of various E-glass fabrics and fibers from Burlington Glass Fabrics Co., of Diene 35 NFA from Firestone Tire and Rubber Co. and of dicumyl peroxide, DICUP R, from Hercules, Inc. are gratefully acknowledged. Rubber latexes were kindly supplied by Killian Latex Inc. and Polysar Resins, Inc. Thanks are given to Janine Rizer for help with tensile strength measurements.

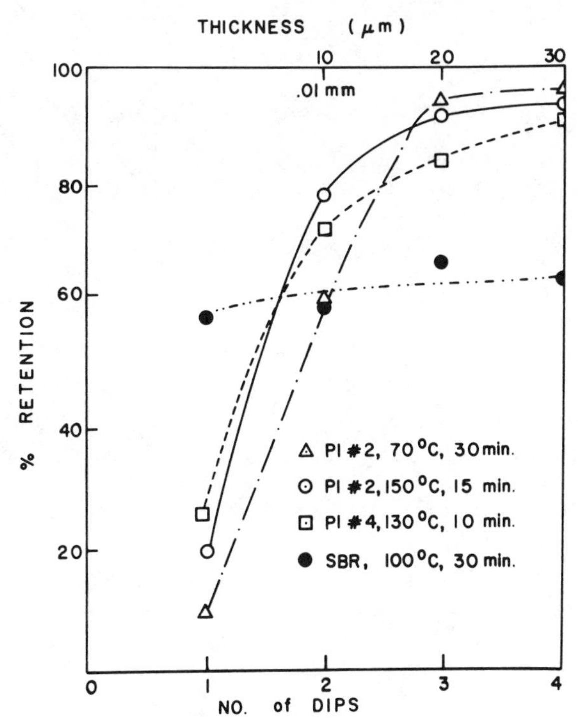

Figure 5. Percent retention of strength of latex coated #3701 fabric with 3-aminopropylpolysiloxane coating, after immersion in alkali for 7 days at 80°C., as a function of number of dips in latex, kind of latex, and curing conditions.
PI#2 is Killian 500 clear polyisoprene latex - Cure 2;
PI#4 is Killian 500 clear polyisoprene latex - Cure 4;
PI#2 and PI#4 were received from Killian as partially cured resins. SBR is Polysar XE-432 styrene-butadiene latex.
The temperature and time of cure are cited after each latex.

Literature Cited

1. Boyd, D. C.; Thompson, D. A. Encyclopedia Chem. Technol. 1980, 11, p. 843.
2. Plueddemann, E. P. "Silane Coupling Agents"; Plenum Press:New York, 1982.
3. Eckstein, Y.; Dreyfuss, P. J. Adhesion 1983, 15, 193-202.
4. Ahagon, A.; Gent, A. N. J. Polym. Sci.:Polym. Phys. Ed. 1975, 13, 1285-1300.
5. Liang, F.; Dreyfuss, P. J. Appl. Polym. Sci. 1984, 29, 3147-3159.
6. Anledter, H. F. Encyclopedia Polym. Sci. and Technol. 1967, 6, p. 634.
7. Mack, Jr., E.; Garrett, A. B.; Haskins, J. F.; Verhoek, F. H. "Textbook of Chemistry"; Ginn and Co.:New York, 1949; p. 760.
8. Hannant, D. J. "Fibre Cements and Fibre Concretes"; John Wiley & Sons:New York, 1978; p. 100.

INDEXES

Author Index

Subject Index

P

R

S

Production by Joan C. Cook
Indexing by Keith B. Belton
Jacket design by Pamela Lewis

Elements typeset by Hot Type Ltd., Washington, DC
Printed and bound by Maple Press Co., York, PA